P9-BJT-901

Combustibility of Plastics

Frank L. Fire

Vice President, Marketing
Americhem, Inc.

Library of Congress Catalog Card Number 91-23836
ISBN 0-442-23801-0

Manufactured in the United States of America

Published by Van Nostrand Reinhold
115 Fifth Avenue
New York, New York 10003

Chapman and Hall
2-6 Boundary Row
London, SE1 8HN, England

Thomas Nelson Australia
102 Dodds Street
South Melbourne 3205
Victoria, Australia

Nelson Canada
1120 Birchmount Road
Scarborough, Ontario M1K 5G4, Canada

16 15 14 13 12 11 10 9 8 7 6 5 4 3 2 1

Library of Congress Cataloging-in-Publication Data

Fire, Frank L., 1937–
Combustibility of plastics / Frank L. Fire.
p. cm.
Includes index.
ISBN 0-442-23801-0
1. Plastics—Flammability. I. Title.
TH9446.5.P45F47 1991
628.9'222—dc20 91-23836
CIP

Dedication

This book is dedicated to my wife, Marlene, and my four sons, Frank, Kevin, Eric and Christopher—but especially to my first son, Frank, without whose gift it would not have been possible, and to Kevin, who taught me how to use it.

Contents

Preface		vii
How to Use This Book		1
Chapter One	The Chemistry of Plastics: A Primer	4
Chapter Two	Principles of Combustion	21
Chapter Three	Plastics: An Overview	41
Chapter Four	Descriptions of Individual Plastics	50
Chapter Five	Additives: Other Chemicals That Might be Found in Plastics	110
Chapter Six	The Combustion of Plastics	122
Chapter Seven	Flame Retardants and Flame Retarded Plastics	156
Chapter Eight	Fire Tests	180
Chapter Nine	Toxicity Testing	221

Appendixes

A. Abbreviations for Plastics 236

B. Plastics Trade Names 243

C. Glossary 267

D. Fire Tests and Miscellaneous Tables 289

E. References 310

F. Addresses of International Code and Standards Setting Organizations 314

G. Abbreviations for Code Bodies and Standards Setting Organizations 317

Index 319

Preface

Although many people may want to know how plastics burn, the question of how plastics burn is only a part of a much larger question: "How does *anything* burn?" To examine the burning process of plastics in isolation does not make any sense and may lend credence to the idea that plastics are a special material with a special hazard. An analogy would be to study how "carbonless" carbon paper burns and then to relate the burning characteristics and combustion products of this single type of paper to all other types of paper, and even to other wood products and to wood itself.

The testing of one type of product from a very large family of products that are physically similar, and the relation of these test results to all other members of that family, has no basis in the scientific method. Those who do so are ignorant of science—unless they have some hidden reason for wanting to ascribe certain properties to all related materials.

Differences among the many kinds of plastics are as large as or larger than the differences found among other groups of materials such as glass, wood, wood products, and metals. Then, consider the myriad of additives and processing techniques employed by the plastics industry and you will soon realize that an almost inifinite number of properties can be built into certain plastics. When polymers are blended, or alloyed, the combinations and permutations of different compounds with different properties becomes mind-boggling. And accompanying these different compositions and different properties are differing combustion properties and combustion products related not only to the composition of the plastic blend or alloy but to its shape, its use, and the many other factors that determine the manner in which any material burns. Anyone who simply burns an article made of plastic and then claims to know all there is to know about how it burns and what all of its products of combustion are deserves an Olympic gold medal in conclusion jumping.

In other words, "Which burns faster, a pound of plastic or a pound of wood?" is a simplistic question. One cannot begin to answer it until the type of plastic *and* the type of wood have been determined. Then one needs to know the polymerization process or the molecular weight of the plasic or establish its proper place within the plastics family. Next, it must be determined whether the plastic material from which it was made was a single polymer or was copolymerized. Then come more questions. Is the plastic an alloy or blend of polymers? Does it contain additives that would speed up or slow down combustion? How large is the object? What is its shape? Both attributes influence how *any* material burns. In actual use, will the plastic part stand alone? Or will it be laminated with or in some other way be in contact with another object made from the same or different material?

Then, the type of ignition source must be known. Are additional materials going to be involved in the fire (in the real world, they usually are)? Additional properties of the plastic would then have to be defined: ease of ignition, flame spread, rate of heat release, mass loss, or gas generation—all the many factors related to the combustion of any material.

Then most of the same questions would have to be addressed to the pound of wood. Is it a single kind (like lumber) or a blend, like plywood or composition board? What type is it: pine? oak? walnut? Is it "green" or aged? Is it high or low in water content? What are the dimensions and shape?

Once all these variables have been determined *scientifically,* by an accepted, standardized procedure, a large question remains. *Did the tests reproduce what would happen in a real, accidental, unwanted fire?* Or only what happens in a contrived fire, large or small? Which burns faster—wood or plastic? It seems unlikely that a correct, scientific answer will ever be found, because the simple question raises enormously complicated issues. It *has* no simple answer. Those who argue that it does usually have an emotional or economic (or both) reason for their untenable position.

This book examines the polymers, blends, alloys, copolymers, and compounded plastics known to the author at the time of its writing. The plastics industry is complex and subject to constantly changing demands for materials engineered to do specific tasks. Some products will become obsolete and new products will be introduced almost daily. The area of combustion and toxicity testing is also evolving. Every effort has been made to include all tests currently in use or being developed and to avoid editorializing about their practicality.

The author's goal was to report accurately the current state of affairs in the study of the combustibility of plastics, while bearing in mind the constant change occurring in the marketplace. To do this I have focused on the fundamentals of plastics composition and the principles of ignition and combustion. These subjects are presented in such a way that they can be applied to all foreseeable combinations of polymers and additives.

My intent was not to present a scholarly or scientific tome, but rather to present the material as straightforwardly as possible, eliminating many of the chemical, theoretical, or highly scientific aspects of the polymer industry. I have simplified (perhaps in some cases oversimplified) an extremely complex subject, taking some shortcuts, regrouping materials, and otherwise departing from the way information about plastics is usually presented. This book was not written for polymer chemists but rather for those who have no (or only a very slight) working knowledge of plastics, and little or no knowledge of chemistry: architects, builders, building code writers and inspectors, fire personnel (including fire investigators, fire marshals, fire prevention officers, and firefighters), legislators, and anyone else who has an interest in plastics. This book is about how plastics burn. The information is presented with that purpose continually in view.

How to Use This Book

Plastics are arguably the most engineered group of materials ever made by humans. Each polymer, each resin, each blend, each compound, and each alloy is designed to do a specific job and to perform a specific function. There is very little serendipity in the field of plastics. Polymer engineering is a recognized and respected profession, and it is practiced for one reason only: to design and produce a specific polymer, resin, blend, compound, or alloy that will fill a specific need in the marketplace, as demanded by that marketplace.

As a subject to be discussed in a small space or relatively short time, plastics is not a simple topic. To address this complex and rapidly changing field properly, it will be necessary to break it down into simple, easy-to-manage pieces. We will start with a brief introduction to the subject of chemistry. Sometimes the very mention of this word turns people off, especially those who have chosen to stay away from this subject all of their lives. The material is presented as simply as possible, however, and readers who pay attention will absorb enough chemistry to understand the terms used in the rest of the book.

Before the combustion of any particular material can be discussed, the process of combustion itself must be discussed. This subject is covered next in Chapter 2. Like the subject of plastics, the subject of combustion is a complicated one. Literally thousands of chemical reactions occur spontaneously and simultaneously in a fire, especially a large one. We will break the subject of combustion into simple parts, and two simple, straightforward theories will be presented and explained. It is important to understand the factors determining the combustibility of *all* materials. Once those factors have been introduced and explained, their effect on plastics will be compared with their effects on other "traditional" materials. The subject of pyrolysis and how plastics pyrolyze is introduced in Chapter 2. This topic is important because most solid materials (excluding metals) undergo pyrolysis, and although plastics pyrolyze, they behave somewhat differently than do other solid (nonmetal) combustible materials.

Chapter 3 presents an overview of the main subject, plastics, introducing and explaining the concept of polymers. Some terminology may be new to some readers, but each term is defined as it is introduced and defined again in a glossary presented at the end of the book.

Chapter 4 describes each major, commercially important polymer and briefly discusses its properties and uses. It is followed by a short chapter describing the different chemicals that are added to plastics to give the resulting compound the properties desired by the customer.

Chapter 6 introduces the factors that determine the combustibility of plastics and presents the combustion characteristics of each polymer. The chemistry of each polymer is simply presented in a such a manner that *no* prior knowledge of chemistry is required to understand it. The chemical makeup of the materials that are used to make the plastics, the materials that are added to them later, and the manner in which the finished part is used and subsequently exposed to fire are extremely important to know. If you want to see how an individual plastic burns, obtain an article made from that plastic *only* and burn it—either outdoors or in a laboratory smoke hood. But, as is emphasized over and over in this book, this is not a scientific test and has little or no bearing on how the plastic will behave in a real fire. The reasons are discussed elsewhere in this book.

Once the material has been burned, the concern usually shifts to the combustion products of the burned product: smoke and gases. These are discussed together with the manner in which each individual plastic burns. Information about chemistry and combustion from the first two chapters can be used to make rather accurate predictions about the composition of the smoke and fire gases released during combustion of an individual plastic material. Such predictions are of vital importance to occupants of a burning structure or vehicle, who are exposed to the products of combustion of *all* the materials (both natural and synthetic) being consumed in the fire. It is also of vital importance to emergency personnel who respond to the fire and do not wear protective gear.

Chapter 7 discusses the use of flame retardants to reduce the combustibility of plastics and what this means in terms of how these materials can then be used. The most important flame-retarded plastics are briefly described.

Chapter 8 concerns what is probably the most complicated and sensitive subject in the entire field of plastics: testing for combustibility. Arguments abound concerning the relevance of *any* small-scale test involving only the plastic to be tested, as compared with a full-scale test using finished parts and systems that will more closely approximate the conditions found in an actual accidental or unwanted fire. The arguments for both sides are presented.

The subject of testing leads naturally to the topic of the toxicity of combustion products, which is presented in Chapter 9. The tests used to determine plastics toxicity are introduced and discussed.

Appendices A through G provide information on uses of plastics, abbreviations and trade names used in the plastics industry, and the glossaries, as well as tables containing results of tests for combustibility and other thermal properties of plastics.

From place to place within this book, you will notice what appears to be repetition of certain statements, facts, or warnings. This repetition is deliberate, for two reasons. First, in approaching such an emotional subject as the combustion of plastics, it is critical that certain concepts be repeated wherever certain topics are discussed. Second, many readers will read this text selectively in terms of their own interests and needs. Readers who use the text as a reference may not see a fact or warning presented earlier. The author hopes that this repetition will not offend readers who encounter it but will simply reinforce its importance.

1

The Chemistry of Plastics: A Primer

You need not become a polymer chemist to be able to understand the structure of polymers and what they are made of. To be able to look at a molecular or structural formula of a polymer and to understand what elements make up the polymer chain does not even require you be any kind of chemist. In fact, to understand what happens when plastics burn you need not even have taken a course in chemistry at any time in your educational career. Just read the material in this and the next chapter carefully, and you will find everything you need to understand the rest of this book.

One warning, however. The material will *not* be presented as if you were enrolled in a formal chemistry class. This book is intended as a primer in plastics and how they burn, and it will *not* delve deeply into the intricate chemistry that is required of the polymer chemist. I apologize here, up front, to chemists and others who may find the material oversimplified. Some liberties were taken in the name of clarification and understanding for the nontechnical readers of this book.

Elsewhere in this book, I question the work being done by some individuals in studying the combustion properties of plastics. My argument is that too many steps and processes have been simplified only because studying all aspects of the phenomenon of fire as it involves plastics would require impossibly complicated experiments. Yet here, the author appears to be arguing from an opposing point of view: that the presentation of certain aspects of chemistry should be simplified and even modified somewhat for the readers of this book.

There is a great difference between simplifying the presentation of information concerning a *science* such as chemistry and simplifying the presentation of information surrounding the *art* of combustion testing. That difference is simply that the science of chemistry (*science* being defined as a systemized body of knowledge, or, as the *American Heritage Dictionary* has it, "the observation, identification, description, experimental investigation and theoretical explanation of natural phenomena") is well established and accepted, particularly the basic principles covered here. Chemical reactions and the periodic table of the elements have been studied for many years by tens of thousands of accomplished scientists. What is presented in the following pages is not a bending of the facts to prove an argument but an attempt to make what *is* fact understandable to individuals who may have had little or no formal scientific education. There is no hidden agenda here, and the method of presenting the material has been proved successful in teaching chemistry to many students with little or no chemistry background.

On the other hand, the decomposition and combustion processes through which plastics pass are extremely complicated, and experimenters are still trying to develop tests that not only make sense but correctly identify what really happens when these processes occur. This makes the activity an art rather than a science. There is no well-established set of directions to determine the correct combustion characteristics of plastics through which experimenters can travel to be sure they have captured what actually happens—and captured it correctly. There are, nevertheless, experimenters who say they can leave out this step or that measurement and still come up with correct information. This simply cannot happen in the study of chemistry or the study of complex chemical reactions such as fire.

The subject of chemistry is very large and complex. However, if one can conquer the basic principles of any subject, that person will be able to understand any part of that subject that relates to those basic principles. If the discussion of any complicated subject does not go beyond that first, basic level, the student cannot get lost. This is only common sense, and common sense is the foundation upon which the following is built.

One must grasp only a few basic principles to be able to understand a simplified chemistry of polymers, and therefore the simplified chemistry of plastics. These principles will be accompanied by definitions, and each will be tied to the others before the next step is taken.

Chemistry is the science of matter, energy, and reactions. All matter is made up of certain building blocks, and these chemical building blocks react (hence the word "reaction") in certain ways when energy is applied to or taken from them. Chemical reactions occur because of the manner in which a particular building block is structured and the amount of energy involved in the desired reaction. Very specific things happen to very specific chemicals when they are brought together in a very specific way, and the way the building block (as a raw material in the reaction) is constructed is fundamental to these phenomena. That construction is the focus of this chapter.

ELEMENTS: THE BUILDING BLOCKS OF MATTER

All matter is composed of three things: pure substances (the basic building blocks which are called elements), compounds made from those building blocks (also pure substances), and mixtures of those pure substances. It is now time to identify these substances and show how and why they become **reactants** (raw materials that undergo chemical reactions).

Elements

The basic pure substances in the universe are called **elements**. Elements are the basic building blocks of everything that exists, everywhere on Earth and everywhere outside it. An element is defined as a pure substance that cannot be broken down into simpler pure substances by *chemical* means. The emphasis on chemical procedures suggests that there are other methods by which elements are broken down, but only the chemical approach is important in defining elements and their chemistry. Many of the elements have names that are familiar to most of us: as oxygen, hydrogen, nitrogen, carbon, gold, silver, copper, and so on. There are 103 elements that have been identified (more, depending on the chemistry book one might have), and everything in the universe as we understand it is made up of one or more of these basic materials—and only these basic materials. Nothing can be found that is not made up of elements—even if a new element must be identified to solve the mysterious composition of a particular strange material.

You can find the names of all the elements listed in any textbook on chemistry, and you can find such a book at your local library or at any high school or college library. The text will list the elements in alphabetical order or in some other grouping that the author of that book feels will make sense to the student of chemistry. However, one thing that is common to all chemistry textbooks is the presentation of the elements in a device called the periodic table of the elements. The periodic table (for short) is a document that displays all the known elements according to a particular property known as atomic number (defined later). When arranged according to electronic structure (also defined later), the elements line up in groups or "families" that all have somewhat similar chemical properties. You need not be familiar with the entire periodic table. For now, if you accept the presence of elements in the universe, and their role as the basic building blocks, all chemical substances (not just polymers) will be more understandable to you.

THE PERIODIC TABLE OF THE ELEMENTS

In a periodic table, each element is listed in a box that contains certain information about that element. In the simplest of periodic tables (presented here as Figure 1.1), the only information contained in the box is the name of the element (on some tables, even the name is omitted), its atomic number, its atomic weight (defined later), and its chemical symbol. The chemical symbol is a kind of shorthand derived from the element's English name or the original Latin or Greek name. Examples of chemical symbols are C for carbon, H for hydrogen, N for nitrogen, O for oxygen, Cl for chlorine, and Fe for iron. The symbol stands for the name of the element and also for one atom of that element. The symbols for all the elements are given in Table 1.1. The periodic table is broken down into columns, each of which represents a "family" or group of elements. There are some other divisions of the periodic table, but that is beyond the scope of this book.

Elements are classified as metals or nonmetals on the basis of the chemical reactions they undergo. By far the greater number of elements are metals, so it is much simpler to name the 22 nonmetallic elements. They are: argon, arsenic, astatine, boron, bromine, carbon, chlorine, fluorine, helium, hydrogen, iodine, krypton, neon, nitrogen, oxygen, phosphorus, radon, selenium, silicon, sulfur, tellurium, and xenon. All the other 81 elements are metals.

Besides being metallic or nometallic, elements may be gases, liquids, or solids. The gaseous elements (all nonmetals) are hydrogen, nitrogen, oxygen, fluorine, chlorine, helium, neon, argon, krypton, xenon, and radon. The liquid elements are mercury, a metal, and bromine, a nonmetal. All the other elements are solids.

ATOMS AND SUBATOMIC PARTICLES

As stated above, the symbol for an element also stands for one atom of that element. An atom is defined as the smallest particle of an element that can still be identified as the element. Therefore, the atom is the unit particle, the simplest particle of an element. Elements are pure substances, therefore an atom is a pure substance. An element is made up of its own atoms and nothing else. An atom, which is made up of even smaller particles, cannot be broken down into those smaller parts chemically. To break down, or "smash" an atom, physical means rather than chemical means must be employed.

The atom of each element is unique to that element. That is, all atoms of an element may be identified as being an atom of that particular element. This uniqueness is due to the presence of certain particles within the nucleus of an atom. The nucleus of an atom is the subatomic particle that is located at the

Group IA	IIA	III B	IV B	V B	VI B	VII B	VIII	VIII	VIII	I B	II B	III A	IV A	V A	VI A	VII A	VIII A
1 H 1.008																	2 He 4.0026
3 Li 6.941	4 Be 9.01											5 B 10.81	6 C 12.01	7 N 14.007	8 O 15.999	9 F 18.998	10 Ne 20.179
11 Na 22.99	12 Mg 24.305											13 Al 26.98	14 Si 28.086	15 P 30.974	16 S 32.06	17 Cl 35.453	18 Ar 39.95
19 K 39.102	20 Ca 40.08	21 Sc 44.956	22 Ti 47.9	23 V 50.941	24 Cr 51.996	25 Mn 54.938	26 Fe 55.847	27 Co 58.933	28 Ni 58.71	29 Cu 63.546	30 Zn 65.37	31 Ga 69.72	32 Ge 75.59	33 As 74.92	34 Se 78.96	35 Br 79.904	36 Kr 83.80
37 Rb 85.468	38 Sr 87.62	39 Y 88.906	40 Zr 91.22	41 Nb 92.906	42 Mo 95.94	43 Tc 98.906	44 Ru 101.07	45 Rh 102.91	46 Pd 106.4	47 Ag 107.87	48 Cd 112.40	49 In 114.82	50 Sn 118.69	51 Sb 121.75	52 Te 127.6	53 I 126.90	54 Xe 131.3
55 Cs 132.91	56 Ba 137.34	57 La 138.91	72 Hf 178.49	73 Ta 180.95	74 W 183.85	75 Re 186.2	76 Os 190.2	77 Ir 192.22	78 Pt 195.09	79 Au 196.97	80 Hg 200.59	81 Tl 204.37	82 Pb 207.2	83 Bi 208.98	84 Po (210)	85 At (210)	86 Rn (222)
87 Fr (223)	88 Ra 226.03	89 Ac (227)	104														

Lanthanide series	58 Ce 140.12	59 Pr 140.91	60 Nd 144.24	61 Pm (147)	62 Sm 150.4	63 Eu 151.96	64 Gd 157.25	65 Tb 158.93	66 Dy 162.50	67 Ho 164.93	68 Er 167.26	69 Tm 168.93	70 Yb 173.04	71 Lu 174.97
Actinide series	90 Th 232.04	91 Pa 231.04	92 U 238.03	93 Np 237.05	94 Pu 239.05	95 Am (243)	96 Cm (247)	97 Bk (245)	98 Cf (248)	99 Es (254)	100 Fm (253)	101 Md (256)	102 No (254)	103 Lw (257)

Figure 1.1 The periodic table of the elements

TABLE 1.1 The Elements and Their Symbols

Name	Symbol	Name	Symbol	Name	Symbol
Actinium	Ac	Hafnium	Hf	Praseodymium	Pr
Aluminum	Al	Helium	He	Promethium	Pm
Americium	Am	Holmium	Ho	Protactinium	Pa
Antimony	Sb	Hydrogen	H	Radium	Ra
Argon	Ar	Indium	In	Radon	Rn
Arsenic	As	Iodine	I	Rhenium	Re
Astatine	At	Iridium	Ir	Rhodium	Rh
Barium	Ba	Iron	Fe	Rubidium	Rb
Berkelium	Bk	Krypton	Kr	Ruthenium	Ru
Beryllium	Be	Lanthanum	La	Samarium	Sm
Bismuth	Bi	Lawrencium	Lr	Scandium	Sc
Boron	Bo	Lead	Pb	Selenium	Se
Bromine	Br	Lithium	Li	Silicon	Si
Cadmium	Cd	Lutetium	Lu	Silver	Ag
Calcium	Ca	Magnesium	Mg	Sodium	Na
Californium	Cf	Manganese	Mn	Strontium	Sr
Carbon	C	Mendelevium	Md	Sulfur	S
Cerium	Ce	Mercury	Hg	Tantalum	Ta
Cesium	Cs	Molybdenum	Mo	Technetium	Tc
Chlorine	Cl	Neodymium	Nd	Tellurium	Te
Chromium	Cr	Neon	Ne	Terbium	Tb
Cobalt	Co	Neptunium	Np	Thallium	Tl
Copper	Cu	Nickel	Ni	Thorium	Th
Curium	Cm	Niobium	Nb	Thulium	Tm
Dysprosium	Dy	Nitrogen	N	Tin	Sn
Einsteinium	Es	Nobelium	No	Titanium	Ti
Erbium	Er	Osmium	Os	Tungsten	W
Europium	Eu	Oxygen	O	Uranium	U
Fermium	Fm	Palladium	Pd	Vanadium	V
Fluorine	F	Phosphorus	P	Xenon	Xe
Francium	Fr	Platinum	Pt	Ytterbium	Yb
Gadolinium	Gd	Plutonium	Pu	Yttrium	Y
Gallium	Ga	Polonium	Po	Zinc	Zn
Germanium	Ge	Potassium	K	Zirconium	Zr
Gold	Au				

center of the atom and contains almost all the weight of the atom. Within this nucleus of the atom are nuclear particles called protons. They have a positive electrical charge and a weight defined as one "atomic mass unit" (a.m.u. or atomic weight). The number of protons in the nucleus of the atoms of an element give that element its unique atomic number. That is, the atomic number of an element is defined as the number of protons in the nucleus of each of its atoms.

Also within the nucleus (of every atom of every element except atoms of nonradioactive hydrogen) is another nuclear particle called a neutron. Neutrons are defined as nuclear particles with an atomic weight of 1 (equal to the weight or mass of the proton) but with no electrical charge (they are electrically neutral). The atomic weight of an element is defined as equal to the total amount of protons and neutrons in one of its atoms. There is one proton and no neutrons in the nucleus of the stable isotope (defined below) of hydrogen, so its atomic number is 1 and its atomic weight is 1. Fluorine, on the other hand, has nine protons in its nucleus (hence its atomic number of 9) and ten neutrons (giving it an atomic weight of 19).

There are certain atoms of certain elements that have more neutrons in the nucleus than the "stable" atoms of that element. These are the radioactive isotopes of the element. An **isotope** is defined as an atom of the same element (that is, it has the same number of protons as do all the other atoms of that element) that has a different number of neutrons in the nucleus. The number of neutrons in the nucleus of atoms of the same element may vary (producing isotopes), but the number of protons in *all* the atoms of the same element is always the same. If somehow the number of protons in the nucleus of an atom changes, the atom changes and becomes the atom of another element. Which element it becomes depends upon the new number of protons in the nucleus, and the element is determined by the new atomic number.

The existence of isotopes is what causes the listed atomic weight of the elements to be other than whole numbers. The calculated atomic weight that is listed for each element is a weighted average of all the atoms of that element, which will include stable and radioactive isotopes, each with a different atomic weight. For ease of calculation, the atomic weight of each element considered to be its closest whole number (with the exception of chlorine, whose atomic weight is usually considered to be 35.5). Therefore, atomic weight of oxygen is considered to be 16, the weight of nitrogen is 14, hydrogen is 1, and carbon is 12.

Orbiting around the nucleus of each atom are particles called **electrons,** the number of which is equal to the number of protons in the nucleus. The electron is so tiny that it has almost no appreciable weight. However, it does have an electrical charge equal to but opposite in sign (negative) to the proton. Therefore, since there are eight protons in the nuclei of all the atoms of oxygen (oxygen's atomic number is 8), there are eight electrons in orbit around the nucleus of each of oxygen's atoms. There are also eight neutrons in the nucleus of most of oxygen's atoms, so its atomic weight is 16 (8 protons plus 8 neutrons equals 16).

Since protons have a positive charge of 1, the nucleus of all atoms is always positively charged, the size of such charge being equal to the number of protons in its nucleus. Since the electron has a negative charge of 1, and there are the same number of electrons in orbit as there are protons in the nucleus, these charges attract and balance each other out, and the net electrical charge on all atoms is zero. An unbreakable law of nature is that all atoms are electrically neutral.

Figure 1.2 is a drawing of the oxygen atom, looking like some miniature solar system, with the nucleus (containing eight protons and eight neutrons) in the center and eight electrons in orbit. Remember, no one has ever seen an atom of oxygen, so the drawing is presented in a manner that will clearly separate the various particles. The current theory is that electrons are not distinct, individual particles but exist as electron "clouds." For our purposes, we will consider electrons as separate and distinct particles, leaving other theories for more complicated, formal chemistry classes.

Each atom has its own electronic configuration, or electronic structure, defined as the manner in which the electrons arrange themselves in orbit around its nucleus. There are specific rules about how many electrons are in each orbit (or orbital, shown as "rings" around the nucleus). It is this electronic configuration, particularly the electrons in the outer ring, that determines the chemistry of each element. Even more specifically, those electrons in the outer ring that are unpaired (electrons have a tendency to pair up with each other, and when they do, they are not available for chemical reactions) are the electrons that determine what the element will do in a chemical reaction. Again, to go any further into

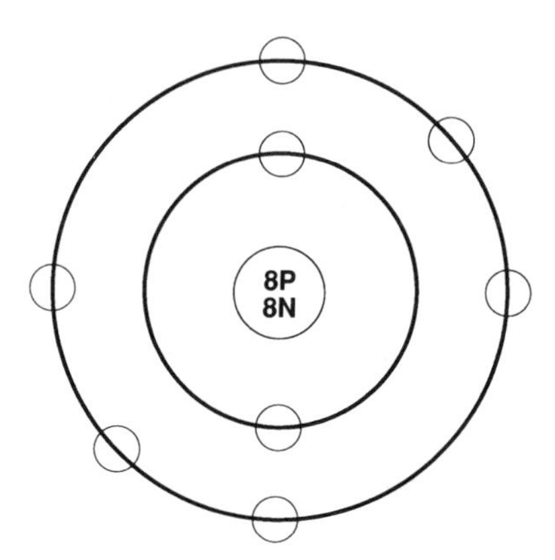

Figure 1.2 The oxygen atom

electronic configuration would be to move beyond the intended scope of this book. However, if you are interested in learning how to predict what reactions occur and what determines the chemical formula for any simple compound, a simplified explanation can be found in chapters 1, 2, and 3 of *The Common Sense Approach to Hazardous Materials,* by this author. It is published by Fire Engineering, a division of PennWell Publishing Company, 1421 South Sheridan, Tulsa, OK 74112.

COMPOUNDS

The word "compound" will appear many times in this book. It has two meanings, and they have to be differentiated here.

A **chemical compound** is a pure substance made from a chemical reaction that started with atoms of elements, other chemical compounds, or a combination of both elements and compounds. A chemical compound may be either organic (covalent) or inorganic (ionic). If it is organic, or covalent, it is made up of two or more atoms of the same or different elements. An example is atmospheric oxygen, whose chemical formula (a chemical or molecular formula shows, in writing, what atoms or ions are present and in what quantity) is O_2. The chemical *symbol* for the element oxygen is O, while the chemical *formula* for the covalent compound is O_2. This means that two atoms of oxygen have combined (with two covalent bonds between them) chemically with each other to form a chemical compound. The identifying characteristic of the organic or covalent compound is that it is made up of atoms that are connected to other atoms by the sharing of one or more pairs of electrons (each shared pair of electrons between the atoms is a **covalent bond**). This method of bonding is what sets it apart from the next example, an inorganic, or ionic, chemical compound.

The chemical formula for sodium chloride (common table salt) is NaCl. It is not obvious from the formula that the Na (the chemical symbol for sodium) and the Cl (the chemical symbol for chlorine) do not represent one atom of sodium and one atom of chlorine. Those readers who have had previous education in chemistry are aware of what metal and nonmetal elements are, and they know that when metals and non-metals react with each other to form a compound, it is always ionic in form. Therefore, the sodium really exists as Na^{+1}, which is the chemical formula for the sodium ion, and the chlorine really exists as Cl^{-1}, which is the chemical formula for the chloride ion. The chloride ion has a negative (–) electrical charge of 1 because the chlorine atom has gained one electron (1 negative charge) from the sodium atom. This "taking" of one electron from the sodium atom converts it to a sodium ion, which has a positive (+) electrical charge of 1 because it has lost one electron (1 negative charge) to the chlorine atom.

The difference in bonding between the two types of chemical compounds is the difference between ionic and covalent bonding, but they are both chemical compounds. In this book, when the compound under discussion is a chemical compound, the words "chemical compound" or "ionic compound" or "covalent compound" will be used as appropriate.

Again, it is important to remember that chemical compounds are pure substances, just as elements are. That is, chemical compounds are substances created by a chemical reaction, and, if they are covalent chemical compounds, they contain only molecules of the chemical compound. Each molecule of that covalent chemical compound may always be identified as belonging to that chemical compound. In fact, a molecule is defined as the smallest particle of a covalent chemical compound that may still be identified as being part of that chemical compound. If the chemical compound is ionic, *ions,* not molecules, will be present. The only ions present will be those making up that chemical compound.

This discussion of chemical compounds as pure substances is necessary to distinguish them from **mixtures,** which are physical mixtures or blends of different chemical compounds, different elements, or chemical compounds and elements or other mixtures. Examples of chemical compounds are water, atmospheric oxygen, carbon monoxide, carbon dioxide, and sodium chloride. Examples of mixtures are air, wood, milk, blood, gasoline, and kerosene. If the mixture is one of metals only, that have been blended together in the molten state, it is called an alloy.

The other kind of compound that will be discussed in this book is a **plastic compound.** In this context, a compound is a resin or polymer to which another substance has been added. The resin or polymer may have begun as a chemical compound—for example, polyethylene—but some other substance was added to it to give it specific properties. That other substance might have been a pigment to give it color, or it may have been an ultraviolet absorber to allow the plastic article made from polyethylene to be used outside without rapid degradation. In either case, a pure material, polyethylene, has been modified by the addition of another substance. The polyethylene is now referred to as a polyethylene (or other plastic) compound. Plastic compounds are always mixtures and are never pure substances.

In some cases, two or more different polymers may be blended and melted together. These physical mixtures are called **alloys** (just as in the physical mixing of metals mentioned above). Alloys differ from **copolymers** which are polymers made from copolymerizing (polymerizing together) two different monomers. Polymerization, explained further in the next chapter, is a special type of chemical reaction (as opposed to the physical reaction of melting and mixing), in which a very special type of chemical compound is used. The chemical compound, called a **monomer,** is special because, under the proper conditions, it can react with itself. Out of over ten million chemicals that have been isolated, identified, or synthesized and named, probably less than three hundred chemical compounds classify as monomers. Other chemical compounds, when subjected to the heat and/or pressure to which monomers are subjected, will either break down into decomposition products, re-form into new chemical compounds, or not react at all.

These special chemical compounds or monomers, usually containing a double covalent bond (two covalent bonds existing between the same two atoms), with the right amount of heat and pressure, will react with and add onto themselves to form a much larger chemical compound called a **polymer.** The chemical reaction in which this takes place is called a polymerization reaction. It takes place in a large, sealed pressurized container (called a polymerization reactor) that can be heated or cooled, as necessary. Polymerization is a very specialized type of chemical reaction and will be discussed further in Chapter 4.

To summarize the rather complicated subject of compounds, remember that we will be discussing two kinds of compounds, chemical and plastic. Chemical compounds are always pure substances, made up only of atoms or ions. Plastic compounds are always mixtures, because chemical compounds are added to the pure polymer (a chemical compound) to alter it in some way. What makes it confusing is that the starting chemical, the monomer, is a chemical compound (a pure substance). When it is polymerized, it forms a new chemical compound, a polymer, which, since it is a chemical compound, is a pure substance. This pure substance, this polymer, then has some additive mixed in with it. When this happens, the resulting mixture (no longer a pure substance) becomes a plastic compound.

ORGANIC VERSUS INORGANIC

The science of chemistry has two great divisions, generally called organic and inorganic chemistry. Organic chemistry originally was concerned with the chemistry of chemical compounds that came from living things, and was once defined as the chemistry of living things. We humans, however, have not only learned to synthesize (produce in the laboratory) those chemical compounds once produced only in nature, but we have moved a great deal beyond that. Once early chemists discovered that carbon has the unique capability of combining with itself in long chains, it was not difficult for them to learn how to duplicate nature's compounds and then to construct chemicals not known in nature at all.

Today, organic chemistry is better described as chemistry based on the chemical compounds of the element carbon. All organic compounds contain carbon. (A tiny percentage of nonorganic chemical compounds also contain carbon.) With the exception of some silicone polymers that are based on the element silicon rather than on carbon, polymers are organic: They are based on carbon and chemical compounds of carbon. The force that holds these chemical compounds (made up of atoms) together is the covalent bond. A **covalent bond** is the pairing up of electrons *between* atoms (rather than pairing up in the orbit of one atom).

Inorganic chemistry, on the other hand, is considered to be the chemistry of minerals and salts, including chemical compounds made from them. These chemical compounds are often referred to as ionic, because they are made up of ions rather than atoms. Ions are atoms or groups of atoms bound together chemically that have gained or lost one or more electrons and therefore possess an electrical charge. The force that holds these ionic chemical compounds together is the electrostatic attraction of opposite electrical charges created by the loss (resulting in a positive charge equal to the number of electrons lost) or gain (resulting in a corresponding negative charge) of electrons.

Since polymers are organic (again, silicones being the only exception), our discussions will focus on covalent chemistry and the covalent bonds that hold these compounds together. These bonds are sources of stored-up energy, since energy had to be used to form them. This is true for natural chemical compounds as well as synthetic chemical compounds. It is the release of energy from these bonds when they break that produces the heat and light (two forms of energy) in a fire.

CHEMICAL BONDS

To be sure you understand chemical bonding, a quick review is in order. Chemical bonds are the forces that hold chemical compounds together. There are two types the casual student of simplified chemistry must be familiar with: ionic bonds and covalent bonds.

The **ionic bond** is simply the attraction of opposite charges of electricity. An ionic compound contains: two types of ions, metallic (positively charged, or cations) and nonmetallic (negatively charged, or anions). These ions are held together by the electrostatic attraction of the opposite charges, always in the proper ratio to make the final chemical compound electrically neutral. The ionic bond is very strong, but it is overcome if the ionic compound dissolves in water (forming a solution that conducts electricity). The water that dissolved this ionic chemical now contains the positive and negative ions of that chemical compound moving about freely in the solution. If the water is evaporated, the positive and negative ions will rejoin to re-form the solid ionic compound.

On the other hand, the covalent bond is totally different. There are no electrically charged particles, and no parting or relaxing of the bonds if the chemical compound dissolves in water. The **covalent bond** is formed by the pairing up of two electrons between the atoms involved (only nonmetallic elements reacting with each other can form covalent bonds). There may be one pair of electrons (a single bond), two pairs (a double bond), or three pairs (a triple bond) between atoms. Certain elements may form only one covalent bond with another element, some may form single or double bonds, and some may form single, double, or triple bonds. These will be enumerated below.

MOLECULES

A **molecule** is defined as the smallest unit of a covalent chemical compound that may still be identified as that chemical compound. It is a chemical combination of two or more atoms of the same or different elements bound together chemically by the covalent bond—already defined as the sharing of a pair of electrons between two atoms. Since all atoms are electrically neutral, and molecules are made up of atoms, all molecules must then be electrically neutral.

Molecules, unlike atoms, may be broken down into simpler substances (the atoms that make up the molecule) by chemical means. Molecules may be changed in a chemical reaction, releasing the atoms from their original chemical combination to form new molecules with different combinations of the same atoms. Thus, molecules of carbon monoxide may be burned (combined with oxygen or oxidized) to form new molecules of carbon dioxide. Methane, which contains carbon and hydrogen and whose molecular formula is CH_4, when burned in air will form molecules of water (H_2O), carbon dioxide (CO_2), and carbon monoxide (CO). If the flame is hot enough, nitrogen, N_2, in the air will be oxidized to nitrogen dioxide (NO_2). If combustion is carried out in an oxygen-poor atmosphere, free carbon (C) will be liberated. The point is, these molecules (all except for carbon, which is an atom), are made up of atoms that are joined together, and they may be broken apart and the atoms may be rearranged into new molecules.

Molecules may be very small (smallness is relative; all molecules are small, with only the largest recently being viewed through very powerful electron scanning microscopes), like the hydrogen molecule (H_2) or the ethylene molecule (C_2H_4), or very large, like the polyethylene molecule, made up of thousands of repeating —C_2H_4— units. Molecules are the building blocks of polymers, so most of the attention in this book is directed toward chemical reactions involving the molecules of monomers and polymers. It is important to know what elements are represented in each molecule of a monomer, since these will then be the only elements present in the polymer. It will be from the knowledge of what elements are present in the molecules of the polymer that one will be able to conclude what possible combustion products are formed when a material burns. Other chemical compounds may be introduced into the polymer through additives, but they usually will be present in very small quantities. None of the additives that are ionic in composition will burn. Ionic compounds may be identified by the presence of a metal in the chemical compound.

ELEMENTS INVOLVED IN POLYMERIZATION

Only a few of the 103 known elements can be involved in the polymerization process. This is because only the nonmetals can form covalent compounds, and all the monomers are covalent chemical compounds. The nonmetallic elements (with their chemical symbols in parentheses) that are found most often in monomers and therefore the resulting polymers are carbon (C), hydrogen (H), oxygen (O), nitrogen (N), chlorine (Cl), fluorine (F), sulfur (S), and to very small extent, silicon (Si). Other elements make up the additives that go into making plastics compounds; these are discussed in another section. The additive chemical compounds that are covalent chemical compounds are mostly carbon, hydrogen, and oxygen, so they contribute very few additional elements (by weight) to the plastic compound. The mineral (ionic) additives are nonflammable, so they add no fuel in a fire.

The main backbone of the polymer, the carbon chain, *is* the fuel, however. All polymers are made up of repeating units of carbon-to-carbon linkages, with other atoms bonded to the carbon atoms. The carbon atoms are capable of forming (and will always form) four covalent bonds with other atoms (including other carbon atoms). These bonds will be either single, double, or triple covalent bonds, depending on how many bonds the atom to which carbon is bonded can form. Silicon, the element directly below (and therefore in the same "family" as carbon) on the periodic table can also form four covalent bonds, and it does so in exactly the same manner as carbon. Hydrogen, fluorine, and chlorine can form only one covalent bond, so there will always be a single bond between carbon and any of these three elements, or between these three elements and the atoms of any other element. Oxygen and sulfur can and will form two covalent bonds, so there will be either a single or double bond between them and carbon. The bond will be a double bond if there are no other atoms bonded to the oxygen or sulfur, and it will be a single bond if there is another atom bonded to these atoms. Nitrogen can and will form three covalent bonds, so there can be a single, double, or triple bond between it and carbon, again depending on whether or not there are other atoms bound to the nitrogen.

It takes energy to form these covalent bonds, and that energy is stored within the bonds until they are broken by the further input of energy. That is what happens in a fire. An input of energy (from a flame or other ignition or energy source) impinges upon the organic material, and the covalent bonds begin to break. The speed at which they break depends on the intensity of the energy source, the size of the area affected by the energy input, and the nature of the material upon which the energy is impinging.

Since energy may take the form of heat, light, or both, heat and light will both affect the breakdown of bonds. If the energy input is slow, the breakdown is almost imperceptible in the amount of energy released by the breaking bonds and the resulting effect on the material itself. A very slow breakdown of an organic material (not including slow oxidation) cause the material to discolor or the surface to break down slowly. In some cases, input of heat and the subsequent breakdown of covalent bonds may be more rapid than this yet too slow to cause ignition. In this situation, **pyrolysis** will occur. Pyrolysis is the breakdown by heat of covalent bonds, usually in the absence of oxygen. Pyrolysis of thermosetting polymers resembles that of wood, whereas thermoplastics usually melt first, or at least soften to a great degree before bond breakage (pyrolysis) begins to occur.

In the case of a rapid input of a large amount of heat energy, however, the covalent bonds will break more rapidly, and the energy they release is added to the input energy. Pyrolysis is still taking place. As the amount of energy involved in the reaction increases, the speed of the chemical reaction taking place increases, further adding to the energy released. A rule of thumb is that the speed of a chemical reaction will double for every 10°C (18°F) increase in the temperature of the reactants involved. Finally, the reactants absorb enough energy to reach ignition temperature, and combustion begins. The combustion of the material adds still more energy to the system, and the fire builds and keeps building until all available fuel has been consumed. Combustion is defined as the rapid oxidation of a substance, accompanied by the release of energy. A more detailed description of the combustion process is found in Chapter 2.

All of this chemical reactivity depends upon the breaking of the bonds that hold covalent compounds together, and the release of the energy contained therein. This energy can be calculated precisely, since the energies of each type of bond are already known. Table 1.2 lists the bond that involve the type of covalent bonds found in plastics. These bond energies represent the amount of energy contained in (and therefore released in bond breakage) the types of covalent bonds listed. Calories per mole is defined as the number of calories of energy released per atomic weight (expressed in grams) of the compound. The greater the number of bonds, the greater the amount of energy contained in the compound, and therefore the greater the amount of energy released as the material burns. This does not necessarily mean that the material with the greatest amount of energy bound up in covalent bonds will burn hotter than another material. It simply means that the *total heat released* will be higher.

In some situations, a plastic article will be made only from the pure polymer, with no other chemicals added. However, if the color or physical or chemical

TABLE 1.2 Energies of Covalent Bonds

Type of Covalent Bond	Average Bond Energy in Calories/Mole*
Carbon—Carbon (single)	82,600
Carbon=Carbon (double)	145,800
Carbon≡Carbon (triple)	199,600
Carbon—Hydrogen	98,700
Carbon—Fluorine	116,000
Carbon—Chlorine	81,000
Carbon—Oxygen (single)	85,500
Carbon=Oxygen (double)	177,500
Carbon—Sulfur (single)	65,000
Carbon=Sulfur (double)	165,000
Carbon—Nitrogen (single)	72,800
Carbon=Nitrogen (double)	147,000
Carbon≡Nitrogen (triple)	212,600
Carbon—Silicon	78,000

*The definition of a mole is the atomic weight of the material expressed in grams. Actually, this is an outdated definition, but it is more understandable than the current one, which defines a mole as the amount of pure substance containing the same number of chemical units as there are atoms in exactly 12 grams of carbon=12 (i.e., 6.023×10^{23}).

properties of the polymer must be altered to make the part perform or look as the end user of the article desires, other chemicals must be added to the polymer, producing a plastic compound. Those materials are called additives, and each type of additive is added to impart a specific property to the plastic compound. A discussion of these materials is contained in Chapter 3.

SUMMARY

- Elements are the building blocks of nature, and everything in the universe is made up of these basic pure substances.
- There are currently 103 known elements, which are arranged on the periodic table of the elements to explain how each one behaves in a chemical reaction: The letters on the periodic table are the symbols for each of the elements, and they also represent one atom of that element.
- The nucleus of atoms (except for hydrogen) is made up of protons and neutrons, and electrons in orbit around that nucleus.
- Electrons are negatively charged and protons are positively charged, and there are equal numbers of them in each atom, so that all atoms are electrically neutral.

- Atoms will join to form chemical compounds, which are also electrically neutral.
- Most substances are made up of chemical compounds, and they are divided into two classes: organic and inorganic.
- Organic compounds, whose smallest parts are molecules, are held together by covalent bonds, which are the sharing of two electrons between two atoms.
- All polymers are organic compounds and are formed from relatively few elements.

2

Principles of Combustion

Combustion is defined as the rapid oxidation of a substance, accompanied by the equally rapid release of energy. One of the definitions of **oxidation** is the chemical combination of a substance with oxygen. Another definition is that oxidation is a chemical reaction involving oxygen in which one or more substances are combined with oxygen. Neither of these definitions is complete as far as a chemist is concerned, but they will suffice here; although oxidation involves much more than a combination with oxygen, it is beyond the scope of this book to delve much deeper. The specific oxidation reaction called combustion is the reaction meant when fires involving plastics are discussed.

A simpler definition of combustion is that it is also known as **fire**, which is the rapid chemical combination of a substance with oxygen, usually producing both heat and light. Flaming (or space burning) and smoldering (or surface burning) are the two types of combustion possible. For combustion to occur, an oxidizer must be present. Since this book deals with the combustibility of plastics, and the fires that usually involve plastics are "normal" or common structure or vehicle fires, the reader needs to be concerned only with oxygen as the oxidizer.

Almost all fires occur with atmospheric oxygen as the oxidizer, but other oxidizing agents do exist. Most of those oxidizers, when subjected to heat, pressure, or both, will liberate oxygen that will subsequently become available to support the combustion of the fuel. There are other oxidizers, such as the halogens (fluorine, chlorine, bromine, and iodine) that will support combustion, but only combustion with oxygen (either from the atmosphere or liberated from oxidizing agents) as the oxidizer will be discussed here. The halogens are mentioned to make you aware that there are a few oxidizing agents that contain no oxygen.

Deflagrations (defined below) and most explosions are exactly the same type of chemical reaction as fires: They are chemical reactions that involve the rapid oxidation of a material. The speed of these reactions, however, may be hundreds of thousands of times faster than a fire. Explosions of plastics dusts are possible, just as the explosion of any finely divided organic material is possible, but they happen much less frequently than other dust explosions because the dust from plastic resins and compounds is usually not so finely ground as dusts from grain, coal, or other organic materials. **Deflagrations** are oxidation reactions that are significantly faster than a fire but slower than an explosion. To complicate matters, some references claim that a deflagration is an explosion that moves (or is propagated through its mass) with subsonic speed, whereas a detonation is an explosion that moves with supersonic speed. Regardless of these precise scientific definitions, you will surely understand what is meant when the word "explosion" is used.

A **fuel** is defined as anything that will burn. In the context of this book, the fuel involved in the combusion process is a plastic material, whether it exists as a pure polymer, a resin, an alloy, a compound, or as a finished article made of plastic. It may also exist in one of the many forms in which the plastic compound exists, that is, in the shape of the finished article. In fact, the overwhelming majority—99 percent—of real, unwanted, unexpected fires involving plastics involve the finished article.

Therefore, outside of the plastics processing plant, where these resins, alloys, blends, and compounds are converted from liquids, powders, pellets, beads, and cubes into a formed part, fires involving plastics will be assumed to involve plastic parts. Furthermore, aside from a warehouse or other storage area where only plastic parts are present, the combustion of plastics will *always* occur in conjunction with another fuel—which it might also do in the warehouse or other storage area. That is, in a structure fire, plastics make up only a part of the fire load, and many other materials will also be acting as fuels. The ways in which these other materials burn has a great deal of influence on the combustibility of plastics, but they will not be discussed at length here.

THEORIES OF FIRE

To understand how materials burn, very simple theories of fire have been proposed. These theories have been developed mainly to help firefighters understand how fires can be suppressed. Plastics follow the same principles of combustion as do other "natural" materials, so two of the theories will be presented here. Anyone with any training in fire suppression knows these theories, but I have "modernized" them to fit with today's understanding of the principles of fire.

Whenever fires are discussed, suppression is usually the central theme. That is, when discussing combustibility of any material, emphasis is usually placed upon suppression techniques. Indeed, many people in the Fire Service use the theories of fire that follow as tools in fire extinguishment, as well they might, since the principles are tried and true. But the true value of the theories should be in *preventing* fires instead of suppressing them.

Fires *are* preventable. But so long as accidents are accepted as being "normal" where humans are involved, and therefore a certain percentage of accidents are considered nonpreventable ("accidents will happen" and "we'll always have fires"), the emphasis shifts to rapid detection and even more rapid extinguishment of fires. This means that an aggressive program of fire prevention is absolutely necessary, backed up by the presence of smoke detectors and fast-acting sprinklers or other fire suppression systems in *every* structure.

The Fire Triangle

This theory is presented as a triangle because there are three parts (or sides) to the theory, and the triangle is a closed figure that represents a closed system. Part of the theory says that the system must be closed for a fire to exist; that is, if any of the sides or legs of the triangle is not touching a side or leg next to it, a fire is not possible. The **fire triangle** is presented in Figure 2.1.

Another way of stating the fire triangle theory is to say that three factors must be present at the same time for a fire to exist—in the right amounts *and* in the proper form. Those three factors are fuel, oxidizer, and energy. The three sides of the original fire triangle are fuel, oxygen, and heat. The modernized version of

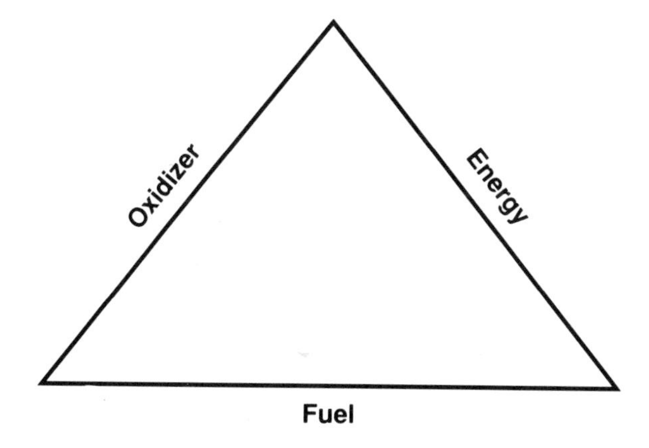

Figure 2.1 The fire triangle

the theory includes *all* oxidizers and states that although atmospheric oxygen is the most common of the oxidizers, oxygen in other forms as well as the halogens must be considered as oxidizers. Similarly, although heat is the most common form of energy to be involved as the ignition source, it must be understood that heat is a form of energy and can be generated and transmitted in more than one manner. It must also be realized that other forms of energy (light, chemical, electrical, mechanical, and nuclear) can and do start fires (if the other "legs" of the triangle are present).

To put it simply, the fire triangle theory says that if fuel, an oxidizer, and energy are brought together in one place in the proper amounts and forms, a fire *will* occur. Not only must a fuel be present, it must be in the proper form; that is, the fuel must be present in a gas or vapor form for it to burn (metallic or organic dusts are very finely divided particles, but a dust explosion involves a different mechanism than a flaming fire). The fuel must also be available in the proper amount, defined as being in the flammable, or explosive, range. The **flammable range** is defined as the percentage of gas or vapor in air between the upper and lower flammable limits. The upper flammable limit is defined as a maximum percentage of gas or vapor in air, *above which* combustion will not occur (the mixture is said to be too rich). The lower flammable limit is a minimum percentage of gas or vapor in air, *below which* combustion will not occur (the mixture is said to be too lean).

The oxidizer portion of the fire triangle must also be in the proper form. This means it must be present as a vapor or gas. It will probably be atmospheric oxygen, but it may also be oxygen released from an oxidizing agent or it may be one of the halogens. The amount of the oxidizer present is also important. The condition whereby the fuel-air mixture is too rich really means there is not enough oxidizer present to support combusion. The normal interpretation of this condition is that there is too much fuel, but what has happened is that the fuel has displaced enough oxygen (or other oxidizer) so that not enough oxidizer is present to support the ignition of the fuel. The oxidizer level can change rapidly, as when a door or window opens or some other opening occurs that allows air to enter the fuel mixture, raising the amount of oxidizer present (bringing the percentage of fuel down below the upper flammable limit into the flammable range). Whenever the fuel is within the flammable range (the fuel and oxidizer are within the "right" percentages), the fuel is ready to burn, lacking only the energy necessary to ignite the mixture.

The energy "leg" of the fire triangle may be defined as the amount of energy required to raise the temperature of the fuel to its ignition temperature. **Ignition temperature** is defined as the minimum temperature to which a fuel must be raised before it will ignite. Although it is true that temperature is interpreted as a method of measuring the amount of energy present in a system, most observers consider heat as the ignition source, rather than as only one form of energy. It is also true that the manifestation of all types of energy input is that the material becomes hotter and hotter until its ignition temperature is reached and a fire begins (or will begin when enough oxidizer is present). This narrow view of heat leads the general public—and unfortunately some firefighters—to overlook other ignition sources. Because heat is the form of energy usually associated with fire, people incorrectly believe that open flames are the only source of ignition.

It is not my intent to discuss further the different types of ignition sources and the methods of transmission of heat energy (conduction, convection, and radiation), but is very important in the study of the combustion of specific materials to understand the relative ease (or difficulty) of ignition of the material under study. When one material is examined against another to compare combustion properties, **ignitability** of the material is very important. Although a material under investigation indeed may burn, it may be so difficult to ignite that another material must be burning nearby to get the first material ignited. In this situation, the ignition temperature of the first material may be well within the range of "ordinary" or common ignition sources, but its oxidizer requirements may be so high that it must be heated by the flames of another material to get it to react with oxygen.

Theories of fire are important in the discussion of the combustibility of any material, because if the theory is correct, it can be used to extinguish or better yet prevent the fire. Recall that the fire triangle states that if the three sides of the triangle "come together" (the three necessary things exist in the proper form and proper amount in the same place), a fire *will* occur. This means that if any one or more of the sides are prevented from existing in the proper form or amount, a fire *will not* occur. On the other hand, if a fire *has* occurred, removal of one or more of the three sides will extinguish it. Since plastics are organic in composition (like wood, paper, cotton, wool, or silk), they are fuels. Since these fuels are used in everyday life, they exist in an atmosphere of air where the oxidizer necessary for combustion, oxygen, is present. Of course, this means that in our everyday lives, we live in a world where fuels and oxidizers exist in the same place, and the only "leg" missing is the energy leg. Fire prevention, therefore, necessarily becomes mainly a matter of preventing the existence of sources of energy that will become ignition sources.

In terms of fire extinguishment, however, if one or more of the sides of the triangle are removed, the fire will go out. There are different methods of accomplishing this, and the extinguishment technique is determined by the type of fire; its size; its location; the fuel involved; the type of oxidizer involved; the threat to human life; the threat to the environment, to property, or to systems (highways, water supply, communications); and the labor, equipment, and extinguishing agents available.

The type of fire may be broken down into what material might be burning in terms of its physical location (is it a house that is burning? a car? a field of grass?) and the class of fuel into which it has been categorized for extinguishment purposes. The classes of fuel or fire are Class A, B, C, and D. **Class A fires** involve materials that are "ordinary" combustibles, such as wood and wood products, paper and paper products, cloth made from natural or synthetic fibers, and rubber and plastics. **Class B fires** involve materials that are liquids; **Class C fires** are those that involve charged electrical wires, equipment, or both; and **Class D fires** are fires involving metals.

The most common method of extinguishing fires is by removing the energy side of the triangle. In most instances this means removing the heat of the fire by cooling it down below the ignition temperature of the fuel by applying water. There are other methods of "cooling" a fire, and in some cases water is definitely not the extinguishing agent of choice, such as in a Class C fire (charged electrical wires or equipment), a fire involving water-reactive materials, or a fire involving metals. The technique of dilution, which is the addition of water to a water-soluble liquid, will extinguish a fire by cooling the liquid below its **flash point**, which is defined as the minimum temperature of the fuel at which it will produce vapors sufficient to form an ignitable mixture near the surface of the liquid or container. Energy is being removed from the system, but in this technique, the fuel itself is being cooled to the point that it is not producing the gases or vapors in the proper amount to have a fire. Since liquids do not burn, but some liquids (classified as flammable or combustible) produce vapors that do burn, dilution really has the effect of removing the fuel. Whether dilution extinguishes a fire by removing energy or fuel is a minor point, since what is important is that the fire be extinguished.

Plastics are Class A materials, and standard techniques of fire extinguishment for Class A fires may be used to extinguish fires involving plastics. In some cases, however, because of the heat absorbed, the plastic may melt and begin to flow like a liquid, giving the appearance of a Class B fire. The use of finely divided water spray or a water fog will quickly cool the liquid and return it to the solid state, in addition to extinguishing the fire. Other techniques designed for Class B fires may also be used.

A second group of techniques of fire extinguishment is based on the oxidizer leg of the triangle. The application of foams to Class B fires, or the application of carbon dioxide (CO_2) to either Class A, B, or C fires, will exclude atmospheric oxygen from the fire, and the fire will go out. The use of water to blanket a burning liquid (the liquid must be insoluble in water and have a higher specific gravity than water) will also exclude oxygen, as will the placing of a lid or cover over a container of burning fuel, whether liquid or solid. Dropping a burning object into water and causing it to sink below the surface will exclude air and cool the fuel rapidly.

The third method of fire extinguishment based on the fire triangle theory is the removal of the fuel. This may be very simple and quick, as the removal of a burning mattress from a house, as is commonly done in fires caused by smoking in bed. A more complicated example of removal of fuel is the transfer of a liquid involved in a fire from one storage tank to another through a piping network. As the level of liquid in the burning tank is lowered, less and less fuel is available to burn, until the last of the liquid is drawn into the network of pipes and moved to another tank, and the fire goes out because there is nothing left to burn. Deliberately allowing a fire consume itself is another example of removal of fuel, as is the building of backfires or the clearing of wooded areas being threatened by forest fires or wildfires.

The extinguishment of fire by the elimination of one or more legs of the fire triangle is proof that the theory is, indeed, correct. However, the fire triangle theory does not explain how burning covalently bonded or organic materials can be extinguished by using dry chemicals, which clearly do not remove sufficient energy, exclude oxidizer, or remove the fuel from a fire. Therefore, the fire triangle theory of fires may be correct, but it is not complete, at least for covalently bonded materials, which make up almost all the materials that burn in fires. This shortcoming led to the formulation of a new theory to explain the mechanism by which most fires are extinguished.

The Tetrahedron of Fire

A new side has been added to the fire triangle, making it a four-sided solid figure called a tetrahedron (rather than a two-dimensional rectangle). The same sort of rationale exists for the **tetrahedron of fire** as for the fire triangle: All sides of the figure must be in contact with each other for a fire to exist. That is, in this theory, four things (instead of three) must be present in the proper amount and proper form for a fire to exist. Indeed, just as in the triangle, if all four "sides" of the tetrahedron are present (touching each other or "closed"), there *will* be a fire.

The three entities that must be present in the fire triangle for a fire to exist must also be present in the tetrahedron theory, plus one more. The fourth side of the tetrahedron is the reaction that takes place when materials burn that are covalently bonded. This reaction is called "the chain reaction of burning," or the formation and reaction of free radicals in a fire. The tetrahedron of fire is presented in Figure 2.2.

Chapter 1 covered the subject of covalent bonding, and the nature of those covalent bonds. That material will be summarized rather quickly here.

A covalent bond is formed by the pairing up of two electrons between the atoms involved (only nonmetallic elements reacting with each other can form covalent bonds). There may be one pair of electrons (a single bond), two pairs (a double bond), or three pairs (a triple bond) between atoms. Certain elements may

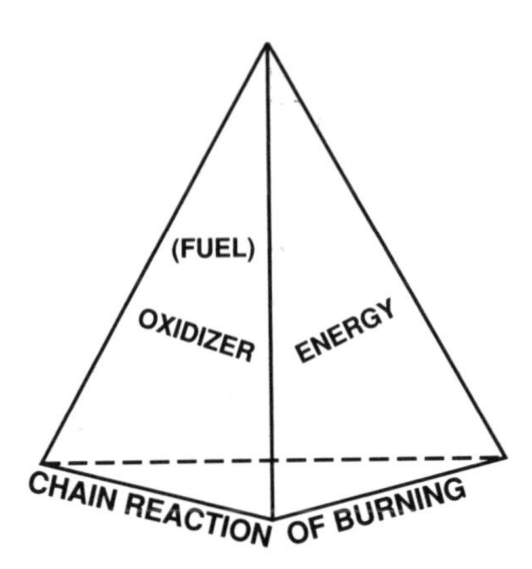

Figure 2.2 The tetrahedron of fire

form only one covalent bond with another element; some may form single or double bonds; and some may form single, double, or triple bonds. No matter whether the nonmetallic elements can form one, two, three, or four covalent bonds, the bonds will be formed only with another nonmetallic element.

There are 22 nonmetallic elements: hydrogen, boron, carbon, silicon, nitrogen, phosphorus, arsenic, oxygen, sulfur, selenium, tellurium, fluorine, chlorine, bromine, iodine, astatine, helium, neon, argon, krypton, xenon, and radon. The nonmetallic elements (with their chemical symbols in parentheses) that are found most often in monomers and therefore in the resulting polymers are carbon (C), chlorine (Cl), fluorine (F), hydrogen (H), nitrogen (N), oxygen (O), sulfur (S), and to a very small extent, silicon (Si).

Free Radical Formation

Each covalent bond between two nonmetal atoms (either from the same or different elements) is a pair of electrons shared between the two atoms. Each one of these bonds has a bond energy (which was stored within the bond when it was formed), which represents the energy released when the bonds are broken. When covalent bonds are broken, one electron per broken bond will remain with the atom, producing a molecular fragment that is extremely reactive. That is, the atom (or molecule) that was previously "satisfied" in its bonded state, now is very "unsatisfied" and unstable with its unbalanced state. It now has at least one unpaired electron that represents one half of a covalent bond, and Mother Nature does not allow this "excited" or unstable particle to exist for very long. It must seek out another "unsatisfied" or "excited" fragment with which to combine, thus satisfying (and stabilizing) each of the fragments and forming a new particle (with new bonding) that is satisfied with its new configuration.

What has been described is the breaking of covalent bonds within a molecule, forming two or more **free radicals**, which are the "excited" molecular fragments that each now has at least one unpaired electron forming one half (or a "dangling" bond) of a covalent bond. Those free radicals are very reactive. That is, they cannot exist in the excited state for long, but must react with other free radicals to form new molecules that contain all paired electrons.

This free radical formation is what occurs in a fire. For example, a material such as methane, the simplest of all hydrocarbons, is mixed in the proper proportions in air (that is, within methane's flammable range), and an ignition source with high enough energy to reach the ignition temperature of methane is present. The immediate result is an explosive fire, but the chemist will describe the reaction between methane and the oxygen in the air in the rather sterile manner shown in Figure 2.3.

$$CH_4 \quad + \quad 2O_2 \longrightarrow CO_2 \quad + \quad 2H_2O$$

methane plus oxygen yields carbon dioxide plus water

Figure 2.3 The combustion of methane

The chemical equation in Figure 2.3 shows the end result of the chemical reaction (fire) that occurs when methane reacts with oxygen. One molecule of methane will react with two molecules of oxygen to form one molecule of carbon dioxide and two molecules of water. Notice that only carbon dioxide and water are formed in this reaction, yet it is a commonly known fact that whenever a substance containing carbon is burned, carbon monoxide is formed. The reaction in Figure 2.3 represents the laboratory reaction, in which combustion is always complete or, for the sake of writing balanced chemical equations, the reaction is *considered* complete. As the reaction is broken down and examined, the formation of carbon monoxide will become evident. Since carbon monoxide is a flammable gas, it undergoes an oxidation reaction itself to form carbon dioxide.

In the reaction just described, the fire is represented by the reaction of methane with oxygen, and there is an implied input energy (heat) to start the chemical reaction. In this instance, the chemical reaction is one of oxidation, and it is so very rapid that there is a manifestation of heat and light energy being liberated. We know this reaction as fire or combustion. The simple chemical equation presented above represents the total reaction. In Figures 2.4 through 2.14, the individual steps are separated. Just as the CH_4, O_2, CO_2, and H_2O

represent methane, oxygen, carbon dioxide, and water, respectively, the symbols and dashes in the figures represent the *structural* formulas for the compounds and free radicals in the reaction. A structural formula shows the symbols for the elements involved, and the long dashes shown between two atoms represent a covalent bond. When the dash is attached to just one atom, it represents an unpaired electron, and the structure thus shown represents a free radical.

When heat energy is applied to the mixture of methane and oxygen in Figure 2.3, the covalent bonds in both molecules are stressed, and they eventually break. The oxygen molecule, O_2, simply breaks in the middle. Since there is a double bond (the sharing of two pairs of electrons) between the two oxygen atoms, its structure is shown in Figure 2.4.

$$O{=}O$$

Figure 2.4 Structural formula of molecular oxygen

When free radicals of oxygen are formed, they contain two unpaired electrons, one on each side of each of the two oxygen atoms, as represented by the "dashes" and shown in Figure 2.5.

$$O{=}O + \text{Heat} \longrightarrow 2\ {-}O{-}$$

Figure 2.5 Creation of oxygen radicals

This free radical is very reactive, and it will seek out other free radicals with which to combine. It may recombine with other oxygen radicals to reform molecular oxygen, or it may react with other radicals available, such as carbon or hydrogen, to form carbon monoxide, carbon dioxide, and water.

The methane molecule has a structure shown in Figure 2.6.

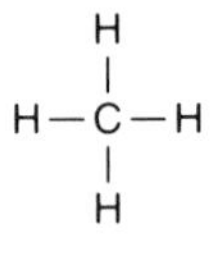

```
    H
    |
H — C — H
    |
    H
```

Figure 2.6 Structure of methane

As energy attacks the methane molecule, hydrogen is broken off the structure and becomes a hydrogen radical as shown in Figure 2.7. The remaining portion of the original methane molecule becomes a methyl radical (the radical of every different hydrocarbon has its own name). The methyl radical results from the methane molecule's losing one atom of hydrogen.

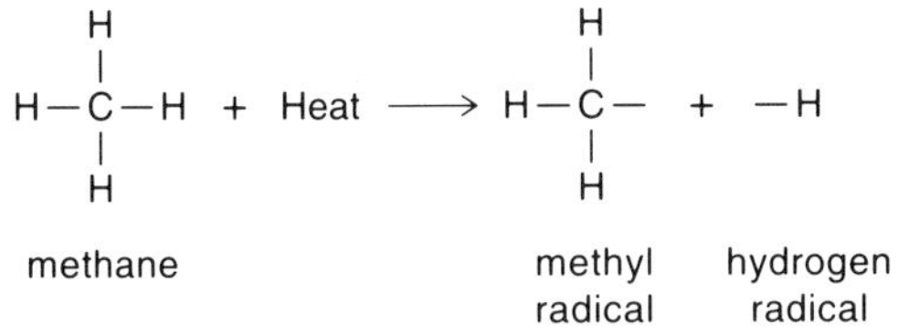

```
      H                           H
      |                           |
  H—C—H  +  Heat  ——>  H—C—    +   —H
      |                           |
      H                           H

  methane                      methyl     hydrogen
                               radical    radical
```

Figure 2.7 Creation of the methyl radical

The methyl radical undergoes further decomposition with the continued application of energy, forming radicals with more than one unpaired electron until the carbon radical is formed. The more unpaired electrons the radical has, the more "active" it is—that is, the more quickly it will seek to react and become a balanced, stable compound. This is demonstrated in steps in Figures 2.8, 2.9, and 2.10. Do not be concerned with the names of the radicals formed with each subsequent splitting off of a hydrogen radical. Each does have a different name, but those names are not important here.

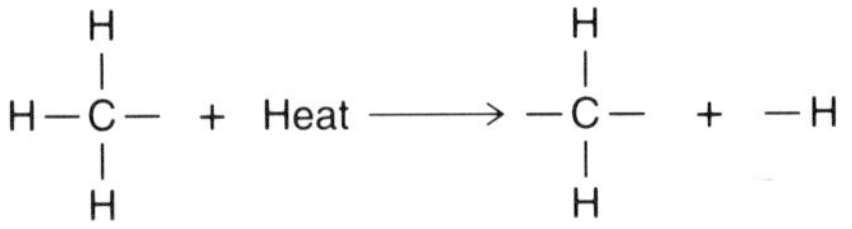

```
      H                    H
      |                    |
  H—C—  +  Heat  ——>  —C—   +  —H
      |                    |
      H                    H
```

Figure 2.8 Creation of additional radicals

```
     H                    H
     |                    |
   —C—  +  Heat  ——>  —C—   +  —H
     |                    |
     H
```

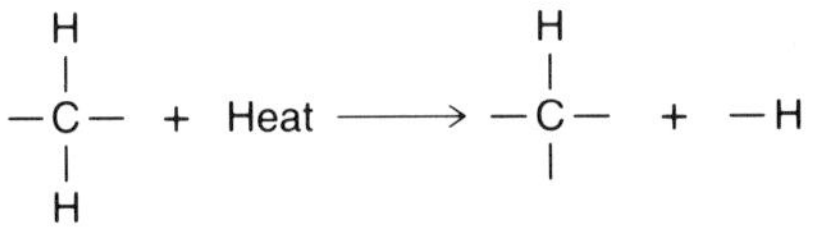

Figure 2.9 Creation of additional radicals

```
     H
     |                    |
   —C—  +  Heat  ——>  —C—   +  —H
     |                    |

                 carbon free radical
```

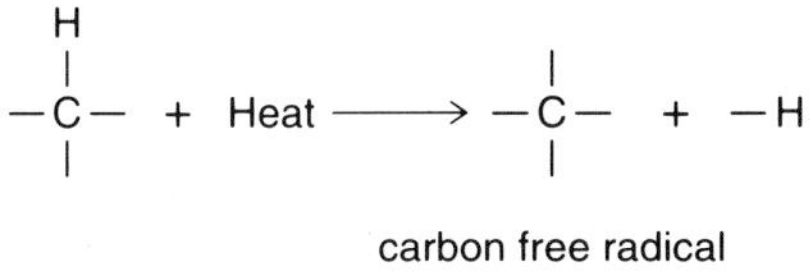

Figure 2.10 Creation of the carbon radical

The end products of this stepwise decomposition of methane combine with the end products of the decomposition of oxygen to form carbon monoxide (Figure 2.11) carbon dioxide (Figure 2.12), and water (Figure 2.13). Any unburned particles of carbon (free radicals) will collect together to form discrete particles of soot, which gives smoke its black color. Carbon monoxide is not only toxic, but it is also a flammable gas. When carbon monoxide burns, it "hooks up" with an oxygen radical and forms carbon dioxide, as shown in Figure 2.14.

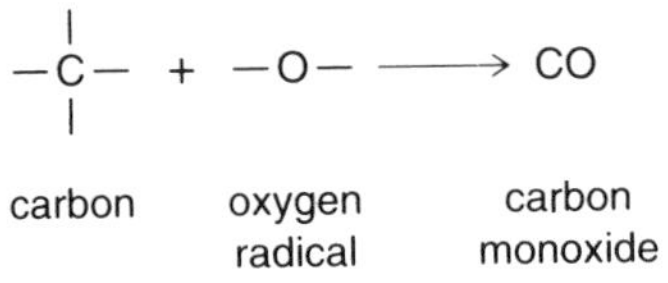

$$\begin{array}{ccccc} -\overset{|}{\underset{|}{C}}- & + & -O- & \longrightarrow & CO \\ \text{carbon} & & \text{oxygen radical} & & \text{carbon monoxide} \end{array}$$

Figure 2.11 Oxidation of carbon to carbon monoxide

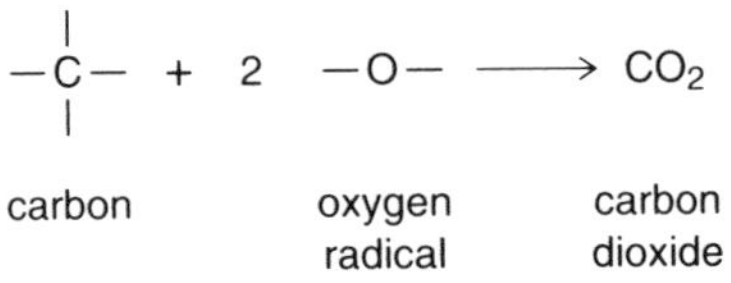

$$\begin{array}{ccccc} -\overset{|}{\underset{|}{C}}- & + & 2\ -O- & \longrightarrow & CO_2 \\ \text{carbon} & & \text{oxygen radical} & & \text{carbon dioxide} \end{array}$$

Figure 2.12 Oxidation of carbon to carbon dioxide

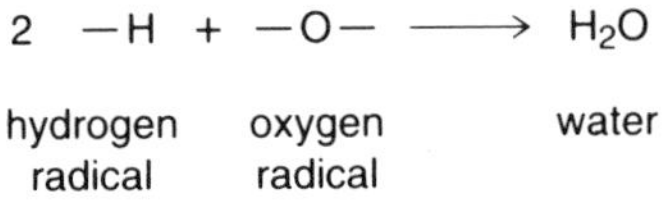

$$\begin{array}{ccccc} 2\ -H & + & -O- & \longrightarrow & H_2O \\ \text{hydrogen radical} & & \text{oxygen radical} & & \text{water} \end{array}$$

Figure 2.13 Formation of water

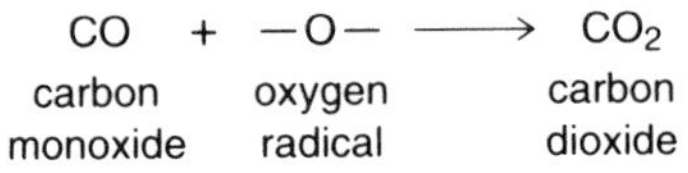

$$\begin{array}{ccccc} CO & + & -O- & \longrightarrow & CO_2 \\ \text{carbon monoxide} & & \text{oxygen radical} & & \text{carbon dioxide} \end{array}$$

Figure 2.14 The oxidation of carbon monoxide

Thus the tetrahedron of fire theory states that in a fire involving covalently bonded materials, there will be a continual formation of free radicals, which must react with other free radicals during the combustion process to become balanced and stable compounds. The discussion of pyrolysis that will follow shortly will complete the picture of how free radicals form and the subsequent combustion reaction. But since free radicals do form in the combustion process involving covalently bonded materials, if the theory is correct, the fire will be extinguished if the formation of these free radicals is prevented. Indeed, that is exactly the process by which dry chemicals and halogenated substances work in the extinguishment of a fire on which they are used.

As the heat from the ignition source is applied and the covalent bonds break, the temperature of the fuel rises to its ignition temperature and combustion begins with the rapid chain reaction in which free radicals form and subsequently combine with other free radicals. When a dry chemical or halogenated material is applied to the fire, a breakdown also occurs in the fire extinguishing agent. The particles (free radicals) formed from the fire extinguishing agent will preferentially react with free radicals formed from the fuel, thereby preventing the fuel free radicals from reacting with each other or with oxygen free radicals. As the free radicals from the fuel are used up by the extinguishing agent, the chain reaction is broken, and the formation of additional radicals is stopped. As fewer and fewer free radicals are formed, the fire goes out. The fire extinguishing agents that work this way are also known as *free radical quenchants* or *free radical traps*. Since preventing the formation of free radicals extinguishes fires, the tetrahedron theory of fire is proved to be correct.

All **hydrocarbons** (covalently bonded compounds containing only hydrogen and carbon) will react in the same manner, with the molecular covalent bonds of any hydrocarbon compound other than methane breaking in the same manner as they do in methane. The only difference between methane and other hydrocarbons is that methane is the simplest of all the hydrocarbons, and only one carbon atom is present. In all other hydrocarbons, since there is more than one carbon atom present, there are carbon-to-carbon bonds that will also break. And since there are more carbons in these molecules than in methane, there usually will be more hydrogens also (there are a few exceptions). The longer the hydrocarbon chain (or the more carbon present), the more fuel will be available to burn. That is, more carbon-to-carbon and carbon-to-hydrogen bonds will be broken, therefore the heat of combustion (the total amount of heat released in the complete combustion of the molecule) of hydrocarbon compounds increases as the number of carbon atoms in the molecule increases. However, since there are more covalent bonds to be broken, more energy must be input to these longer-chain molecules to provide the energy to create the free radicals that will react. Because of the length and complexity of the molecular structure of some fuels, many intermediate products of combustion will be formed. These intermediate products, chemical compounds in their own right, will then undergo further bond breakage until they reach the simplest stages, producing carbon monoxide, carbon dioxide, free carbon, and water.

Not all fuels are hydrocarbons. Indeed, there are larger classes of fuels known as hydrocarbon derivatives which are, literally, covalently bonded compounds derived (or made) from hydrocarbons. Some are made in nature, but the vast majority of these compounds are made (or synthesized) by humans. These materials include the classes of chemical compounds known as alcohols, ketones, aldehydes, amines, organic acids, organic peroxides, ethers, and esters, plus some halogenated hydrocarbons. These compounds act much like hydrocarbons when they burn, since bond breakage and free radical formation must occur before a fire will begin. Almost all of these compounds contain only carbon, hydrogen, and oxygen. The amines contain carbon, hydrogen, and nitrogen, and there are some other classes of hydrocarbon derivatives that contain sulfur.

Additional theories of fire exist, including scientific models that predict growth, spread, and smoke generation and movement. However, the person interest in the basics of combustion need understand only the triangle and tetrahedron theories. These enduring theories have been well tested by extinguishment techniques.

A warning: You have just read a step-by-step representation of what happens in a fire involving covalently bonded fuels. The formation of free radicals takes several minutes to describe but only milliseconds and/or microseconds to happen. What is observed when a gas or vaporized fuel is mixed in the "proper" amount of air is an *explosion,* not just a fire. The fire usually occurs after the explosion. It is one thing to read about chemical reactions and the mechanisms by which they take place, but in the real world where these reactions take place in uncontrolled surroundings, the reaction may be instantaneous.

Pyrolysis

The word "pyrolysis" has its roots in two words: the Greek *pyro,* meaning fire, and the word *lysis,* which means breakdown. Pyrolysis may be defined very simply as the breakdown of a molecule by heat. The process of pyrolysis is what causes fuel to be in the proper form to burn, and the process may be rapid or slow, depending mainly on the energy source providing the input heat. When a material pyrolyzes, the covalent bonds within the molecule are broken, usually generating more heat, but it is the breakdown of the fuel into simpler substances that allows the fire to exist. "Classic" pyrolysis occurs when solid substances like wood and other cellulose-containing materials are heated, often in the absence of air. Even in the presence of air, wood is pyrolyzed, and the pyrolysis products are the products that participate in the combustion reaction. However, pyrolysis is the breakdown of any covalent compound by heat, and flammable and combustible liquids and gases are covalently bonded compounds.

Liquids, as already stated, do not burn. The terms "flammable" and "combustible," when used in conjunction with liquids are used only to differentiate between two classes of liquids with different flash point ranges. Many people believe that a flammable liquid burns, when the truth is that it is only the *vapors* of the liquid that burn. The same, of course, is true for combustible liquids. A **vapor** is defined as the gaslike state of a liquid or solid (the three states of matter are gas, liquid, and solid), that occurs when the material has been heated to the point of evaporation. Evaporation is defined as the conversion of a liquid to a vapor. Therefore, **flammable** or **combustible liquids** are liquids that produce vapors that burn. Whether a substance is a liquid or a gas depends on a scientific definition of its physical state under certain conditions of temperature and pressure, and is of no real concern here. The point is that liquids do not burn, and to become a fuel they must first be converted into a vapor.

Flammable liquids, are liquids with flash points below 100°F, and combustible liquids are liquids with flash points at 100°F or higher. **Flash point** is defined as the minimum temperature of the liquid at which it produces vapors sufficient to form an ignitable mixture at the surface of the liquid or near the container.

All liquids evaporate at some rate below their boiling point and reach their maximum evaporation rate at their boiling point. The vapor mixed in air is said to be ready to burn, but if the tetrahedron of fire theory is correct (and it is), another step must occur: the formation of free radicals. The ignition source is the cause of the breakdown of the molecule into simpler substances (free radicals). When the fuel is mixed in the proper proportion in air (within its flammable range), the mixture will be raised to the fuel's ignition temperature and the fire will begin, usually with an explosion of the vapors. Although most students of fire consider that pyrolysis occurs only with solid substances such as wood and wood products, the molecules of the vapor are also broken down by the application of heat, and that fits the definition of pyrolysis. Any time that a covalently bonded material is broken down by heat into simpler compounds (and free radicals), pyrolysis has taken place.

The pyrolysis of solid covalently bonded materials is really a two-step process, and the materials most often pyrolyzed are wood and wood products (including paper and paper products). First, the energy source is exposed to the wood, and the absorption of energy by the wood causes the breakdown of bonds in the atmospheric oxygen molecules and in cellulose, which is the major constituent of wood and all wood products. Cellulose is a naturally occurring polymer containing carbon, hydrogen, and oxygen, and it is found in every growing plant in nature. It is a hydrocarbon derivative, very long-chain in its makeup, and therefore contains a tremendous amount of carbon-to-carbon, carbon-to-hydrogen, carbon-to-oxygen, and oxygen-to-hydrogen bonds. In an extremely hot fire, atmospheric nitrogen may be broken down into nitrogen radicals that cause the formation of a group of gases called the nitrogen oxides. Normally, an ignition source that breaks oxygen–oxygen bonds in air will not break down nitrogen–nitrogen bonds. The energy to do that comes either from a very hot fire or the heat of an internal combustion engine (which, of course, qualifies as a very hot fire).

As the cellulose in the wood, in other products, or in both breaks down when heated, it forms free radicals, usually hydrogen, hydrocarbon radicals, and other radicals containing carbon, hydrogen, and oxygen. The hydrogen radicals are ready to react with the oxygen radicals in the air, but the hydrogen is being grabbed by other hydrocarbon radicals to form short-chain hydrocarbons. The short-chain hydrocarbon compounds thus formed come off the wood as gases, such as methane, ethane, propane, and butane, and vapors of liquid hydrocarbons (pentane, hexane, and so on) and hydrocarbon derivatives (such as acrolein, acetaldehyde, and so on). These in turn are further broken down into smaller and smaller hydrocarbons until a breakdown such as depicted in Figures 2.7 through 2.10 occurs. The final combustion products (assuming complete oxidation) of any compound containing carbon and hydrogen or carbon, hydrogen, and oxygen will always be carbon dioxide, water, and free carbon. Experience tells us that complete combustion of a substance occurs only in controlled conditions in a laboratory or in controlled industrial processes, and that carbon monoxide is always formed in any real, unwanted, and unexpected (and sometimes wanted and expected, as in cooking and heating) fire involving a carbon-containing organic material. In a real fire, other intermediate combustion products of hydrocarbons and hydrocarbon derivatives may also be formed, depending on the fuel and the conditions of the fire. Acrolein, one of the hydrocarbon derivatives formed in a fire involving wood and wood products, is a deadly poison. Fortunately, like carbon monoxide, it is flammable and is mostly consumed by the fire, oxidizing to carbon dioxide, as in Figure 2.14.

Conditions of the fire that determine combustion products (assuming that the fuel is identical in shape, amount, and chemical composition) include (1) the amount of oxygen present to support combustion, (2) the location of the fire, (3) the amount of moving air available, (4) the size of the thermal column, and (5) whether heat and combustion products are trapped or free to leave the immediate vicinity of the fire. The hotter the fire—that is, if heat and combustion products from the fire cannot be ventilated—a different set (or at least a different percentage mix of the same combustion products) of final combustion products will be found than if the fire is cooler and well ventilated. A higher percentage of the nitrogen oxides will be formed in a very hot fire than in a relatively "cooler" one. A large influx of air and a corresponding large thermal column will produce more aldehydes and hydrocarbons as combustion products because of the turbulence present, and that will disrupt the breakdown of larger compounds into simpler compounds for combustion. A shortage of oxygen will dramatically increase the amount of carbon monoxide generated, since CO is always a product of incomplete combustion. A very difficult-to-burn material will also produce large quantities of free carbon, which also takes place whenever large amounts of carbon monoxide are being generated.

An interior structure fire is said to be oxygen regulated, since sufficient fuel is usually present and the fire usually can draw from the oxygen in the entire building in order to grow. A fire "in the open" is said to be fuel regulated, since the atmosphere in which the fire is burning contains (theoretically) all the oxygen needed for combustion, and the only barrier to the growth and size of the fire is the amount and type of fuel present.

Knowing the type of fuel involved in the fire will help determine the final combustion products, since it is the chemical composition of the fuel that contributes most to the final combustion products (aside from the oxygen in the atmosphere). If the fuel is a long-chain hydrocarbon, it will burn hotter (assuming the time of the combustion reaction is the same for all fuels) since it contains more fuel in its molecules than a shorter-chain hydrocarbon. If the fuel is an alcohol, it will burn cooler than hydrocarbons of the same carbon content since it is already partly oxidized. This is because all alcohols contain the hydroxyl radical, which is made up of one atom of hydrogen and one atom of oxygen. Other fuels containing multiple carbon–carbon bonds will produce more heat (and usually more quickly) because of the additional energy stored within the double and triple covalent bonds between some of the carbon atoms. Halogens (fluorine, chlorine, bromine, or iodine), if present in the fuel molecules, will always act to retard the combustion reaction by functioning as free radical traps.

The most common and most visible pyrolysis is that of wood burning. In a fireplace, where the fuel is a log of a particular type of wood, pyrolysis is induced by starting a smaller fire to produce enough energy to get the log to begin to pyrolyze, at least on the surface. A match is struck, with the heat produced by the burning phosphorus beginning the pyrolysis of the match. As the cellulose in the paper or wood of the match breaks down into simpler and simpler gases, these gases are further broken down into free radicals that react with other free radicals, and the match begins to burn. In **space burning**, which is what occurs when a match is lit, the flame is nothing more than the manifestation of energy as light (as well as heat), the energy coming from the reaction of the free radicals. The energy from this match flame is then used to begin pyrolysis and combustion in more paper, whose subsequent burning initiates pyrolysis and fire in small pieces of wood (kindling), whose release of heat in their combustion begins the pyrolysis of the log and subsequent burning of its pyrolysis products.

The log itself really doesn't burn, at least at first. As more and more cellulose is decomposed, and more and more vapors are broken down and flames appear, the radiated heat from these flames increase the pyrolytic and burning actions. This continues until all the covalently bonded material is consumed. What is left is solid material of a carbonaceous nature, much like charcoal, which begins to glow. This glowing is **surface burning**, which is the direct combination of oxygen with the fuel on the surface of the material. By now pyrolysis is complete, and no flames will be visible.

Pyrolysis needs only heat and the fuel to take place. Indeed, in a smoldering fire involving cellulose-containing material, the heat will break down the cellulose and re-form shorter-chain hydrocarbons, since there is not enough oxygen for the fire to burn with a flame. As more and more of these highly flammable (easily broken-down short-chain compounds) gases fill the room, three sides of the tetrahedron are present: free radical formation, energy, and fuel. These three components usually are present at temperatures above the ignition temperature of the gases present. Consequently, when a room is ventilated by a door or window opening, the oxygen is carried into the reaction by the onrushing air, and the four sides of the tetrahedron "slam shut," producing a "backdraft" explosion. The explosion is caused by the rapid introduction of oxygen into the super-heated gases, and the majority of the fuel ignites simultaneously, producing the explosion. All that was missing from either the triangle or the tetrahedron was the oxygen, and as soon as it was introduced into the fuel- and energy-charged atmosphere, its covalent bonds were broken and oxygen free radicals were available for the reaction. The instant the other free radicals had oxygen radicals available, the reaction, an explosion, took place.

When there is sufficient air within a fire room, the flames will build and send superheated combustion products to the upper regions of the room. The heat radiated from these hot gases and from the flames themselves will cause the simultaneous breakdown of other articles in the room, producing large additional amounts of fuel in the form of short-chain gases and free radicals. These short-chain gases will further pyrolyze, and the unburned area of the fire room will be filled with free radicals. When this happens, free radicals of fuel and oxygen are being produced simultaneously. What is missing from the tetrahedron here is the heat energy required to raise all the fuel to its ignition temperature. What is happening during the buildup of heat and the generation of free radicals is that other fuels (objects such as furniture and carpeting) in the room are absorbing the heat, acting as a sort of heat sink. As they absorb more and more heat from the fire, these objects begin pyrolyzing. This process produces more and more free radicals and more heat, which brings about a very unstable condition in the fire room. Under proper conditions, this unstable condition will continue until the four sides of the tetrahedron slam together, at which time the entire room bursts into flame simultaneously, producing a condition called flashover.

Pyrolysis may be as rapid, as in a fire, or it may be very slow and unnoticed. In the classic case of the "oily" rags (the oil really has to be a specific type of animal or vegetable oil containing double covalent bonds) in an area exposed to the atmosphere, but with very little air movement, a slow combination of oxygen with the oil can take place. As the oil oxidizes, the heat released from the breaking of the double covalent bonds is absorbed by the rest of the pile of rags, and the temperature of the material rises slowly. This increase in temperature encourages increased bond breakage, and more heat is released and then absorbed by the remaining fuel. This process continues breaking covalent bonds and producing heat until the ignition temperature of the oil (or rags) is reached, and the entire mass appears to spontaneously break into flames. This, of course, is the definition of the process called **spontaneous ignition** (formerly called spontaneous combustion).

This process, in a slightly different manner, may also occur with freshly washed coal or wet charcoal that is drying out. In this case, free electrons ("dangling" bonds) are present on the carbon atoms on the surface of the coal or charcoal, and are therefore available for direct combination with oxygen from the air. Although some may consider this process spontaneous ignition because it can begin surface burning after the direct oxidation has raised the temperature of the fuel to its ignition temperature, it is not the same set of fuel and reactions as the classic case of spontaneous combustion.

The Pyrolysis of Plastics

This discussion of the way combustion occurs (including pyrolysis) covers all covalently bonded materials. Plastics, even though they are a member of this class and undergo similar reactions, will not react exactly as described above, especially the class known as thermoplastics. These materials will actually resist pyrolysis at first, by absorbing the original amounts of energy and melting. In many cases, the molten plastic will then flow away from the source of heat. As it flows away, contacting cooler surfaces in the process, the plastic will give up its heat to the cooler surface and resolidify. Continued absorption of heat will cause this process to continue until the plastic flows away from the source of heat, or the fire is extinguished. This absorption of heat by the melting plastic material serves as an effective **heat sink**, slowing or preventing the rising of the temperature of the plastic material to its ignition temperature.

If the plastic is subjected to sufficient energy and prevented from leaving the source of heat, it will still accept the original energy and melt. It will eventually pyrolyze, however, producing many of the same pyrolysis products as wood and wood products, depending upon the chemical makeup of the plastic. The pyrolysis products will then react, and the plastic material will burn, assuming enough heat energy is available.

Thermosetting plastics, being different chemically and structurally from the thermoplastics, will not melt when subjected to heat. They will more closely follow the classic pyrolysis reactions of wood and wood products. The burning of individual plastics and plastics compounds is discussed in Chapter 6.

SUMMARY

- Fire is a chemical reaction called oxidation, and certain things must be present in the right amount before it can occur.
- One theory (the fire triangle) says that fuel, an oxidizer, and energy in the form of heat, when present in the proper amount, will always produce a fire.
- A second theory (the tetrahedron of fire), adds that the fuel must be in the proper form (free radicals), and that the oxidation reaction will proceed in the manner of a chain reaction.
- Pyrolysis is the process by which the fuel is put into the proper form for burning, and it is nothing more than the breaking of covalent bonds.
- Knowing the chemical composition of the fuel allows one to predict the composition of the combustion products, and understanding the theories of fire provides methods of preventing or extinguishing the fire.
- Plastics are Class A materials, and burning plastics may be extinguished by the same methods used on other Class A materials.

3

Plastics: An Overview

Plastics are a family of materials that belong to a much larger group of materials called polymers. A **polymer** is defined as a giant molecule made up of thousands of tiny molecules that have the unique property of being able to react with themselves to form larger molecules. Of course, all molecules are tiny in relation to what we can see, but tiny in this context means in relation to the size of other molecules. The process by which this takes place is called polymerization, and each compound whose molecules can react with themselves to form a polymer is called a **monomer**. This ability for molecules to combine with themselves is very rare and is possessed by relatively few chemical compounds. Each compound that can react with itself to form a polymer is called a monomer.

Both words, "monomer" and "polymer," come from Greek roots: *mono-* meaning one, *poly-*, meaning many, and *-mer* meaning part. Therefore, monomer means one part and polymer means many parts. Sometimes a polymer is made up of many tens of thousands of the monomer molecules (called repeating units), depending on the type of polymer the manufacturer (the company carrying out the polymerization) wants to make. But before we get into the different types of plastics, we have to examine this much larger group of materials that are called polymers.

There are really two large groups of polymers. The first and much larger group of polymers consists of the natural (existing in and created by nature) polymers, and the other, smaller group is known as synthetic, or artificial, polymers. If one combines all the manufacturing capabilities of all the companies in the world that make synthetic polymers (and there are quite a few companies making quite a large amount of synthetic polymers), the total volume of polymers produced would seem like a mere pittance compared to the amount of polymer that is made in nature on a regular basis. The natural polymers include such familiar and abundant materials as cellulose (the major ingredient in wood and woody plants), cotton (which is nearly pure cellulose), leather, silk, wool, and human skin. The cellulose from wood is also the major ingredient in paper and all paper products, including cardboard and cardboard products. Humans have taken advantage of the volume of polymers and polymeric products that nature manufactures to build shelters, make clothing, and to create the other comforts that are demanded by society.

At some point, however, we humans ourselves decided that society wants certain other creature comforts that can be made from other materials that can be created to duplicate or improve upon those made by nature. It was in this manner that the first plastics and the first plastics products were made. Plastics were first developed as "replacement" or substitute materials. Today, plastics are so spread throughout our society and accepted by its users that anything else used in place of plastics might today be considered a substitute material.

The first use of a commercially feasible plastic was to make billiard balls that had originally been made of ivory. Most people over 40 recall that vinyl (or polyvinyl chloride) was originally called "artificial leather." As new polymers were introduced as substitute materials, they replaced more and more so-called "traditional" or natural materials. Soon, manufacturers began to make products that could be made only from plastics, and plastic began to lose its image as a replacement material and began to be known as an "original" material. However, as medical science has discovered that certain plastics are the only satisfactory replacement for arteries and joints, and doctors have begun the development of plastic parts for the human body, it seems that plastics have come full cycle, and will always be called upon to duplicate or improve upon something natural that might have become defective or is simply not acceptable by society in its natural form.

Precisely because of the manner in which the polymerization of plastics has developed, plastics are by far the most specifically engineered group of materials in existence. The plastics industry has grown up around the marketing concept; that is, it is necessary to find out what problems the customer has, and gather together the resources of the organization and focus them on the solution of that problem. The quest for new products by the consumer and new processes by the manufacturer has been the driving force for the plastics industry to continue to engineer new products and to refine old ones. No new plastic is developed unless there is a specific need (demand) for it, and the chemical engineers and polymerizers are constantly striving to produce only those materials that will satisfy those demands.

The stated objective of this book is to present the manner in which plastics burn, so the topic of natural polymers will be excluded. However, during discussions of the behavior of plastics under certain conditions, the properties of natural polymers such as wood may be compared, especially as they relate to combustion.

TYPES OF SYNTHETIC POLYMERS

Synthetic polymers can be divided into the two large groups known as rubber and plastics. Many may claim that rubber is really a thermosetting (defined later) plastic, and therefore is not a separate group of materials. This is true, but the distinctions need not be discussed here. Most consumers probably consider rubber and plastics to be two distinct types of materials. One should be aware, however, that there are more and more plastics that have **elastomeric** properties (the ability to be stretched to twice its length and to snap back to approximately its original length), and more and more of these plastics materials resemble rubber. There are also more and more rubberlike materials that more closely resemble what most people consider plastics. Additionally, there are many materials that possess a set of properties that make it very difficult to describe them as either rubber or plastic. These include both solid and liquid systems (such as paints and coatings). However, all these materials are polymeric in composition and are artificial. Therefore, for the purposes of this book, they will be called plastics with the knowledge that many scientists and plastics experts will argue that not all synthetic polymers are considered to be plastics.

CLASSES OF PLASTICS

For purposes of simplification, plastics are divided into two great divisions called thermoplastics and thermosets. A **thermoplastic** is a plastic that can be formed in a machine or processed by heat and pressure and, once formed, may be "chopped up" (or otherwise reduced in size) and re-formed in the same machine or process by the use of additional heat and pressure. That is, the processor, or the company that changes plastic resins and compounds from a liquid, powder, pellet, or cube into a finished part by heating and applying pressure in a particular process, is not happy with the part it has just made, may grind it into small pieces and put it back through the equipment and make a new part. This may be done once, or it may be done several times, depending on the particular thermoplastic and its unique properties.

A **thermosetting** plastic, on the other hand, is defined as a plastic that may be formed by heat and pressure just once. Any further attempts to reheat the thermosetting plastic to the point where it might be re-formed will result in the thermal degradation (called decomposition, or breakdown) of the product. Whether a particular plastic is thermoplastic or thermosetting depends upon the chemistry of its monomer(s) or other "feedstock" chemicals, and the polymerization process. However, a plastic may be either thermoplastic or a thermoset, and cannot be both (although there are groups of chemically similar plastics, like the polyesters and polyurethanes, which include both thermoplastic and thermosetting plastics). The thermoplastics include, but are not limited to, polyethylene, polypropylene, polybutylene, polystyrene, polyvinyl chloride, polyurethane (there are also thermosetting polyurethanes), acrylics, polyamides, polyesters (thermosetting polyesters also exist), cellulosics, and fluoropolymers. Thermosets include the alkyds, melamines, phenolics, polyurethanes, polyesters, epoxies, and the urea-formaldehyde plastics.

POLYMERIZATION

The chemical reaction known as **polymerization** is a very special type of reaction that is undergone by only very few chemical compounds, called monomers. This special chemical reaction allows these special molecules to react with themselves to form new molecules, made up of "repeating units" that resemble the original monomer molecule. The best example to examine is the most common of all the monomers, ethylene (also known as ethene). The molecular formula for ethylene is C_2H_4, and its structural formula is shown in Figure 3.1.

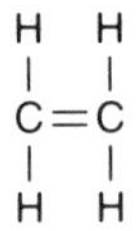

```
H   H
|   |
C = C
|   |
H   H
```

Figure 3.1 The structure of ethylene

Recall from chapters 1 and 2 that the structural formula of a compound shows the individual atoms in the molecule and how they are bound to each other by covalent bonds. The symbols C and H stand for atoms of the elements carbon and hydrogen, respectively, and the dashes between two atoms stand for the covalent bonds that hold those atoms together.

What makes ethylene and other monomers different from other compounds is what happens under special conditions of temperature, pressure, and the addition of special chemicals that begin and control the rate of reaction. When these monomers polymerize, the double bond that is characteristic of the alkenes (the family of hydrocarbons to which ethylene belongs) undergoes a remarkable change. Instead of both bonds breaking as they do in the combustion reaction (which, of course, the monomer will still undergo under "normal" conditions outside the reactor), only one bond breaks. Both electrons of the broken bond are transferred to the outermost carbon atoms, giving them each an unshared (or unpaired) electron. This happens to all of the monomer molecules in the reactor, and a molecule with unshared electrons is in a very unstable condition (they are all considered to be free radicals) that nature will not permit except for an extremely short time (see Chapter 2). Since these molecules have all been prepared to react, the only other materials with which they *can* react are the other free radicals around them, and so they do react with them by joining together in long chainlike molecules.

In the polymerization reactor, this chemical reaction is very carefully controlled, and the monomer molecules are not turned into free radicals all at once, but rather at a controlled rate. This is necessary because covalent bonds are breaking, which causes the release of energy. If this release were allowed to proceed at an uncontrolled rate, the result would be a very rapid release of energy in the form of a tremendous amount of heat, which could produce a catastrophic explosion called "runaway polymerization." Since the reaction is carefully controlled, however, it proceeds at the rate called for by the chemical engineer in charge of the process, so that the polymer with the desired properties will be produced.

With the double bond now broken in the reactor under controlled conditions, and the two unpaired electrons now attached to the "other side" of the carbon atoms, the monomer now looks like that in Figure 3.2.

```
  H   H
  |   |
—C — C—
  |   |
  H   H
```

Figure 3.2 Radical formed by breaking a bond in ethylene

Each of the unconnected (or "dangling") bonds shown on the outside of each carbon atom represents an unpaired electron. This is a very reactive state for any chemical compound or part of a compound, and nature will not allow this "unit" to exist very long in this unattached and extremely reactive and unstable state. To become stable, and therefore "acceptable" in nature, it must react with something, and that something will be another reactive "unit" close to it in the reactor.

To keep the runaway polymerization explosion from occurring before the material is charged into the reactor—that is, during the transportation or subsequent storage of the monomer—an **inhibitor** is mixed in with the monomer. This chemical prohibits the process called polymerization from beginning prematurely. If the inhibitor is not added to the monomer, or if, after being added, it is allowed to escape by evaporation during an emergency of some kind, the protection against the instantaneous total polymerization of all the monomer in the tank truck or rail car is gone. In this situation, an extremely dangerous condition will now exist, and the predicament in which emergency responders will find themselves is a very serious hazardous materials incident.

The monomer will begin to react with itself, breaking the double covalent bonds and liberating heat energy. This increase in energy will cause the chemical reaction speed to increase, allowing it quickly to get out of control (the speed of chemical reactions generally doubles with every 18°F rise in temperature). The resulting explosion will resemble a BLEVE (a Boiling Liquid, Expanding Vapor Explosion). Many monomers are easily liquefiable flammable gases, so these materials are subject to the BLEVE anyway. Regardless, the resulting explosion, no matter how it is classified, will kill anyone within the danger zone and cause tremendous property damage.

Therefore, it is imperative that the polymerization process be tightly controlled, not just for quality control reasons, but also for safety reasons.

Following through with the ethylene example, if the polymerization process has been successful, the resulting polyethylene molecule will look like Figure 3.3, with a continuation of the "chain" on both ends of the molecule where an unpaired electron is shown as an "open" bond. Because of the lack of space, the entire molecule of several thousand repeating units cannot be shown, so the reader must imagine the continuation of the carbon chain on both sides of the portion of the molecule shown. The real molecule is not as straight as in Figure 3.3, but may be twisted and bent. Some polyethylene chains may be branched, but that is a special polymer. Using accepted shorthand, the polyethylene molecule is usually written as shown in Figure 3.4.

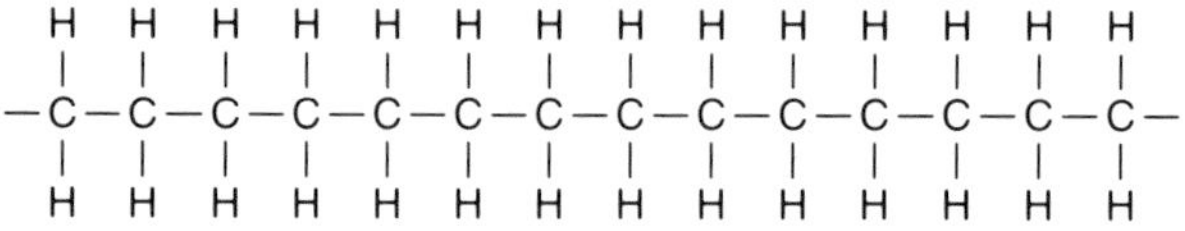

```
  H   H   H   H   H   H   H   H   H   H   H   H   H   H
  |   |   |   |   |   |   |   |   |   |   |   |   |   |
— C — C — C — C — C — C — C — C — C — C — C — C — C — C —
  |   |   |   |   |   |   |   |   |   |   |   |   |   |
  H   H   H   H   H   H   H   H   H   H   H   H   H   H
```

Figure 3.3 Portion of the polyethylene molecule

```
 ┌ H   H ┐
 | |   | |
—┼ C — C ┼—
 | |   | |
 └ H   H ┘n
```

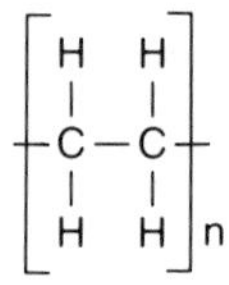

Figure 3.4 Shorthand method for drawing the structure of polyethylene

The brackets around the molecule indicate that this is the repeating unit of the polymer, and the *n* stands for the number of repeating units in the molecule.

Depending on the particular polymer to be made, n can be in the hundreds, thousands, or tens of thousands. The last units on each end of the chain do not have the free electrons, and the chain is "capped off" in the reaction. This sort of reaction occurs with each type of thermoplastic polymer produced, so long as only one monomer is required. Some types of thermoplastics are made from more than one monomer, but the reaction, for our purposes, can be considered essentially the same.

THERMOPLASTICS

By far the largest group of plastics in volume is the thermoplastics. This group contains all the common plastics you are likely to know, plus all of the engineering plastics. Engineering plastics are so named because they are called upon to perform some task other than packaging or being decorative. They may be loadbearing as in the case of certain thermoplastics that are wheels, gears, slides, and other working and functional parts in many applications, or they may perform some other difficult function.

To understand the thermoplastics, one must know a little bit about their chemistry. It is not necessary to become a polymer chemist or a plastics engineer, but it is extremely helpful to know what starting materials these polymers are made of. Basically, the starting raw materials for the thermoplastics are monomers, as in the polyethylene example. Monomers are those small molecules that can react with themselves, "hooking up" with each other in extremely long chains. For the most part, monomers are very reactive chemicals, and often they are very unstable, hazardous materials that must be handled very carefully. The polymer chemist or the chemical engineer knows that under the proper conditions of temperature and pressure in a reactor (really a large, sealed vat), and in the presence of an **initiator** (a substance that will overcome the inhibitor that has been added to the monomer to prevent premature polymerization, and begin the polymerization process at the proper time), plus, in some cases, a **catalyst** (a substance added to the reaction that speeds it up but is not used up in the process), the molecules of monomer can be coaxed to "hook up" with each other to begin to form a long-chain molecule. The chemist or engineer also knows that by varying the temperature, the pressure, and the length of time the reactants are subjected to such conditions, and in some cases, the medium in which the polymer may be dispersed or suspended, different variations on the same polymer can be produced. An example would be a polymer plant that with variations on the same equipment and processes, can produce linear low density polyethylene (LLDPE), low density polyethylene (LDPE), high density polyethylene (HDPE), high molecular weight high density polyethylene (HMWHDPE), and ultrahigh molecular weight polyethylene (UHMWPE). All of these polymers would have started with the same monomer, ethylene (ethene), but ended up as quite radically different plastics with widely varying properties. These different types of polyethylenes (PEs) can be used for many different purposes, with each use having a completely different set of requirements for the plastic selected.

All these types of polyethylene had the same starting monomer (ethylene). That is, the only thermoplastic one can make by starting with ethylene as the only monomer is polyethylene. In other words, each thermoplastic has its own starting monomer. For polypropylene, propylene (propene) is the monomer; for polybutylene, butylene (butene) is the monomer; and for polystyrene, styrene is the monomer. To produce polyvinyl chloride, one must start with vinyl chloride monomer (VCM), and to make acrylonitrile–butadiene–styrene (ABS) plastics, three monomers are needed: acrylonitrile, butadiene, and styrene. Some other thermoplastics require more than one monomer, which may not even be monomers in the manner in which we defined them. For instance, some thermoplastic polyurethanes need an isocyanate and a polyol (a type of alcohol) to react to form the polymer, and some thermoplastic polyesters may require polyethylene glycol and terephthallic acid to form PET (polyethylene terephthalate). Do not concern yourself with all the different chemicals needed to make the different thermoplastics. Whatever the method and monomers or other chemicals used, if the polymer has the ability to be reheated and re-formed without decomposing, it is a thermoplastic.

THERMOSETS

As already stated, a thermoset (or thermosetting plastic) is a plastic that can be processed by heat and pressure only once. If the part made using a thermosetting polymer does not meet specifications, it must be discarded. Any further attempt to reprocess the part will cause it to decompose. The "normal" decomposition of any plastic during compounding or processing is usually referred to as "burning up," even though no actual combustion is occurring. This decomposition may be pyrolytic in nature, or it may be oxidative. This is not to ignore the fact that if enough heat is applied in the presence of enough oxygen, the material can begin actual combustion.

Thermosetting plastics often have two raw materials that must be mixed together to promote polymerization. In most cases, both the resin and the catalyst (hardener) are liquids, and the **exothermic** reaction produced when the materials react provides the heat necessary to "set up" or cure the resin. In some cases the thermosetting material must be heated, and in other cases pressure might have to be added to effect a satisfactory "cure." The difference between thermoplastic polymerization and thermosetting polymerization is that crosslinking of the polymer chains occurs in the latter process. This crosslinking is what produces the stiffness and hardness associated with most thermosetting plastics (there are some thermoplastics that involve some crosslinking, but they are relatively few). **Crosslinking** is defined as the formation of covalent bonds *between* polymer chains, in addition to those formed *along* the chain.

Thermosets involve a chemical reaction between two or more materials, in contrast to the monomers of thermoplastics that react with themselves in a different type of chemical reaction. The properties of thermosets are usually not as variable as those of thermoplastics. The properties of the final part made from thermosetting plastics are usually predetermined by the nature of the uncured resin, whereas the properties of thermoplastics are determined by the polymerization process and substances that may be added later in a separate compounding process. Once again, thermosetting plastics include the alkyds, epoxies, melamines, phenolics, polyesters, polyurethanes, and the urea-formaldehyde plastics.

This very short discussion of plastics in general must be accepted for what it is: a very brief overview. However, for someone to understand how these materials burn, not very much additional information is necessary, beyond the chemical makeup of each polymer, alloy, blend, or compound. That information is provided in the next chapter.

SUMMARY

- Plastics are a family of materials that belong to a much larger group of materials called polymers.
- A polymer is a giant molecule made up of thousands of tiny molecules called monomers.
- Monomers are tiny, very reactive molecules that can react with themselves to form polymers.
- Polymerization is the chemical reaction in which molecules of monomers react with themselves to form long-chain molecules called polymers.
- Thermoplastics are plastics that can be processed with heat, pressure, or both, more than once without degrading.
- Thermosets are plastics that cannot be processed with heat, pressure, or both, more than once.
- Synthetic polymers are artificial polymers.
- Natural polymers are polymers that exist in nature.

4

Descriptions of Individual Plastics

This chapter describes individual plastics, their processing, and their applications. Not all plastics are described, but some time and space is devoted to each of the important plastic polymers. No time is devoted to alloys and blends, simply because it would be impossible to try to describe all the possible combinations. The physical properties of an alloy are usually a combination of the best properties of each plastic in it, in terms of the application for which it is intended. That is, if a strong, flexible, chemical-resistant material can be produced by alloying two thermoplastics, the individual polymers and copolymers will be chosen for the blend on the basis of those properties.

The descriptions of the individual plastics are of necessity very general. To cover all possible polymers, resins, compounds, alloys, and blends would be an impossible task, if only because these materials are constantly changing to meet the needs of the marketplace. New uses for these materials are being discovered every day, and variations on the existing materials, compounds, and their processing are occurring equally as fast. Consider what follows as a snapshot in time of the plastics industry, and the materials it makes. Some materials that are necessarily omitted here may be included in some future edition.

ACETALS

Acetals (the first was made in 1960) belong to the class of polymers known as engineering plastics, so-called because they can perform some loadbearing task that once required a metal part or a part made from some other "traditional" material. They are available as homopolymers and copolymers, with homopolymers having somewhat higher strength properties and higher (short-term) heat deflection properties. Acetals are chosen because of their high loadbearing characteristics, high rigidity, good mechanical strength, good dimensional stability (resistance to creep, the tendency to deform permanently under prolonged stress or pressure), high fatigue strength, low coefficient of friction against themselves and many other materials, light weight, good dielectric properties, and good chemical and corrosion resistance. They are tough at both low and high temperatures. Acetals are grease resistant and are resistant to most hydrocarbons and their derivatives. They have a glossy surface and a low coefficient of friction owing to the "waxy" surface of the finished part.

Some copolymer acetals resist attack from strong alkalies (the homopolymer is attacked), whereas both homopolymer and copolymer acetals are attacked by strong acids and oxidizing agents. Both have low water absorption, but the copolymers are more resistant to hydrolysis produced by exposure to hot water. Acetals are also attacked and made brittle by ultraviolet radiation and must be stabilized against them if the final part is to be used outdoors.

The monomer for acetals is formaldehyde, and it is the $—CH_2O—$, or oxymethylene unit that is the repeating unit. For this reason, acetal polymers are also known as polyoxymethylene (POM) or polyformaldehyde. Formaldehyde's molecular formula is HCHO. Its structural formula is shown in Figure 4.1. The structure of acetal is shown in Figure 4.2.

$$\begin{array}{c} H \\ | \\ C{=}O \\ | \\ H \end{array}$$

Figure 4.1 Stucture of formaldehyde

$$\left[\begin{array}{ccccccccccc} & H & & H & & H & & H & & H & \\ & | & & | & & | & & | & & | & \\ -O- & C & -O- & C & -O- & C & -O- & C & -O- & C & -O- \\ & | & & | & & | & & | & & | & \\ & H & & H & & H & & H & & H & \end{array} \right]_n$$

Figure 4.2 Structure of acetal

Resistance to heat and many chemicals can be increased by attaching certain hydrocarbon derivatives to the terminal group of either structure.

Since no side chains are formed in the polymerization of acetal, the chain takes the structure of a densely packed crystal, and this is what gives acetal its hardness, toughness, strength, and resiliency. The finished parts may also be painted or plated to resemble a metal part, and indeed acetals *have* replaced many metals in certain applications.

Additives

As in most other thermoplastics, additives may be compounded into acetal resins to impart particular desired qualities. Acetal's flame resistance, ultraviolet radiation resistance, impact strength, coefficient of friction, and flow rate may all be improved by the addition of such materials as flame retardants, ultraviolet light stabilizers, glass fibers or beads, elastomers, minerals, and lubricants. The polymer must be stabilized to prevent depolymerization at processing temperatures.

Processing

Whether a particular acetal is a homopolymer or copolymer dictates the process to be used in shaping the final part. However, depending on the grade, acetals may be processed in all the normal thermoplastic methods. They may be extruded, injection molded, and blow molded, which is a fairly difficult process. Copolymers are, in general, easier to process than the homopolymers. Work has been done recently to improve homopolymer processing by eliminating much of the gasing (producing strong odors) and mold deposit associated with these grades.

Applications

Aerosol containers
Appliance parts (loadbearing and functional)
Audiocassette reels and guide rolls
Automotive parts (loadbearing and functional)
Ball sockets
Bearings
Bolts
Bushings
Butane lighter bodies
Cams
Chains
Cigarette lighter bodies
Combs
Conveyor chains
Electrical switches
Electronic parts
Fans
Fasteners
Friction pads
Garden hose nozzles
Gears
Housings
Levers
Linkages
Meat hooks
Nuts
Pen and pencil barrels and tips
Plumbing fittings
Pulleys
Pump parts

Applications (Continued)

Rails
Rollers
Shower heads
Sinks
Soap dispensers
Sprayers
Springs
Toys
Valves
Videocassette reels and guide rolls
Water faucets

ACRYLICS

The term "acrylic" covers a very wide family of plastics used in an even wider range of uses. Generally acrylics are selected wherever properites such as crystallike clarity and superior weathering are required. Acrylic plastics also possess high surface hardness characteristics and resistance to many chemicals.

The most common acrylic polymers begin with the monomer methyl methacrylate (MMA) but may also be copolymerized with methyl acrylate or ethyl acrylate to produce resins having properties desired by particular processors. Whenever the impact strength must be increased, butadiene, butyl acrylate, or vinyl acrylate may be added.

A second group of "acrylic" plastics are included here that are not grouped with acrylics in other references. These materials are often referred to as "branched polyethylenes," since the backbone of the polymer molecules are polyethylene, with side branches of other polymers copolymerized onto the main chain. Technically, these materials are closer to polyethylene than to acrylics in their physical properites and uses, but they are grouped here because they all contain an "acrylate" portion. That is, even though their softness, low melting points, and adhesive properties will put them into uses sometimes reserved for polyethylene, they will burn like acrylics. Since this text is concerned with the flammability of plastics rather than a strict classification, any polymer containing an acrylate portion of the molecule will be classified an acrylic.

In the world of polymer chemistry, however, this classification is not acceptable. Also in the real world, there will never be such an accumulation of articles made from these branched polyethylenes in their final form as to constitute a fire hazard. Nevertheless, at the processing plants where the final form is produced, there *will* be large accumulations of these resins. Powdered or pelletized or cubed resin in large containers will certainly burn in a different manner than if they were in film form. It is the condition under which they will most likely be burning in large quantities that concern emergency responders, and this condition is present in the unprocessed form. Therefore, the classification as acrylics for combustion (and extinguishment) purposes is valid.

However, polymethylmethacrylate, in its processed form, is very different from these other materials, and the applications for PMMA are always in a rigid form, rather than a flexible form that the soft-branched polyethylenes take. This must be remembered when looking at the list of applications for all acrylic polymers and copolymers at the end of this section.

The molecular formula for methyl methacrylate is $CH_2{=}CCH_3COOCH_3$ and its polymer is polymethylmethacrylate (PMMA). The structure of the polymer is simply the repeating of the molecule with the end hydrogens removed.

Properties

Acrylics are very light stable and therefore are well suited to outdoor applications because of their superior weatherability. This ability to resist sunlight without the deterioration of their optical properties makes acrylics the material of choice in transparent outdoor signs, glazing materials (their transmission of visible light makes them equal to glass as window material), and other enclosing applications. In fact, whenever there is an outdoor requirement that also calls for clarity, such as automobile tail lights, acrylics are selected. Acrylics are also very popular in the design and manufacture of light diffusers used in lighting fixtures, both indoors and out.

Acrylics are resistant to solutions of inorganic alkalies and acids, as well as aliphatic (straight-chain or branched) hydrocarbons but will be attacked by aromatic (unsaturated cyclic) hydrocarbons.

Electrical properites of the acrylics are characterized by low electrical conductivity and excellent arc resistance.

Impact properties of acrylics can be improved somewhat by the addition of certain materials, but the levels of any additive must be kept low to prevent loss of clarity in the final part.

Ethylene Acrylic Acid

Ethylene acrylic acid (EAA) copolymers are branched polyethylene thermoplastics that are stronger and tougher than low density polyethylene, and they have better adhesion properties and higher melt strength. The structural formula for ethylene acrylic acid is $CH_3(CH_2)_2CHCOOHCH_3$. The structure for the polymer is complicated, but it contains the same elements as the other acrylics.

The polymer of ethylene acrylic acid is similar to polyethylene in its water and chemical resistance and in its gas transmission properties. It can be processed by extrusion. The use of ethylene acrylic acid films includes food packaging, medical item packaging, paperboard coating, blister packs, and as a laminating film used with aluminum foil.

Ethylene Butyl Acrylate

Ethylene butyl acrylate is a branched polyethylene thermoplastic that is used when a very tough, low melt-index film is required. However, the clarity of the film is very poor, and its major use is as a wrapping film for frozen foods. The molecular formula for ethylene butyl acrylate is $CH_3(CH_2)_2CHCOO(CH_2)CH_3$. Its polymer has a rather large structure, but contains only carbon, hydrogen, and oxygen, just as do the other acrylic plastics.

Ethylene–Ethyl Acrylate

Ethylene–ethyl acrylate (EEA), another branched polyethylene, differs from ethylene methacrylate only in having a slightly different structural formula. but it has strikingly different properties. The molecular formula for ethylene–ethyl acrylate is $CH_3(CH_2)_2CHCOO(CH_2)CH_3$.

Ethylene–ethyl acrylate has better heat resistance than ethylene methacrylate and low temperature flexibility, and it is not as transparent. It is among the toughest of the polyolefin films, and because of its toughness it is used to make specialty hoses and tubing. Ethylene–ethyl acrylate can be injection molded and blow molded in addition to being extruded. The impact resistance of some engineering resins can be improved by blending ethylene-ethyl acrylate with them.

Ethylene Methacrylate

Ethylene methacrylate (EMA) copolymer is a branched polyethylene that is softer and has better heat stability than ethylene–vinyl acetate. The molecular formula for ethylene methacrylate is $CH_3(CH_2)_2CHCOOCH_3CH_3$, showing that it contains only carbon, hydrogen, and oxygen in its molecule.

It is extruded principally as a film, and its major uses include food packaging, medical packaging, aseptic cartons, paperboard coating, binders for nonwoven fibers, adhesion laminating, disposable gloves, and composite cans.

Ethylene Methacrylic Acid

Ethylene methacrylic acid (EMAA) is a branched polythylene thermoplastic very similar to ethylene acrylic acid in its uses, and it has better adhesion properties than ethylene acrylic acid. It has higher melt strength than polyethylene. The molecular formula for ethylene methacrylic acid is $CH_3(CH_2)_2CCH_3COOHCH_3$.

Ethylene methacrylic acid has water resistance, chemical resistance, and gas transmission properties similar to those of low density polyethylene. It can be processed by extrusion.

Processing

Some acrylics are available in the liquid monomeric form, since the processor may wish to coat a material or to produce sheets, rods, or tubes by casting (pouring) the liquid into forms that will produce the desired shape. Care must be taken when acrylic is used in conjunction with another material because of its high coefficient of thermal expansion, which is higher than that of most other thermoplastics. The liquid monomer may also be cast onto a moving belt that carries the monomer through a heated area where polymerization takes place, producing cast acrylic sheet in a continuous manner. The sheet may then be cut to the customer's specification.

Acrylics are also available in bead or pellet form for processes such as extrusion and injection molding. Sheet, as well as specific profiles, may be produced by extrusion. Sheet that is cast or extruded may be thermoformed into parts by heating the sheet and "drawing it down" over a male or female mold by using vacuum to form the parts. The sheet may also be compression molded into a finished part by heating and then pressing the sheet into a mold. Extruded sheet may be thermoformed more easily than cast sheet.

Additives

Even though one of acrylics' best properties is its outstanding clarity, the sheet is often made opaque by adding pigmentation. If the transparency of the polymer is to be maintained, transparent organic pigments or dyes may be added to color the "clear," uncolored sheet. Acrylics can be colored rather easily because of this inherent transparency of the polymer. Its clarity and ease of pigmentation make it very useful in the manufacture of transparent tinted glazing, where much of the harmful (and hot) radiation of sunlight may be removed while still allowing visible light to pass through.

Applications for All Acrylics

Adhesives
Aircraft enclosures
Appliance panels
Automobile tail lights
Blister packs
Building enclosures
Display cases
Food packaging
Handle grips
Heat-seal films
Knobs
Light diffusers
Medical packaging
Sanitary ware
Signs
Skylights

Applications for All Acrylics (Continued)

Spas	Telephone buttons
Swimming pools	Toys
Toothpaste tubes	Windows
Telephone switchhooks	

ACRYLONITRILE–BUTADIENE–STYRENE

Acrylonitrile–butadiene–styrene (ABS) is a thermoplastic that has three monomers as its building blocks. It began its life as an engineering thermoplastic but now ranks as one of the highest volume plastics in the United States, even though it is still considered an engineering thermoplastic. Since it is made up of three "building block" materials, the final resin is usually referred to as a compound.

Acrylonitrile–butadiene–styrene is not strictly a terpolymer (three monomers polymerized together) but is really an alloy of a copolymer of styrene–acrylonitrile and butadiene. The three monomers are selected for the properties they bring to the final compound. Acrylonitrile provides heat stability and good chemical resistance to the alloy; styrene adds stiffness and processibility; whereas butadiene provides toughness and impact strength. The relative amounts of each of the monomers in the alloy determine the actual properties for the particular compound required by the end user. Because of the presence of styrene in the alloy, ABS is often called a *styrenic* resin. Since it is such a high-volume material, ABS is discussed here rather than as part of the styrenics covered in another section.

The molecular formulas for the three monomers that make up acrylonitrile–butadiene–styrene are:

Acrylonitrile	C_2H_3CN
Butadiene	C_4H_6
Styrene	$C_2H_3C_6H_5$

The structural formulas for the three monomers are shown in Figures 4.3, 4.4, and 4.5.

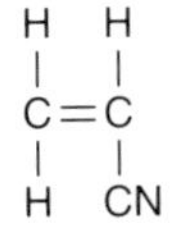

Figure 4.3 Structure of acrylonitrile

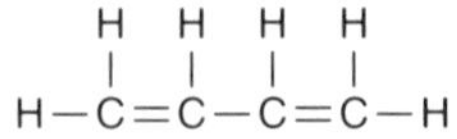

```
    H   H   H   H
    |   |   |   |
H — C = C — C = C — H
```

Figure 4.4 Structure of butadiene

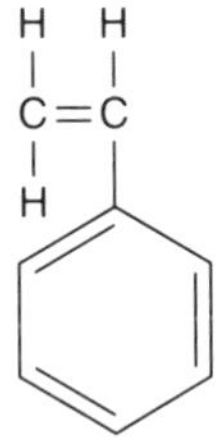

```
H   H
|   |
C = C
|   |
H
```

Figure 4.5 Structure of styrene

The structural formula for acrylonitrile–butadiene–styrene is a combination of the structural formulas for acrylonitrile, butadiene, and styrene, brought together in the polymerization process.

Acrylonitrile–butadiene–styrene is a very tough and abrasion-resistant material; therefore it is used in areas where treatment of the final part can be expected to be rough on certain occasions. It has good mechanical strength, hence its designation as an engineering resin. Lengthening the polymer chain adds to the strength characteristics of acrylonitrile–butadiene–styrene.

It has relatively poor weatherability, so when it is to be used to make parts that will be exposed to the environment (that is, sun and weather), it is usually cap-coated (laminated) with a thin layer of another plastic that does have good weathering properties.

In some specific applications, the required properties specified by an end user may not be attainable with an existing ABS resin. In those situations, acrylonitrile–butadiene–styrene can be further modified through additional alloying. This is done by taking an existing ABS resin and blending it with other thermoplastics such as polycarbonate (PC), polysulfone (PSO), styrene–maleic anhydride (SMA), or polyvinyl chloride (PVC) to produce a material that will have the properties required.

Additives

Although acrylonitrile–butadiene–styrene is top-coated or cap-coated when it must be used outdoors, light stabilizers (ultraviolet absorbers) are often added to the compound if it will be used where sunlight may pass through a window onto it, such as in automobile parts and other articles where color retention is necessary. Constant work is being done by chemical companies to improve the UV (ultraviolet) stability of acrylonitrile–butadiene–styrene.

Colorants are almost always added to acrylonitrile–butadiene–styrene, since the natural color (a milky white to yellowish white) is not acceptable for most applications. Both organic and inorganic pigments as well as special-effect colorants may be used to color acrylonitrile–butadiene–styrene.

Other additives, such as antioxidants, antistatic agents, fillers, flame retardants, foaming agents, impact modifiers, and lubricants may be used to zero in on the exact properties required by the end user of the article.

Processing

Acrylonitrile–butadiene–styrene is processed by extrusion, injection molding, blow molding, rotomolding, and thermoforming. The resin may have to be dried before processing since it is mildly hygroscopic. It softens over a relatively wide spread of temperatures rather than melting rapidly. ABS parts are able to accept chrome plating, so metallic-looking parts may be processed and plated much more cheaply than using real metal for the parts. Flame retardant acrylonitrile–butadiene–styrene is ideal for business machine housings and housings for tools and household appliances.

Applications

Aircraft parts
Appliance housings and parts
Automotive parts
Bathtubs
Battery cases
Boat parts
Business machine housings and parts
Cabinets
Camera housings
Camper bodies
Canoes
Chrome-plated parts
Conduit
Crates
Fans
Fume hoods and ducts
Football helmets
Furniture parts
Highway safety devices
Housewares
Lawn and garden instruments
Luggage
Medical disposables
Mobile home parts
Office equipment
Packaging
Pickup truck bed liners
Picture and mirror frames
Pipe and fittings
Power tool housings
Radio housings
Recreational vehicle parts
Refrigerator inner liners and parts
Safety equipment
Sanitary ware
Seatbacks
Spas
Swimming pool filter housings
Telephones
Tools
Toys
Vacuum cleaners
Wall surrounds

CELLULOSICS

The polymer that exists in the largest volume in the world today (and in all of recorded history) is the naturally occuring polymer cellulose. Cellulose is the major ingredient in wood and woody plants, and cotton is almost pure cellulose. Actually, cellulose is the major constituent of *all* plant tissue and fibers, and if one can visualize the volume of plant life on earth, it becomes apparent rather quickly that Mother Nature carries out the largest polymerization operation imaginable. Naturally occurring cellulose has the chemical formula $(C_6H_{10}O_5)_n$. The *n* in the formula represents the number of cellulose molecules in the polymer. It is used whenever a chemical formula for any polymer is given.

"Cellulosics" is a general name for a family of plastics containing (or using as a beginning raw material) cellulose. The properties of this family of plastics are very wide ranging but can be summarized by clarity, glossiness, toughness, and versatility in processing. The designer of finished parts has a very wide spectrum of materials from which to choose, since the properties of individual cellulosics may be modified by the use of additives.

Cellulose Nitrate

Since the purpose of producing synthetic polymers was to replace or improve upon natural polymers, it was logical to start by emulating nature. The very first polymer made successfully to solve a particular commercial problem was a cellulosic to specifically cellulose nitrate. It was developed in answer to a problem in the manufacture of billiard balls, which were made of ivory from the tusks of elephants.

Cellulose nitrate had excellent physical properties, such as good toughness, good dimensional stability, and low water absorption. But one specific property made it very hazardous to use: it was very combustible. It burned not only readily but very rapidly. Its use as motion picture film led to many large movie house fires that took many lives, and because of these tragedies, cellulose nitrate fell into disfavor. It was not used extensively again until the development of additives that allowed it to be used safely. When it returned as motion picture film, it was dubbed "Safety Film." Even in its safer formulation, however, cellulose nitrate has very poor weatherability. It degrades quickly in sunlight or high heat conditions, turning yellow, then amber, then losing its toughness and becoming brittle and useless. Its most common use today is not as a plastic from which to make final parts but as a dissolved base for lacquers, cements, and adhesives, and it is also used as a base for explosives.

Other Cellulosics

The three major cellulosic plastics today are cellulose acetate (CA), cellulose acetate butyrate (CAB), and cellulose acetate propionate (CAP) (three of the so-called cellulose ester polymers). Other variations use cellulose as the central building block, and the final choice of the proper cellulosic to use depends on the specifications for the final, shaped article.

Cellophane

Cellophane is a film made from the natural cellulose extracted from wood pulp. It is also known as regenerated cellulose. It is not technically a plastic, although it is made from a naturally occurring polymer. The cellulose is treated and converted to an ester which *is* a synthetic polymer, extruded into fibers or film, and then treated with an acid to reconvert it to cellulose. Its major use is as a packaging film.

Celluloid

Celluloid is a trademarked name for a cellulosic plastic made by dissolving cellulose nitrate in camphor or another plasticizer and adding a flame retardant. The resulting solution is solid. It is also known as pyroxylin.

Cellulose Acetate (CA)

Cellulose acetate has good chemical resistance and good hardness and stiffness properties. Sheet made from cellulose acetate can be thermoformed by conventional methods, and it can be finished in many forms. It may be colored (opaque or translucent), or it may be made in its natural transparent stage. It has good electrical properties. Cellulose acetate also has good electrical insulating properties.

Cellulose Acetate Butyrate (CAB)

Cellulose acetate butyrate is superior to cellulose acetate in weatherability, low temperature impact strength, dimensional and color stability, and toughness. It has faster cycle times in injection molding and extrusion than cellulose acetate but equal cycles to cellulose acetate propionate. Both cellulose acetate butyrate and cellulose acetate propionate also thermoform well.

Cellulose Acetate Phthalate

Cellulose acetate phthalate is made by reacting cellulose acetate with phthalic anhydride. It is used as a coating for medicinal tablets and capsules.

Cellulose Acetate Propionate (CAP)

Cellulose acetate propionate is harder and stiffer than butyrate or cellulose acetate and has a higher tensile stength. The properties of the three cellulose ester plastics can be changed by the addition of plasticizers, which will reduce hardness and stiffness but increase the impact strength of the finished parts.

Cellulose Triacetate

Cellulose triacetate is another cellulosic ester, but it is different from the three major ester plastics (CA, CAB, and CAP) in that plasticizers do not soften it very much, and its softening temperature is very high. Sheeting and film are made by dissolving the cellulose triacetate, casting it onto a surface, and removing the solvent by evaporation. Fibers are made by extruding a very thick solution of the dissolved polymer and evaporating the solvent. Products made from cellulose triacetate have good clarity, dimensional stability, and water resistance, and they are resistant to most common solvents.

Ethyl Cellulose

Ethyl cellulose is another cellulosic plastic. It does not have the exceptional clarity of some of the other cellulosics, but it processes well, can be made somewhat heat- and impact-resistant, and can be used in contact with food. It is very tough at low temperatures and is better than some other cellulosics in this property. Ethyl cellulose is sometimes referred to as the impact grade of cellulosics.

Rayon

Rayon is made in much the same way as is cellophane, but an extra chemical reaction occurs, and the reconverting to cellulose does not take place. Rayon, therefore, is a synthetic polymer. This regenerated cellulose fiber is known as "viscose," "viscose rayon," or just rayon. It is used to make blankets, clothing, coated fabrics, felts, home furnishings (when blended with cotton), nonwoven fabrics, and surgical dressings.

Miscellaneous Cellulosics

Other cellulosics include methyl cellulose, carboxymethyl cellulose, hydroxyalkyl cellulose, cellulose propionate, and cellulose xanthate. The last two are cellulose ester polymers, whereas the others are cellulose ethers. The ether plastics are very often made into coatings, some of which are soluble in water.

Processing

Most cellulosics may be processed on all conventional plastics processing equipment. Because of their instability to heat, cellulose nitrate resins are usually processed as liquid solutions.

Additives

The major additive for injection molded cellulosics is glass fiber, added to give strength to the final part.

Applications

- Appliance parts
- Audiotape
- Automotive trim
- Blister packaging
- Buttons
- Cigarette filters
- Combs
- Engineering/drafting tools
- Eyeglass frames
- Face shields
- Fibers
- Flashlight cases
- Gold-stamped foils
- Hairbrushes
- Handles
- Knobs
- Lacquers
- Magnetic tapes
- Osmotic cell membranes
- Packaging
- Pens
- Photographic film
- Pipe
- Pressure-sensitive tape
- Signs
- Steering wheels
- Strippable coatings
- Tags
- Tail-light lenses
- Telephones
- Toothbrushes
- Tool handles
- Toys
- Tubular clothing packages
- Venetian blind wands

CHLORINATED POLYETHYLENE

Chlorinated polyethylene is a polymer that is halogenated after polymerization. It is considered a partially halogenated polymer and is softer than polyethylene, is more rubberlike, and more soluble. It may be used alone, extruded, or calandered into sheets, which are very effective as roofing material. A higher-volume use, however, is as an additive to other plastics. When added to rigid polyvinyl chloride, it acts as an impact modifier, giving the final product more impact strength and avoiding the use of a plasticizer. When added to a polyolefin or another polymer, it will act as a flame retardant because of the presence of chlorine in its molecule.

ENGINEERING PLASTICS

The plastics described briefly here are not common materials that one would encounter in everyday uses, as one would find polyvinyl chloride, polyethylene, or polystyrene. Rather, these polymers were developed and engineered to solve very specific problems, and they have very specific, limited uses. They are produced and sold in such small quantities (relatively speaking), and the amount of material in use in any one place at any given time is so small, that they do not constitute a significant contribution to any fire load.

Many polymers have earned the name "engineering plastics" but have become so versatile and so popular that their sheer volume and number of uses have placed them in a high-volume category. Acrylonitrile–butadiene–styrene (ABS) is the leading material in this category. The following polymers are not among those commonly used, and some of their names are strange sounding. These are true engineering polymers, and their rather demanding specifications and subsequent high prices will always limit their usage.

Engineering Thermoplastic Elastomers

Engineering thermoplastic elastomers (ETE) are thermoplastic copolymers that incorporate the versatility of the "normal" nonengineering thermoplastic elastomers (TPE) with increased strength, creep resistance, and higher service temperatures (the maximum temperature at which the part made from the polymer will satisfactorily perform its function). These materials are mostly copolyester elastomers.

Processing

Engineering thermoplastic elastomers may be processed readily be extrusion, injection molding, blow molding and rotational molding.

Applications

Applications for engineering thermoplastic elastomers include use in the automotive, electrical, business machine, communications, and medical instrumentation industries.

Additives

For such engineering thermoplastics as polyphenylene oxide and polyphenylene sulfide, typical additives include fillers such as calcium carbonate, clay, colloidal silica, glass, graphite, carbon fibers, additive inorganic and organic flame retardants, and organic and inorganic pigments.

Polyarylates

Polyarylate polymers are aromatic polyesters with noncrystalline structures. They are very tough, heat-resistant, warp-resistant polymers with good weatherability. They have good electrical properties, low moisture absorbance, and good creep resistance.

Processing

Polyarylate polymers may be extruded, injection molded, blow molded, and thermoformed.

Applications

Polyarylate polymers find use as door handles, brake light housings, headlight housings, windshield washers and window trim in automobiles, wire insulation, and circuit boards in the electrical/electronics industry, solar energy collectors, tinted glazing, and construction components.

Polyarylsulfone

Polyarylsulfone (polyphenylsulfone) is a thermoplastic engineering resin with high heat resistance, good electrical properties, and good mechanical properties similar to those of other aromatic sulfones. It has been sanctioned by the FDA (Food and Drug Administration) for use in food contact.

Processing
Polyarylsulfone can be processed on conventional extruders and injection molders.

Applications
Applications for polyarylsulfone include aircraft parts, chemical process piping, circuit boards, electrical parts, electroplated parts, lamp housings, and motor parts.

Polyetheretherketone

Polyetheretherketone (PEEK) is a strong, tough, rigid aromatic ketone–based thermoplastic that has good loadbearing properties, good abrasion and wear resistance, and good chemical resistance. It will withstand attack from most solvents and all but the most concentrated acids.

Processing
Polyetheretherketone can be extruded and injection molded.

Applications
Polyetheretherketone finds applications as wire and cable insulation, automotive engine parts, aerospace composites, and monofilaments.

Polyetherketone

Polyetherketone (PEK) is a tough, strong, rigid, aromatic ketone–based thermoplastic with good loadbearing properties, good wear and abrasion resistance, and good chemical resistance. It has all the resistance to solvents and acids that polyetheretherketone has and may be somewhat easier to process.

Processing
Polyetherketone can be extruded, injection molded, spun, and cold formed.

Applications
Polyetherketone has applications in printed circuit boards, automotive engine parts, and very demanding aerospace composites.

Polyethersulfone

Polyethersulfone (PES) is a high performance engineering thermoplastic with high heat resistance, good electrical properties, high strength and rigidity, good water resistance, and good transparency. It is resistant to most inorganic chemicals and most hydrocarbons but can be attacked by esters, ketones, some chlorinated hydrocarbons, and polar aromatic solvents. It is a member of the polysulphone family.

Processing

Polyethersulfone can be processed on conventional extrusion, injection molding, and blow molding equipment.

Applications

Polyethersulfone is used both in glass-filled and unfilled forms, and its applications include use in aircraft components, electrical parts, gearboxes, lamp sockets, pump housings, small appliance parts, water meters, and medical applications.

Polyimides

Thermoplastic polyimides are aromatic polymers with fairly complicated chemistry, containing the imide group (—CONCO—) in the polymer molecule. It is a very tough polymer, with good electrical properties and excellent heat resistance.

Processing

Thermoplastic polyimides are not easily processed, and any forming must be done before the resin is fully reacted. They can be cast into film by dissolving the resin, pouring it, and evaporating the solvent, and the powder may be compression molded.

Applications

Uses for parts made from polyimide are in applications where high temperatures are present and the part must bear a load. This makes them useful as gaskets, compression bearings, piston rings, and nonlubricated seals. Their low coefficient of friction allows them to replace polytetrafluoroethylene (PTFE) in some applications.

Polyamide–Imides

Polyamide–imides (PAI) are high temperature thermoplastics with better strength properties than most engineering resins. They have superior thermal stability and high strength ranging from –32°F to 450°F. They have good chemical properties, resisting attack from most acid and base solutions, halogenated solvents, alkanes, alkenes, and aromatic solvents. They have low coefficients of friction and high impact strength. Compounding with reinforcing additives produces even stronger materials. Disadvantages include a low resistance to some acids, hot caustic solutions, and steam, and they are somewhat hygroscopic.

Processing
Polyamide–imides may be processed on extruders, injection molders, and compression molders. Soluble powders are available for coating applications.

Applicatons
Applications for polyamide–imides include the replacement of metal parts in compressors, generators, and jet engines. Metal replacement parts are also made for automobiles, business machines, and heavy equipment.

Polyetherimide

Polyetherimide is a thermoplastic engineering resin that displays excellent physical and chemical properties over a very wide range of demanding applications. It has very broad chemical resistance, good electrical properties, high strength and rigidity even at elevated temperatures, and long-term heat resistance. Its chemical resistance includes strong acids and most alcohols, hydrocarbons, and fully halogenated hydrocarbon solvents (it is dissolved by partially halogenated hydrocarbon solvents), but it is attacked by long-term exposure to bases.

Polyetherimide may be reinforced by the addition of glass fibers, thereby significantly increasing the polymer's inherent strength.

Processing
Once the material is dried, polyetherimide may be processed by extrusion, injection molding, blow molding, fiber spinning, thermoforming, melt extrusion, and solvent casting.

Applications
Polyetherimide has one of the widest ranges of application for the very high performance engineering thermoplastics, being used in high temperature applications in automotive under-the-hood parts, jet engine parts, fuel system components, and medical applications where sterilization by autoclaving, ethylene oxide, or gamma radiation techniques are used. It is used in electrical and industrial applications where strength at elevated temperature is required, replacing metal parts. It also has uses in high performance requirements in packaging,

appliance parts, optic fiber components, sporting goods, microwaveable cookware, airplane interior sheeting, foamed structural panels, and impact-resistant composite materials.

Polyphenylene Oxide

Polyphenylene oxide (PPO), also known as polyphenyene ether (PPE) or polyphenylene oxide, modified, is an engineering polymer with a very high heat distortion temperature and good flame-retardant characteristics. However, polyphenylene oxide is usually alloyed with polystyrene to make a wide range of polymer blends with customized properties. The materials sold as polyphenylene oxide are really polyphenylene oxide–polystyrene alloys, and the material is generally known as modified polyphenylene oxide. It has excellent water and corrosion resistance and good chemical resistance. It will accept a wide spectrum of additives that will increase its impact strength and other properties. Polyphenylene oxide is affected by organic solvents, although this can be counteracted by alloying with nylon rather than with polystyrene.

Processing

Modified polyphenylene oxide may be processed on injection molding machines, blow molding machines, and extruders.

Applications

Uses for modified polyphenylene oxide include appliances (coffee brewers and dispensers, hair dryers, portable mixers, power tools, pump housings and impellers, tape cartridge platforms) and automotive uses (connectors, fuse blocks, instrument panels, mirror housings, rear spoilers, seat backs, wheel covers, instrument panels, mirror housings, spoilers and wheel covers). Business machine, telephone, and appliance parts may be made from polyphenylene oxide alloys, as may parts that must be in contact with hot fluids. Other uses include electrical (bus bar insulators, switch housings, terminal blocks, light fixtures), pump and plumbing parts (shower heads, sprinklers), and telecommunications and business machines (card frames, CATV housings, consoles, housings, keyboard bases, printer bases).

Polyphenylene Sulfide

Polyphenylene sulfide (PPS) is a crystalline aromatic polymer whose repeating unit is a benzene ring with a substituted sulfur atom in the link position.

Polyphenylene sulfide is an engineering plastic that has good chemical resistance, good electrical properties, and high temperature resistance. Because of its high crystalline structure, polyphenylene sulfide has good dimensional stability. Its chemical resistance is so good that no material will dissolve it at normal temperatures, and very few chemicals will react with it.

Processing

Polyphenylene sulfide is usually injection molded, but some new applications require extrusion. A newly available compound allows blow molding of the material. In any event, processing temperatures are relatively high.

Applications

Polyphenylene sulfide's high performance properties lend it to uses where a part must resist chemical action, high temperatures, or both. These uses include many appliance parts, under-the-hood automotive parts, boiler sensors, chemical pumps, electrical system parts, lamp housings, office equipment, pumps, valves, monofilaments, motor insulators, and capacitors.

Polysulfone

Polysulfone, or polysulphone (PSO), is a family of high performance thermoplastic engineering polymers with good electrical properties, good heat resistance, good transparency, and good resistance to corrosive chemicals. It is attacked by ketones, chlorinated hydrocarbons, and aromatic hydrocarbons. It is sanctioned by the FDA for use in food contact products. The first commercial polysulfone was polyethersulphone, which is discussed elsewhere.

Processing

Polysulfone may be processed on most common thermoplastic processing equipment, including molding, extrusion, and thermoforming. Polysulfone will absorb some moisture and therefore must be dried before processing through extruders or injection molding, but finished parts may be used in conditions where it is exposed to moisture at high temperatures.

Applications

Polysulfone has applications in appliances and appliance housings, automotive and aerospace components, battery cases, chemical processing equipment, circuit carriers, coil bobbins, electrical equipment, food processing equipment, medical instrumentation, microwave cookware, printed circuit boards, switches, terminal blocks, tube bases, and water purification devices.

FLUOROPOLYMERS

Fluoropolymers are a class of polymers that are derived from alkanes and that have one or more fluorine atoms substituted for hydrogen atoms. They are not necessarily analogs of the chlorine-substituted polymers, since more fluorine may be substituted on the chain. This is possible because the fluorine atom is smaller than the chlorine atom.

The fluoropolymers have properties not possessed by the chlorine-containing polymers such as polyvinyl chloride, chlorinated polyvinyl chloride, chlorinated polyethylene, and polyvinylidene chloride, and they are selected for use because of these differences in properties.

Some of these materials are made by polymerizing two monomers together in a process called copolymerization. This is done to take the best properties of the individual polymers and try to combine them, or to compensate for the lack of a desired property in one of the polymers. The resulting material is called a **copolymer**.

Ethylene–Chlorotrifluoroethylene

Ethylene–chlorotrifluoroethylene (ECTFE) is made by copolymerizing ethylene and chlorotrifluoroethylene to form a copolymer having a molecular structure that has alternating units of each monomer. Ethylene–chlorotrifluoroethylene benefits from this copolymerization by having certain more desirable properties than many of the other fluoropolymers. It is resistant to most chemicals and solvents, even at elevated temperatures. It has higher strength and wear resistance than most other fluoropolymers, and it can be processed on all conventional plastics processing equipment. It is not only extrudable and injection moldable, it is also available in a foamable grade. Its powders can also be rotation molded, electrostatically deposited as a coating, and used as a fluidized bed coating. Its molecular structure is shown in Figure 4.6.

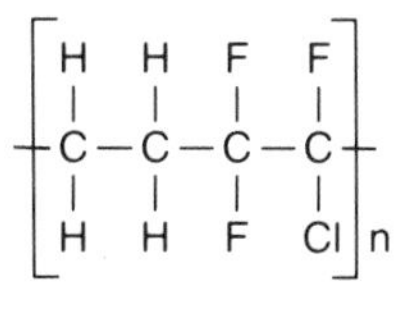

$$\left[\begin{array}{ccccccc} \mathrm{H} & & \mathrm{H} & & \mathrm{F} & & \mathrm{F} \\ | & & | & & | & & | \\ -\mathrm{C} & - & \mathrm{C} & - & \mathrm{C} & - & \mathrm{C}- \\ | & & | & & | & & | \\ \mathrm{H} & & \mathrm{H} & & \mathrm{F} & & \mathrm{Cl} \end{array}\right]_n$$

Figure 4.6 Shorthand method for drawing the structure of ethylene–chlorotrifluoroethylene

Ethylene–chlorotrifluoroethylene finds important uses such as wire and cable insulation, wire connectors, pump valves and other pump components, and coatings in chemical process applications. Other applications include jacketing for optic fiber transmission devices, industrial release applications, and lithium battery construction. Sheet and film made from ethylene–chlorotrifluoroethylene, both supported and unsupported, are used in areas where corrosion resistance must be imparted to pipe, tanks, and other containers.

Ethylene–Tetrafluoroethylene

Ethylene–tetrafluoroethylene (ETFE) is made by copolymerizing ethylene and tetrafluoroethylene to form a copolymer with alternating units of both monomers. Figure 4.7 shows these alternating units.

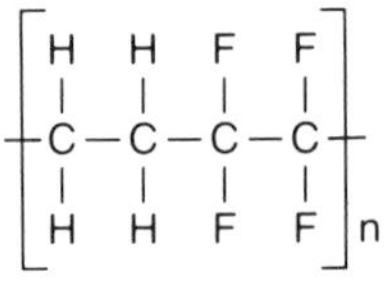

$$\left[-\overset{\displaystyle H}{\underset{\displaystyle H}{\overset{|}{\underset{|}{C}}}} - \overset{\displaystyle H}{\underset{\displaystyle H}{\overset{|}{\underset{|}{C}}}} - \overset{\displaystyle F}{\underset{\displaystyle F}{\overset{|}{\underset{|}{C}}}} - \overset{\displaystyle F}{\underset{\displaystyle F}{\overset{|}{\underset{|}{C}}}} - \right]_n$$

Figure 4.7 Shorthand method for drawing the structure of ethylene–tetrafluoroethylene

Ethylene–tetrafluoroethylene is one of the toughest of the fluoropolymers and has excellent impact and mechanical properties, including chemical resistance and electrical resistance. It can be extruded, injection molded, rotation molded, electrostatically deposited, and used in fluidized bed applications. It is used in wire and cable insulation, chemical process applications, pump components, valves, electrical components, and tie wraps.

Fluorinated Ethylene Propylene

Fluorinated ethylene propylene (FEP) is a copolymer. It has good weathering properties and is very soft. It has a low coefficient of friction and is chemically inert.

Its softness means that fluorinated ethylene propylene has low wear resistance and a low tensile strength, limiting its use in engineering applications. It can be extruded and injection molded into many useful shapes. Its resistance to chemical attack makes it useful in applications involving chemical processing equipment, roll-milling equipment, and transfer equipment such as piping and valve linings. Its high fire resistance makes it useful in a wide variety of wire and cable applications. It is also useful as a glazing material for solar energy collectors and in the manufacture of electronic instruments.

Perfluoroalkoxy Resin

Perfluoroalkoxy (PFA) resin is a relatively new fluoropolymer. Its structural formula is rather complicated and beyond the scope of this book.

Perfluoroalkoxy resin contains a totally fluorine-substituted alkyl radical attached to the oxygen atom. This radical, C_nF_{2n+1}, is similar to the radical of any alkane (C_nH_{2n+2}), except that each hydrogen atom in the molecule has been replaced by a fluorine atom. This particular radical will be selected because of the properties it will impart in the final molecule.

Perfluoroalkoxy resin is very similar to polytetrafluoroethylene and fluorinated ethylene propylene in its properties, except that the wear and creep resistance and tensile strength of perfluoroalkoxy resin are somewhat better, giving it more engineering uses. It can be injection molded and extruded, and in its powder form can be rotation molded into shapes or coated onto other parts. It has good chemical resistance and has many applications in the chemical processing industry as a coating to protect equipment and piping. It can also be extruded into wire coatings in most wire and cable applications requiring the properties of a fluoropolymer.

Additional uses for perfluoroalkoxy resin include the manufacture of chemical tanks, pipes, pump parts, valves, pump bearings and impellers, electrical equipment, and antiadhesive coatings.

Polychlorotrifluoroethylene

Polychlorotrifluoroethylene (PCTFE) is, obviously, a chlorofluoropolymer, since it contains both chlorine and fluorine. Its structural formula is shown in Figure 4.8.

$$\left[\begin{array}{cc} F & F \\ | & | \\ -C- & C- \\ | & | \\ F & Cl \end{array} \right]_n$$

Figure 4.8 Shorthand method for drawing the structure of polychlorotrifluoroethylene

Polychlorotrifluoroethylene is different from most of the other fluoropolymers in that it is soluble in some solvents at elevated temperatures, and chlorinated solvents will penetrate the structure and cause the polymer to swell. However, when products made from polychlorotrifluoroethylene are used at temperatures that do not exceed 72°F, they are inert to most chemicals. It is not processed on normal plastics processing equipment because of its high viscosity: Attempts to process it in extruders and injection molding equipment will cause the polymer to decompose at the temperatures needed to run it through the equipment.

However, polychlorotrifluoroethylene has some outstanding properties that make it the material of choice in some applications. Where water vapor transmission must be prevented in an application calling for a transparent film, polychlorotrifluoroethylene is chosen. Additionally, it has excellent transmission barrier properties for all common gases, and it can be made into transparent sheets in thicknesses up to 125 mils. It is also available in finished forms, such as film, sheeting, rods, and tubes. Its outstanding resistance to chemical attack make it useful in the manufacture of chemical tanks, pumps, pipes, tank linings, connector inserts, valve diaphragms, and insulations.

Polytetrafluoroethylene

Probably the most common fluoropolymer is polytetrafluoroethylene (PTFE). Its trade name is Teflon™.

The structural formula for polytetrafluoroethylene is shown in Figure 4.9. It is a crystalline polymer with a melting point of approximately 620°F, and it has such good mechanical properties that it can be used at temperatures ranging from considerably below 0°F up to 500°F. It has a very good resistance to many chemicals and a high impact strength. Its very low coefficient of friction makes it very valuable in nonstick applications such as in common cookware, and in such high-tech applications such as earthquake-proofing of buildings by insertion of doughnut-shaped pieces of molded polytetrafluoroethylene to lessen the rigidity of crucial joints in the building.

$$\left[\begin{array}{ccc} & F & & F & \\ & | & & | & \\ - & C & - & C & - \\ & | & & | & \\ & F & & F & \end{array} \right]_n$$

Figure 4.9 Shorthand method for drawing the structure of polytetrafluoroethylene

It is considered an engineering resin, but those properties characteristic of good engineering resins, such as tensile strength and wear resistance are very low in polytetrafluoroethylene. When specific mechanical properties are desired, additives such as glass, graphite, and metal fibers may be compounded into the polymer to provide those properties.

Polytetrafluoroethylene cannot be processed (extruded or injection molded) like the volume thermoplastics because of its extremely high melt viscosity. Therefore, methods of processing used for other materials, such as powdered ceramics and powdered metals, must be adapted to produce the desired shapes of articles made from polytetrafluoroethylene.

Additional uses for polytetrafluoroethylene include the manufacture of chemical tanks, pipes, pump parts, valves, pump bearings, and impellers, electrical equipment, and antiadhesive coatings.

Polyvinyl Fluoride

Polyvinyl fluoride (PVF) is the fluorinated analog of polyvinyl chloride (PVC). Its structural formula is shown in Figure 4.10.

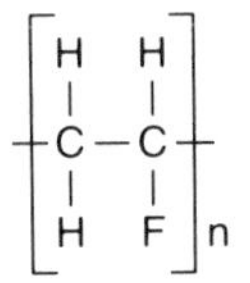

$$\left[-\overset{\displaystyle H}{\underset{\displaystyle H}{\overset{|}{\underset{|}{C}}}} - \overset{\displaystyle H}{\underset{\displaystyle F}{\overset{|}{\underset{|}{C}}}} - \right]_n$$

Figure 4.10 Shorthand method for drawing the structure of polyvinyl flouride

The polymer is not as versatile as PVC, since it is available only as a film. However, that film is very tough and flexible, and has excellent weathering characteristics. It also has excellent abrasion resistance and may be used as a laminate over other materials such as hardboard, metal foils, reinforced polyesters, vinyl, and wood.

Polyvinylidene Fluoride

Polyvinylidene fluoride (PVDF) is the fluorinated analog of polyvinylidene chloride (PVDC). Its molecular formula is $C_2H_2F_2$ and its structural formula is shown in Figure 4.11.

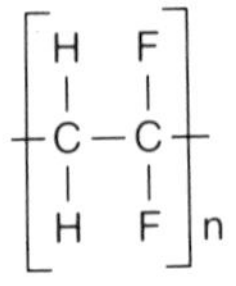

$$\left[\begin{array}{cc} H & F \\ | & | \\ -C- & C- \\ | & | \\ H & F \end{array} \right]_n$$

Figure 4.11 Shorthand method for drawing the structure of polyvinylidene flouride

Polyvinylidene fluoride is a much higher molecular weight polymer than fluorinated ethylene propylene, perfluoroalkoxy resins, and polytetrafluoroethylene, and it has much greater strength and wear resistance than those polymers. It has good chemical solvent, and oxidizer resistance and good weatherability. It can be extruded and injection molded, and coatings can be made by dry powder and dispersion coating techniques.

Applications for polyvinylidene fluoride include film, monofilament, rods and sheeting made for seals, chemical piping and fittings, gaskets, electrical jackets and primary insulation, and several finishes.

MISCELLANEOUS PLASTICS

These plastics are gathered together because there is no one family of plastics into which they could be identified and still be understood by a someone who is neither a polymer chemist nor in the plastics industry. They may be categorized by other reference books into classifications such as "branched polyethylene," "vinyls," or "styrenics," but they are gathered here because I believe that to be less confusing in a text of this type. These materials are not engineering resins, which themselves are gathered into their own classification under that name.

Polybutadiene

Polybutadiene is the polymer of 1,4-butadiene and is considered a rubber rather than a plastic. Butadiene has the same number of carbon atoms as butylene, the monomer of polybutylene, but the bonding is different. This difference carries over into the polymer, and the different properties become evident. Polybutadiene is the second highest volume synthetic rubber, and its blends are used heavily in the manufacture of tires. It is the one of the three materials that is used to make acrylonitrile–butadiene–styrene (ABS), a very versatile plastic. It imparts rubberlike characteristics (impact strength) to the ABS, as well as to high impact polystyrene.

Processing

Polybutadiene may be processed on all conventional rubber and plastics equipment such as extruders, injection molders, and calenders. It is then crosslinked or "cured," as are most conventional rubber polymers.

Applications

Polybutadiene is almost exclusively used to impart its properties to other rubber and plastic compounds and therefore might be considered a modifier of those materials. In styrene–butadiene rubber and in nitrile rubber, however, the percentage of polybutadiene is so high that it is a major portion of the resulting blend. Lesser amounts of polybutadiene may be used in acrylonitrile–butadiene–styrene or in polystyrene compounds, depending upon the degree of impact strength the customer wishes to have in the final compound.

Chlorinated Polyvinyl Chloride

Chlorinated polyvinyl chloride (CPVC) is polyvinyl chloride (PVC) that has had one more chlorine atom substituted for another hydrogen atom on the vinyl chloride "repeating unit." This substitution is not done on the molecule of the monomer, but on the polyvinyl chloride chain itself. The addition of another chlorine atom raises the heat distortion temperature and improves tensile strength above that of polyvinyl chloride, while retaining all the other properties of polyvinyl chloride.

Processing

Although chlorinated polyvinyl chloride can be run on conventional extrusion and injection molding equipment, that equipment might have to be modified to handle chlorinated polyvinyl chloride properly. The modifications could mean chrome plating dies or having stainless steel dies made for extrusion equipment, and special screws for the screw injection molding process.

Applications

Applications for chlorinated polyvinyl chloride include appliance parts, automotive interior parts, business machine covers, high temperature liquid handling in pulp and paper operations, hot and cold potable water piping, piping in chemical processes, pumps, skylight frames, telecommunications equipment, waste disposal devices, and window glazing.

Ethylene–Vinyl Acetate

Ethylene–vinyl acetate (EVA) is classified as a branched polyethylene because the acetate radical is attached to the hydrocarbon chain in a branched position.

The structural formula for EVA is rather complicated and beyond the scope of this book. However, it contains only carbon, hydrogen, and oxygen.

Ethylene–vinyl acetate is really a family of copolymers whose basic properties are determined by the ratio of ethylene to vinyl acetate monomers used in the polymerization process. When extruded into film, ethylene–vinyl acetate exhibits very good clarity, high impact resistance, and the ability to be heat-sealed at a relatively low temperature. The films are used for meat and poultry packaging, stretch film, carpet backing, and extrusion coating. It can also be extruded into hoses and tubing, and injection molded into closures and shoe parts.

Ethylene–vinyl acetate is often blended with other materials called elastomers, and such blends are then very useful as hot melt adhesives.

A major use for ethylene–vinyl acetate resins are as additives to other polymers to alter their properties somewhat. Ethylene–vinyl acetate copolymers are also compatible in so many other resins that EVA is often used as a "carrier" resin in which other materials are compounded at high levels (often called "concentrates"), and this compound is then blended into the compatible resin to add the desired properties of the additives.

Ethylene–Vinyl Alcohol

Ethylene–vinyl alcohol (EVOH) copolymers are branched polyethylene thermoplastics with superior gas transmission properties, moisture resistance, and excellent processability. EVOH copolymers have a high resistance to solvents and oils. They also have good abrasion resistance, good mechanical strength, and good weatherability properties. However, the resin will absorb moisture readily, and this severely affects the gas transmission properties of the film. It contains only carbon, hydrogen, and oxygen in its molecules.

Processing

Ethylene–vinyl alcohol may be extruded as a film that may be laminated onto another film to produce a good gas transmission barrier.

Applications

The ethylene–vinyl alcohol film will then be used to package those materials sensitive to oxidative degradation, such as foods. Ethylene–vinyl alcohol can also be coated onto another substrate to enhance its gas barrier properties.

Ionomers

Ionomers are branched polyethylene thermoplastics that differ from other organic polymers in that they contain a metallic ion as part of their molecule. They form very tough films, having high impact and puncture resistance. Ionomers also exhibit good chemical resistance, good adhesion properties, and good electrical properties. Since they are derivatives of polyethylene, they contain only carbon, hydrogen, and oxygen in their molecules, with the exception of a metal ion being present. The properties of any given ionomer depends largely on the metal ion present in the molecule.

Processing

Ionomers may be extruded, injection molded, blow molded, or foam molded, and films may be cast or blown.

Processing

One of the volume uses for one of the ionomers that takes advantage of its toughness is as the cover for cut-proof golf balls. Other uses for ionomers include both interior and exterior automotive parts, foams, footwear (including ski boots), food packaging, and packaging for electronic and hardware goods.

Polymethylpentene

Polymethylpentene (PMP) is an engineering resin of low density, high melting point, and excellent transparency. Its polymeric structure is beyond the scope of this book, but its monomer contains only carbon and hydrogen in its molecules.

It is available as a homopolymer or copolymer, depending on the properties desired. It is extremely thermally stable, with a melting point of 464°F. Thermal resistance can be enhanced by the addition of glass reinforcement. It has excellent electrical properties, good chemical resistance, and outstanding optical clarity. However, it has poor resistance to sunlight and high energy radiation and will degrade when exposed to such energy sources.

Processing

Polymethylpentene may be processed on all normal plastics processing equipment, such as injection molding, injection blow molding, extrusion, extrusion blow molding, thermoforming equipment, and coating equipment.

Applications

Polymethylpentene finds uses in appliances (connectors, fluid reservoirs, liquid level indicators, microwave oven components, popcorn popper covers), chemical laboratory equipment (diffusion membranes, funnels, petri dishes, test tubes), electrical applications (bobbins, cables, connectors, wire and cable coatings), housewares (food storage containers, microwave cookware), medical products (centrifuge tubes, filter housings, graduated cylinders, syringes), and packaging (bottles, closures, containers, trays).

Nylons

Nylons are a family of polymers that all contain the amide radical (—CONH—) and are usually referred to as polyamides. They get their names, such as nylon—6, nylon—6/6, nylon—11, and nylon 6/12, from their monomers. The numbers refer to the number of carbon atoms in the monomer molecule. If there is only one number, as in nylon—6, there is only one monomer and there are six carbon atoms in that compound. For a nylon like 6/12, there are two monomers, and there are six carbons in the first compound and twelve carbons in the second. The monomers are diamines and dibasic acids, amino acids, and lactams. The monomers and therefore the type of nylon polymerized depends upon the specifications of the end product. Raw materials are then selected, and the desired polymer is made. Therefore, the range of properties of nylons is very wide and may be narrowed by polymerizing specific materials.

Nylon was the first successful engineering resin, being introduced on a commercial scale in 1938. Its growth rate, however, has matched that of the fastest growing commodity resins, and production has surpassed 500 million pounds annually in the United States.

Since there are so many types of nylons, the description of their properties here is general and all-inclusive without reference to the particular polymer that exhibits that particular property. Additives may be used to improve on certain properties of nylons, such as flammability resistance, thermal stability, processibility, and lubricity. Individual chemical compounds or other polymers may be blended with the nylon, or its monomer content may be altered. Glass fiber and mineral reinforcement may be used to increase heat resistance, stiffness, strength, and toughness.

The properties that make nylons the material of choice in so many applications include chemical resistance, resistance to solvents and hydrocarbon fuels, resistance to caustics, high tensile strength, fatigue resistance, gas transmission properties, abrasion resistance, creep resistance, good electrical insulation properties, and a low coefficient of friction.

Nylons are attacked by strong acids, oxidizing agents, and high concentrations of certain salt solutions.

Another disadvantage of nylons is that they are hygroscopic (absorb water from their surroundings). Not all the polyamides absorb water at the same rate or maximum amount, so selection of a particular type of nylon will lessen the problem. Whatever resin is chosen, it is important to control the amount of moisture present, since processing nylon polymers that are either too wet or too dry will cause problems leading to unacceptable parts.

Nylons may be processed by all major methods, including extrusion, injection molding, blow molding, reaction injection molding (RIM), powder coating, and rotational molding.

Applications

- Antifriction parts
- Appliance handles
- Appliance housings
- Automobile fender extensions
- Automobile stone shields
- Automotive trim clips
- Barrier packages
- Bearings
- Brake fluid reservoirs
- Brush bristles
- Bushings
- Business machine parts
- Contacts
- Containers
- Cook-in bags and pouches
- Electrical connectors and plugs
- Electrical insulation
- Engine fans
- Fibers
- Film
- Fishing line
- Fittings
- Fuel tanks
- Gauges
- Gears
- Hammer handles
- Lenses
- Machine parts
- Meters
- Mirror housings
- Monofilaments
- Multilayer packages and pouches
- Power steering fluid reservoirs
- Power tool housings
- Processing equipment housings
- Pump housings and parts
- Radiator headers
- Relays
- Rods
- Rollers
- Screws and other fasteners
- Sewing thread
- Sheet
- Terminal blocks
- Tubing
- Valves
- Vending machine parts
- Wearplates
- Wire and cable insulation

Polyacrylonitrile

Polyacrylonitrile (PAN) is a polymer made from the acrylonitrile monomer and is sometimes referred to as a nitrile resin or a high acrylonitrile polymer. Acrylonitrile's proper chemical name is vinyl cyanide. The structural formula for acrylonitrile, the monomer of polyacrylonitrile, is shown in Figure 4.12.

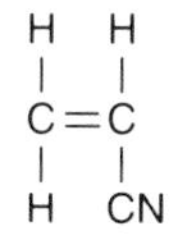

Figure 4.12 Structure of acrylonitrile

The structure is the same as it is for polyvinyl chloride, polyvinyl fluoride, and polystyrene with one exception. Instead of a chlorine atom, fluorine atom, or benzene molecule, respectively, being substituted for one hydrogen atom on the vinyl radical (when one hydrogen is removed from the ethylene molecule, the radical formed is called the vinyl radical), a cyanide radical is attached. This substitution is what provides each of the vinyl-type polymers their characteristics.

Polyacrylonitrile is known primarily for its good gas transmission properties, and it was the first thermoplastic considered for use in plastic soft drink bottles (where polyetheylene terephthalate is used today). However, residual vinyl cyanide was reported in the polymer at levels determined to be unacceptable by the U.S. Food and Drug Administration (FDA), and prior approvals for use in beverage container applications were withdrawn. The polymer has since been reapproved with specifications on the maximum permitted level of residual acrylonitrile.

Polyacrylonitrile has good chemical resistance and fairly high tensile strength. When a rubber-type additive compounded into the product, the resultant material shows good impact strength. It is resistant to dilute corrosives, liquid hydrocarbons, esters, acetates, and chlorinated hydrocarbons.

Polyacrylonitrile may be extruded, injection and blow molded, and thermoformed. It has major applications in the packaging market. Typical uses for finished articles of polyacrylonitrile are as follows.

- Blister packs
- Blow molded bottles and containers
- Caulking compound tubes
- Correction fluid bottles
- Injection molded bottles and containers
- Meat packaging
- Spice containers
- Thermoformed containers

Polybutylene

Polybutylene (PB) is an olefinic polymer because of its monomer, butylene (1-butene). Butylene is the third compound in the alkene series, the first (and smallest molecule) being ethylene (ethene) from which polyethylene is polymerized, and the second being propylene (propene), the monomer of polypropylene. Polybutylene copolymers (with ethylene as the second monomer) are also available.

The structural formula for butylene, the monomer of polybutylene, is shown in Figure 4.13. The actual structure might be branched.

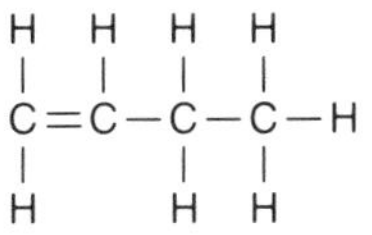

```
H   H   H   H
|   |   |   |
C = C - C - C - H
|       |   |
H       H   H
```

Figure 4.13 Structure of butylene

Polybutylene has good chemical resistance to such diverse materials as acids, bases, detergents, oils, and solvents, so long as the temperature is not elevated, in which case, chemical resistance decreases. However, it does retain its good mechanical properties at higher temperatures. Polybutylene also has good resistance to stress cracking, creep, and abrasion. Films made from polybutylene have good tear resistance and high tensile strength.

Polybutylene can be extruded, injection molded, and blow molded, and film can be made by the blown film method or by casting. Applications include compression wraps, food and meat packaging, hot fill packaging, hot melt adhesives and sealants, piping (PB has been received plumbing code approvals or listings from all the major national and regional code bodies), and tank linings.

Polycarbonate

Polycarbonate (PC) is an engineering thermoplastic that has excellent optical properties (excellent clarity plus colorability), high impact strength, good heat resistance, excellent creep resistance, and good electrical properties. Its original volume use was as a nearly unbreakable glazing material, but it has moved well beyond that, as evidenced by the list of uses below. In its original form, it was easily scratched and tended to turn yellow after prolonged exposure to the sun, but these shortcomings have been overcome by special coatings and stabilizer systems.

Its general overall properties are all good, and they can be improved by additives chosen to enhance specific properties. Alloying with other plastics such as acrylonitrile–butadiene–styrene (ABS), polyethylene terephthalate (PET), and polybutylene terephthalate (PBT) provides specific properties for particular applications not filled by other engineering resins.

Polycarbonate's chemistry is based on bisphenol A as one of the two major reactants in its formation, and its final form is as a polyester of carbonic acid. Even though it does not absorb much water, the resin must be kept dry to maintain its engineering properties after processing.

Polycarbonate can be processed by all the standard methods used for thermoplastics. It can be extruded (both sheet and profile), injection molded, blow molded, rotation molded, and thermoformed. The resin may be blown and extruded into a structural foam.

Applications

- Air conditioner grilles
- Airport runway markers
- Airplane interior parts
- Automobile glazing
- Automobile instrument panels
- Baby bottles
- Beverage pitchers
- Boat windshields
- Business machine housings
- Compact disks
- Computer parts
- Food storage containers
- Football helmets
- Glazing materials for buildings
- Greenhouse glazing
- Headlights
- Health-care products
- Mass transit vehicle glazing
- Medical products
- Microwave cookware
- Milk bottles
- Motorcycle helmets
- Motorcycle windshields
- Optical memory disks
- Outboard motor propellers
- Outdoor lighting devices
- Prescription eyeglass lenses
- Power tool housings
- Safety glasses
- Signs
- Small appliance housings
- Snowmobile helmets
- Structural foam
- Tableware
- Tail-light lenses
- Telephone components
- Tumblers
- Traffic light housings
- Water bottles

Many other plastics that fall into the miscellaneous category in use today, and many more continually are being developed. Some other plastics that may be fairly common but do not exist in quantities sufficient to merit separate descriptions are polyvinyl acetate, polyvinyl butyral, polyvinyl formal, polyvinyl chloride acetate, and ketone-based resins.

POLYESTERS

Polybutylene Terephthalate

Polybutylene terephthalate (PBT) is a thermoplastic polyester also known as polytetramethylene terephthalate. The state of crystallinity of polyesters is important to their properties, and polybutylene terephthalate is known as a semicrystalline material. It is this crystallinity that gives it its resistance to moisture absorption, good chemical resistance, and good electrical properties. Its chemical resistance, which is limited to resistance solvents and to weak acids and bases at room temperature, disappears at elevated temperatures. At temperatures above 140°F, it is attacked by strong acids and bases, oxidizers, aromatic solvents, and ketones.

It has good tensile strengthh, and its low impact strength can be significantly improved through the addition of impact modifiers. Toughness and surface gloss can be modified by the addition of polyethylene terephthalate (PET) or polycarbonate (PC), and mechanical strength and stiffness can be increased by the addition of glass or mineral reinforcement.

Even though polybutylene terephthalate has low moisture absorption, even this small amount of water must be removed in a drying operation before processing may begin. Polybutylene terephthalate may be processed by injection molding, extrusion, blow molding, thermoforming, and structural foam molding.

Applications

Appliances

Coffee maker parts
Hair dryer housings
Handles
Toaster parts
Vacuum cleaner parts

Automotive

Body panels
Brake system parts
Bumpers
Distributor caps
Engine pallets
Fenders
Grilles
Headlight parts
Ignition parts
Mirrors
Racks for parts
Rotors
Water pump parts
Wheel covers
Windshield washer parts

Electrical

Business machine parts
Chip carriers
Computer parts
Connectors
Conveyor components
Fiber optic tubing
Fuse cases
Motor housings
Relays
Switches

Miscellaneous

Bicycle gears
Lawn mower housings
Outboard motor propellers
Power tool housings

Polyethylene Terephthalate

Polyethylene terephthalate (PET) is a thermoplastic polyester that can exist as noncrystalline (amorphous), partially crystalline, and highly crystalline states, each having its own set of properties. Crystallinity is desired because it produces increased strength and a wider range of uses at higher temperatures. In the partially crystalline state, orientation (stress applied in a particular direction) will increase the strength, elongation, and gas transmission properties of polyethylene terephthalate.

Polyethylene terephthalate can be reinforced by the addition of glass fibers or minerals, thereby improving its properties so that it may compete with some engineering resins. It will then be referred to as reinforced polyethylene terephthalate (RPET). Flame retarders and modifiers are also added to allow polyethylene terephthalate to be used in applications not available to the pure polymer.

Polyethylene terephthalate can be processed by extrusion, injection molding, blow molding, film processing, and fiber spinning. Moisture must be removed in a drying operation before any processing can occur. Its ability to be blow molded from preforms has allowed it to dominate the soft drink bottle market, being the material of choice in the 2-liter and 3-liter sizes. It has also captured a great deal of the half-liter and 1-liter market.

Applications

Bottles

Beer
Cosmetics
Household products
Liquor
Mouthwash
Mustard
Nuts
Peanut butter
Pharmaceuticals
Pickled foods
Salad dressing
Soft drink
Syrup
Toiletries
Vegetable oil
Wine

Fibers

Automotive headliners
Carpet
Clothing
Textiles
Tire cord
Upholstery

Miscellaneous

Automobile pump housings
Automotive ignition parts
Bicycle frames
Carburetor parts
Coatings
Cookware
Electrical insulation
Electrical switches and relays

Applications (Continued)

Fiberfill for bedding and thermal clothing
Floor tile
Food and meat packaging
Food trays (microwaveable and conventional oven)
Furniture parts
Magnetic tape
Microwave oven interiors
Photographic film
Plumbing parts
Power tool housings
Strapping
X-ray film

Polyethylene

Polyethylene, in all its forms, is the single largest volume plastic made in the world. Its forms include linear low density polyethylene (LLDPE), ultra low density polyethylene (ULDPE), low density polyethylene (LDPE), medium density polyethylene (MDPE), high density polyethylene (HDPE), high molecular weight polyethylene (HMWPE), and ultra high molecular weight polyethylene (UHMWPE). As the molecular weight (chain length) increases, so do the rigidity and softening-point temperatures. Polyethylene can also be halogenated, as in chlorinated polyethylene (CPE), to attain certain desired properties, and may also be crosslinked to produce different properties.

The desired properties of the final part and the processing requirements dictate which molecular weight polyethylene will be selected, and additional properties can be built in or enhanced by the use of additives.

The monomer of polyetylene is ethylene, whose chemical formula is C_2H_4. Its structural formula is shown in Figure 4.14.

```
H   H
|   |
C = C
|   |
H   H
```

Figure 4.14 Structure of ethylene

Ethylene is a flammable, colorless gas with a sweet odor. It is classified as an alkene, which is defined as a family of unsaturated hydrocarbon compounds containing one and only one double bond. The general formula for the alkenes is C_nH_{2n}.

The alkenes are part of a larger family called olefins. **Olefins** are unsaturated hydrocarbon compounds, all of which have one or more double bonds. Additional members of the olefin family that are monomers are propylene (propene), butylene (1-butene), 1,3-butadiene, and 4-methylpentene. Therefore, all the polymers made from these olefins are called polyolefins.

It is the presence of the double bond that gives the short-chain olefins the reactivity that allows them to polymerize. During the polymerization process, one of the two covalent bonds breaks in the ethylene molecule (a) shown in Figure 4.15, and each of the resulting two unpaired electrons moves to the outside of each of the carbon atoms, producing the very reactive free radical (b).

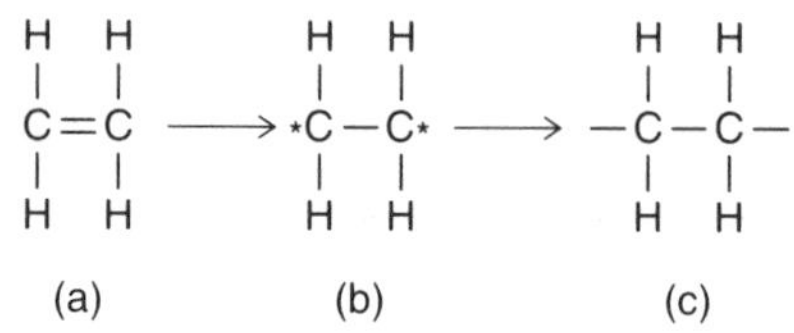

```
H   H          H   H           H   H
|   |          |   |           |   |
C = C  ---->  *C - C*  ---->  -C - C-
|   |          |   |           |   |
H   H          H   H           H   H

 (a)            (b)             (c)
```

Figure 4.15 Radical formed by breaking a bond in ethylene

As this process continues to produce the free radicals—(b) and (c) are exactly the same, but are shown with two different types of shorthand—those free radicals begin to react with each other, producing the very long chain hydrocarbon polymer polyethylene.

Each of the repeating units (b) or (c), or *mers*, will appear in the chain—in some instances, tens of thousands of times. The length of the chain and the degree of branching of each of these giant molecules are two of the variables that determine the physical properties of the polymer. Other variables include the temperature and pressure maintained during the polymerization reaction; the degree of crystallization; and the initiators, catalysts, and solution modifiers used.

The chemical shorthand for writing the molecular formula for polyethylene is shown in Figure 4.16. The bracket drawn through the "extended" bonds designates the unit within the bracket as the repeating unit, or *mer*. The "n" outside the bracket represents the number of mers in the polymer. The "extended" bond as shown (that is, connected to one atom only), stands for an unpaired electron and makes the unit, or free radical, reactive. The fact that there are two unpaired electrons, one on each side of the radical, means that the radical is extremely reactive, which is exactly the way the radical of a monomer is supposed to be.

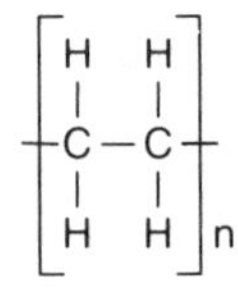

```
 [ H   H ]
 [ |   | ]
-[ C - C ]-
 [ |   | ]
 [ H   H ]n
```

Figure 4.16 Shorthand method for drawing the structure of polyethylene

The uses for polyethylene listed below are not differentiated by type of polyethylene (LDPE, UHMWPE, LLDPE, and so on), except that only polyethylene is considered and not chlorinated polyethylene (CPE) or other substituted materials or copolymers. Polyethylene, in one form or another, may be used as a compounding material to impart certain properties to other polymers or compounds. For example, ultralow density polyethylene (ULDPE) is sometimes used as an impact modifier for polypropylene and high density polyethylene.

Processing

Polyethylene is processed (depending on the type) by extrusion, injection molding, blow molding, film blowing, film casting, extrusion coating, powder coating, and rotational molding.

Uses

Polyethylene is the material of choice in many different applications. The many densities and structures allow for an extremely wide set of specifications.

When the end-use specification calls for such engineering or loadbearing applications such as chemical resistance, low coefficient of friction, self-lubrication, or wear resistance, ultrahigh molecular weight polyethylene is chosen. When large diameter pipe, automotive fuel tanks, or large drums must be made, high molecular weight high density polyethylene is used. This same material, used in such rugged-appearing products, is also the material of choice for tough, thin-film trash bags. When small diameter pipes are extruded, linear low density polyethylene is used. This same material is used to manufacture film for landfills and waste ponds, extruded sheeting, and injection molded articles.

Additives

Although both high density polyethylene and low density polyethylene are usually used in the "natural" (unfilled) form, both can be filled by all the common fillers, including calcium carbonate, glass fibers, mica, silica, and talc. They may be flame retarded by the use of both organic and inorganic materials and may be colored by both organic and inorganic pigments. Polyethylene is a "waxy" material and usually does not need a lubricant. A light stabilizer (ultraviolet absorber) must be added to it if it is to be used outdoors.

- Adhesives
- Automotive parts
- Bottles
  - Beverages
  - Food products
- Cosmetics
- Health-care goods
- Household goods
- Personal care goods
- Pharmaceuticals

- Water
- Bushings
- Canoes
- Chemical process liners
- Chute liners
- Closures
- Communication cable
- Conduit
- Containers
  - Bag-in-box
  - Can liners
  - Cap liners
  - Cartons
  - Crates
  - Cups
  - Drums
  - Dunnage trays
  - Pails
  - Paint cans and lids
  - Tote bins
  - Tumblers
  - Tubs
  - Vats
- Cutting surfaces
- Disposable diapers
- Dinnerware
- Fibers
- Film
  - Agricultural film
  - Blown film
  - Cast and extrusion coating
  - Cast film
  - Construction vapor barrier film
  - Garbage bags
  - Garment bags
  - Grocery bags
  - Laminations
  - Leaf and garden bags
  - Multiwall sacks
  - Packaging
    - Baked goods
    - Food
    - Frozen food
    - Meat
    - Medical products
    - Nonfood products
    - Poultry
    - Produce
    - Seafood
  - Sandwich bags
  - Shipping sacks
  - Shrink-wrap film
  - Trash bags
- Freezer paper
- Furniture parts
- Gaskets
- Gasoline tanks
- Gears
- Gloves
- Handles
- Heat seals
- Hopper liners
- Housewares
- Industrial liners
- Lids
- Luggage
- Novelties
- Pails
- Pen tips
- Pipe
- Pump parts
- Rollers
- Sheet
- Shotgun shell hulls
- Ski sole plates
- Snow plow edges
- Surgical implants
- Toys
- Truck bed liners
- Tubing
- Wastebaskets
- Wire and cable insulation

POLYPROPYLENE

Polypropylene (PP) is another member of the polyolefin family, since its monomer, propylene, is an unsaturated hydrocarbon and belongs to a class of chemical compounds called olefins. Because the monomers of polypropylene and polyethylene (ethylene) are so similar, the polymers have many of the same properties. However, there are several important differences, including density (polypropylene's is lower), service temperature (polypropylene's is higher), rigidity (polypropylene is more rigid), resistance to environmental stress cracking (polypropylene's resistance is higher), and susceptibility to oxidation (polypropylene is worse than polyethylene). Polypropylene is usually chosen for an application that requires toughness, stiffness, impact strength, and the desired melt flow rate (MFR).

The molecular formula for propylene is C_3H_6, and its structural formula is shown in Figure 4.17. During polymerization, different alignments of parts of the molecule may produce a different form of polypropylene. The description below may be complicated, but it explains some of the different possible types of polypropylene. They are all similar in chemical makeup, containing only carbon and hydrogen.

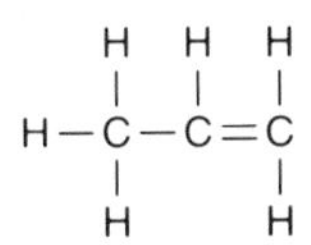

Figure 4.17 Structure of propylene

One of the several different forms of polypropylene is called *isotactic polypropylene*. Isotactic polypropylene has all of the CH_3 groups on the same side of the carbon chain, when the chain is in one plane. This is one of three possible stereotactic arrangements of polypropylene, the other two being atactic (in which the CH_3 groups are arranged randomly on either side of the carbon chain), and syndiotactic (in which the methyl groups are arranged in a regularly alternating arrangement on either side of the carbon chain). Both atactic and syndiotactic polypropylene are more impact resistant than the isotactic, but the atactic polymers are more rubbery and have limited commercial use. Commercially available polypropylene is preponderantly isotactic.

Isotactic polypropylene has good stiffness and tensile properties and is resistant to corrosives and many solvents. It has excellent fatigue resistance, which makes it a perfect choice for packages or containers with an integral hinge. It can be compounded with many additives such as fillers, reinforcements, and other polymers to achieve specific properties.

Polypropylene is available in homopolymer and copolymer grades. The *homopolymer* is used principally for filaments and fibers. Injection molded parts of homopolymer polypropylene are also common. *Copolymer polypropylene* is achieved by copolymerizing propylene and ethylene monomers. Propylene and ethylene may also be polymerized into an ethylene–propylene rubber in a secondary reactor and then used as an impact modifier for the homopolymer. The higher impact strength of copolymer polypropylene is the major property sought by end users.

Another form of polypropylene available is known as *random copolymer polypropylene*. This material is made by randomly attaching ethylene groups to isotactic polypropylene. The appeal of random copolymer polypropylene is due to it clarity, toughness, flexibility, low melting point, and impact resistance, in addition to the chemical resistance and barrier properties of homopolymer polypropylene. These qualities make it desirable for use in packaging all types of products.

Processing

Polypropylene may be processed on extruders, injection molders, and blow molders. It may be further processed by compression and vacuum molding and thermoforming.

Additives

Polypropylene may be filled by all the common fillers, including calcium carbonate, glass fibers, mica, silica, and talc. It may be flame retarded by the use of both organic and inorganic materials and may be colored by both organic and inorganic pigments. It is a "waxy" material and usually does not need a lubricant. It has good exterior properties, so light stabilizers are required less than with polyethylene.

Applications

Appliance parts
Automotive parts
Battery cases
Bottles
Caps and closures
Containers
Dishes
Dishwasher components
Disposable diapers
Fibers
Films
Furniture parts
Housewares
Monofilament
Multilayer coextruded bottles
Office equipment
Packaging
Power tools
Toys
Wire and cable insulation

Polystyrene

Polystyrene (PS) comes in many forms. It is available in grades ranging from crystal to high impact, and in many specialized forms such as foam or flame retardant. Its versatility due to its possible wide range of physical properties and processibility allows it to rank it as one of the most widely used plastics in the world.

The monomer of polystyrene is styrene. Its molecular formula is $C_2H_3C_6H_5$, and its structural formula is repeated in Figure 4.18.

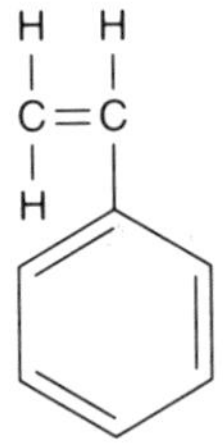

Figure 4.18 Structure of styrene

Because of its name, polystyrene is, of course, classified as a styrenic polymer, a name reserved for a class of polymers that contain styrene. It is also considered a vinyl polymer, because half of the styrene molecule is the vinyl radical, the other half being the benzene structure, or the phenyl radical. Indeed, the proper chemical name for styrene is vinyl benzene.

Crystal polystyrene is really a noncrystalline (amorphous) polymer that is tough, high in stiffness, and low in moisture absorption, has good optical properties, and is a good electrical insulator. Even these properties can be modified and improved with additives, and to improve its resistance to ultraviolet degradation, light stabilizers may be added.

Impact polystyrene is produced by post-blending rubberlike materials into the polystyrene matrix. Styrene–butadiene block copolymers are usually the modifiers of choice, improving impact but reducing transparency. Impact grades are usually designated as medium impact polystyrene (MIPS), high impact polystyrene (HIPS), and extra high impact polystyrene (EHIPS). Flame retardant (FR) grades of high impact polystyrene are available, usually based on halogenated additives blended with antimony oxide.

The styrenes are characterized by a resistance to water, alkaline corrosives, and acids. They are subject to attack by aliphatic and aromatic hydrocarbons and their derivatives. High impact polystyrene has limited resistance to ultraviolet exposure, and therefore is not recommended for outdoor use unless it is protected by ultraviolet stabilization additives or a laminated or coextruded cap coat. It also has low grease resistance and poor high temperatuve performance.

Expandable polystyrene (EPS) is a copolymer of polystyrene that can be processed, by means of a blowing agent, into cellular or foamed shapes. The process by which this is accomplished is slightly different than producing foam polystyrene from a homopolymer. Because of environmental and other problems involving the use of blowing agents or injected gases, however, the entire area of producing expanded or foamed products is undergoing changes in procedures and raw materials.

Processing

Polystyrene may be processed by all conventional thermoplastic methods, including extrusion (film, sheet, and profile), injection molding, blow molding, and structural foam molding.

Additives

Polystyrene will accept antioxidants to protect it from molecular oxidation. It may be colored by most common colorants, both organic and inorganic. Color is important in packaging and other styling uses of polystyrene, whereas polystyrene foam is seldom colored, except perhaps for identification purposes. Polystyrene may be filled or reinforced with a wide variety of fillers and extenders, and flame retardants and smoke suppressants are used to improve the combustion characteristics of the pure polymer. Crystal polystyrene is very brittle and has poor impact resistance, so varying amounts of impact modifiers are used to make the different grades of impact resistant polystyrene. Lubricants are sometimes used to give polystyrene better processing characteristics.

Applications

Appliance parts
Audiocassettes
Automotive parts
Blister packaging
Bottle caps and enclosures
Brush bristles
Business machine and computer housings
Containers
Cosmetic packaging
Dishes
Disposable containers
Egg cartons
Electrical parts
Fast food takeout packages
Flower and other plant pots
Food packaging
Fruit juice packaging
Furniture
Housewares
Insulated cups and containers
Insulation
Lenses
Light diffusers
Lighting fixtures
Model kits
Packing materials

Applications (Continued)

- Picture and mirror frames
- Radio cabinets
- Refrigerator liners and parts
- Smoke detectors
- Toys
- TV cabinets
- Tumblers
- Utensils
- Videocassettes
- Wall tile

POLYURETHANES

Polyurethanes are very versatile materials that owe that versatility to the myriad of ways the materials may be made. Polyurethanes are not made like many other polymers, from a single monomer. Instead, polyurethanes are made by reacting monomers called polyols with other monomers called isocyanates, of which several different types exist. Depending on which isocyanate and which polyol is used, the resulting polyurethane may be thermoplastic, thermosetting, or elastomeric. The polyurethane foams so widely used as cushion material or insulation are usually thermosetting foams. The foams made this way may be flexible, rigid, or semirigid.

Tough, abrasion-resistant foams may be made using polyurethane. Very tough, abrasion-resistant thermoplastic polyurethane film and sheet, may also be made by varying the reactants. If the properties of polyurethane are needed in bar or rod stock, solid thermoplastic polyurethane may also be produced.

Many polyurethane products start as a liquid system, with the liquid monomers (the polyisocyanate and the polyol) being mixed in a special blending chamber of a special mixing machine, and the integrated prepolymer then being injected into a mold. The liquid system reacts in the mold, "setting up" or curing and taking the form of the mold. Polyurethane foams are made this way, as are most other thermosetting polyurethane products.

Thermoplastic polyurethane is prepared in a slightly different manner, after which the solid material is formed into pellets or cubes. The pellets, cubes, or both are then further processed through an injection molding machine to make molded parts or an extruder to make profiles or sheets—or they may be "blown" into film. Blown polyurethane film is extruded as a hollow, sausage-shaped film, which then has air blown into the center of the hollow area to expand it into a larger volume. This film is then slit into the sizes required by customers.

Polyurethane chemistry has evolved to the extent that many of the polymers made today that are called polyurethanes are really something else, but the chemistry is so close to urethane chemistry that it is easier to refer to these newer materials as polyurethanes than their real names. An example of this is the evolution to polyisocyanurates. Rather than polymerization of a polyisocyanate with a polyol, another reaction called trimerization (three parts rather than many parts) is carried out. This results in a much more stable configuration and a much different polymer. But since polyisocyanurates are more rigid and have less abrasion resistance than polyurethanes, the polyisocyanurate is blended with a polyurethane to achieve the desired properties.

Applications

Athletic uniform letters	Mattress cushions
Automobile parts	Pneumatic wheels
Business machine housings	Seat cushions
Diapers	Upholstery cushions
Footwear	Water bags
Insulation	

POLYVINYL CHLORIDE

Polyvinyl chloride (PVC), in all its forms and compounds, is the second largest volume plastic in the world, with only polyethylene being larger in volume. Polyvinyl chloride, however, has more uses than any other plastic in existence. It exists in commercial use only in a compounded form, which means that certain materials have been added to the PVC resin to make it processible and to give it the properties required by the end user. It exists as rigid polyvinyl chloride (RPVC) which, as its name implies, is very stiff, hard, and sometimes brittle. It also exists as flexible polyvinyl chloride (FPVC), which is very soft and pliable. It can be compounded to possess any degree of rigidity or flexibility between these two extremes, depending on the required properties of the final product.

Polyvinyl chloride belongs to a group of polymers commonly referred to as **vinyls** or **polyvinyls**, because the monomer of PVC contains two radicals, the chloride radical and the vinyl radical. The vinyl radical is the radical of ethylene produced by the removal of one hydrogen. When one of the hydrogens is removed from the ethylene molecule (a), the resultant molecular fragment (b) is called the vinyl radical. A different radical or functional group (R) may attach itself to the vinyl radical as in (c). If the radical is chlorine, the resultant molecule (d) is that of vinyl chloride, which is the monomer of polyvinyl chloride. These reactions are shown in Figures 4.19, 4.20, and 4.21.

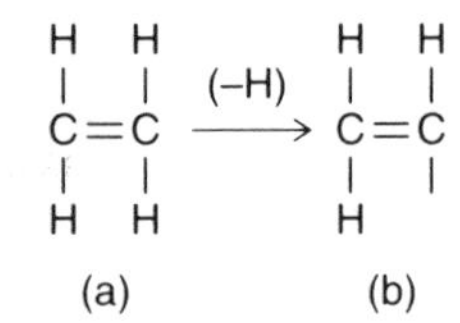

Figure 4.19 Creation of the vinyl radical from ethylene

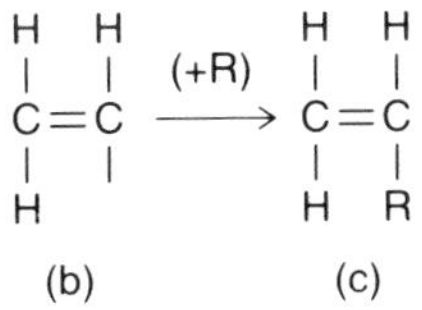

```
H  H          H  H
|  |  (+R)    |  |
C=C  ----->   C=C
|  |          |  |
H             H  R

 (b)           (c)
```

Figure 4.20 Substitution of a radical for an atom of hydrogen

```
H  H             H  H
|  |   (R=Cl)    |  |
C=C  -------->   C=C
|  |             |  |
H  R             H  Cl

 (c)              (d)
```

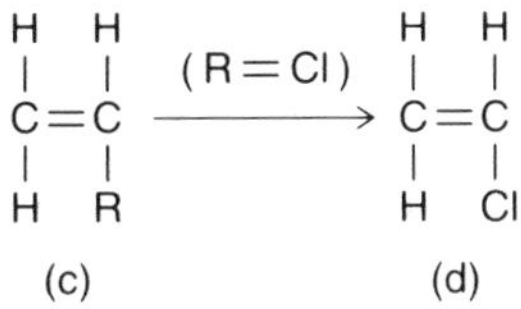

Figure 4.21 Substitution of a radical for an atom of hydrogen

If the radical or functional group (R) is benzene (C_6H_6), the resultant molecule is that of styrene, which is the monomer of polystyrene (PS). Polystyrene is usually referred to as a styrenic material, rather than as a vinyl. If it is the acetate radical (—$COOCH_3$), the resultant is vinyl acetate; the monomer of polyvinyl acetate (PVA), and polyvinyl butyral, and polyvinyl formal are formed in the same manner. Two hydrogens may be removed from the ethylene molecule (a) to make room for the addition of two functional groups (g) as in Figure 4.22.

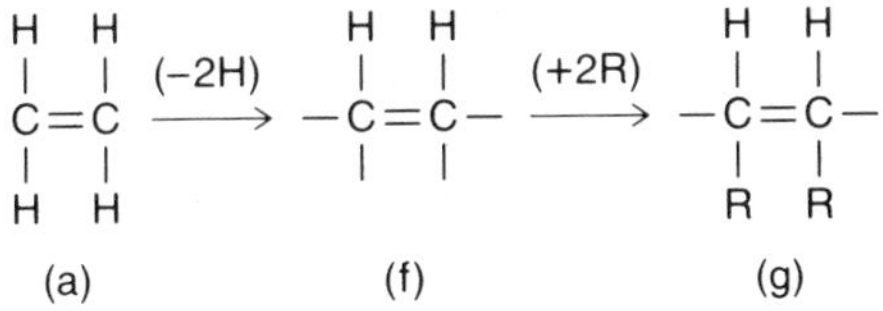

```
H  H            H  H             H  H
|  |  (-2H)     |  |    (+2R)    |  |
C=C  ----->   -C=C-   ----->   -C=C-
|  |            |  |             |  |
H  H                             R  R

 (a)            (f)              (g)
```

Figure 4.22 Substitution of two radicals for two atoms of hydrogen

If both radicals are chlorine, the resultant molecule is that of vinylidene chloride, the monomer of polyvinylidene chloride (PVDC). If the chlorines are on different carbon atoms, the resultant molecule is that of vinyl dichloride, which is the monomer of chlorinated polyvinyl chloride (CPVC), the more recent name for polyvinyl dichloride. If one of the functional groups is the acid radical (—COOH) and the others is the methyl radical (—CH_3), both on the same carbon atom, the resultant molecule is that of acrylic acid, the monomer of one of the many acrylic polymers. There are many more such combinations of the vinyl radical and other functional groups that will form the monomers for many more polymers. They are discussed individually or in tightly related groups of polymers. But by far the most important vinyl polymer commercially is polyvinyl chloride.

Polyvinyl chloride is tough, is a good electrical insulator, has good abrasion resistance, and is resistant to a great many chemicals.

The polymerization of vinyl chloride monomer (VCM) to polyvinyl chloride produces a "giant" molecule, with the repeating vinyl chloride unit producing a long chain of carbon atoms with hydrogen atoms along one side of the chain and chlorine atoms alternating with hydrogen atoms at the other side of the chain.

Each of the monomers named above are individual chemical compounds. Unfortunately, the name "compound" has also been adapted for use when describing a physical mixture of additives with a polymer. The nomenclature is not likely to change, so you will have to make the distinction between a chemical compound and a plastics compound. To try to underscore the distinction, I use the word "compound" in the context of the plastics compound and add the word "chemical" if chemical compound is the intended meaning. Whenever the word "molecule" is used, a chemical compound is the intended meaning. Therefore, one may start with molecules of the chemical compound vinyl chloride, induce the polymerization reaction, and form the new chemical compound, polyvinyl chloride. Then, to make the polyvinyl chloride resin processible, to add desirable properties, or both, other chemical compounds may be added to the polyvinyl chloride resin (in a process called compounding) to make a polyvinyl chloride compound for processing into some useful final product. This polyvinyl chloride compound is not a pure substance, since other materials have been mixed into it, so a polyvinyl chloride (and any other type of plastic) compound is really a mixture of materials.

As implied by the discourse on compounding, polyvinyl chloride resin is commercially useless in whatever form it finds itself immediately after polymerization. For any plastic resin to be made into a useful shape or part, sufficient heat and pressure must be applied to it in one of several mechanical processes. If a processor takes pure polyvinyl chloride and tries to convert it into a useful shape or part, it will begin to degrade (decompose) when heated to a high enough temperature to allow it to soften, liberating hydrogen chloride (HCl) gas as one of its decomposition products. This is an undesirable situation, since with a high enough temperature, enough HCl will be generated to cause discomfort for the human operators of the equipment (HCl gas dissolves in any moisture available to become hydrochloric acid, whose chemical formula is also HCl).

To allow polyvinyl chloride to become processible, that is, to be heated and softened in a piece of processing equipment without decomposing, a **heat**

stabilizer is added to the resin before it is introduced into the equipment that will form it. This heat stabilizer allows the polyvinyl chloride to be processed without degrading and changing color (when PVC decomposes, it turns yellow, amber, brown, and then black as the decomposition reaction progresses), and the resulting shape or part will be very rigid when it cools.

To introduce softness or flexibility in the final product, a **plasticizer** is added to the resin. The addition of the plasticizer may allow the amount of heat stabilizer to be reduced somewhat, since the plasticizer, which in many cases is an oily substance, will have some lubricating properties and may allow the polyvinyl chloride to be processed at slightly lower temperatures.

In some cases, if a rigid finished article is desired but the compound from which it was formed is too brittle, an **impact modifier** is added to give the finished part the ability to resist the impact of a particular force without breaking or deforming too badly.

To lower the cost of the final compound or part, a **filler** or **extender** may be added to the compound. These materials are usually relatively inexpensive minerals (perhaps one-tenth to one-fifth the cost of the resin) that have been ground to a very fine powder. Very often they contribute to the desired properties of the final part.

To aid in the processibility of the polyvinyl chloride compound, a separate lubricant may be added in very small quantities to allow the compound to pass through the equipment more readily, and also to lessen the chance of decomposition occurring.

To allow the final part to have a longer life when exposed to the environment, a **light stabilizer**, or ultraviolet absorber, will be added to the compound. Whenever the finished part will be exposed to the polymer-degrading effects of sunlight, care must be given to selecting the proper stabilizer that will allow the part to keep its chemical and physical properties for as long as possible.

A **flame retardant** may be added to the compound to reduce the flammability of the polyvinyl chloride compound. This usually does not occur with rigid PVC, since it will not ignite easily and, once ignited, burns with some difficulty and only when there is a sufficient heat input to keep the PVC burning. Rigid polyvinyl chloride will not support its own combustion, so there must be a supporting flame from some other burning material before the PVC part will burn. However, flexible polyvinyl chloride is another matter. Most plasticizers are combustible liquids, and the addition of these materials will enhance the combustibility of PVC considerably. Flexible polyvinyl chloride will support its own combustion, and its enhanced combustibility is directly proportional to the amount of plasticizer (if it is a combustible plasticizer) present in the finished part. The compounder may add a flame retardant plasticizer if the processor's customer's specifications allow it. In some cases, polyvinyl chloride itself is added (alloyed) with another polymer to gain some of the particular properties of

PVC that the original polymer may not have. In these cases, the other polymer will burn less readily with the PVC present than if it were not used.

An ultrviolet **brightener** may be added in small quantities to the compound to make the final part "stand out" in sunlight. This material is different than the ultraviolet absorber mentioned above. In this situation the ultraviolet brightener will absorb light at one energy level and emit it at another, giving the material containing it a much brighter appearance than normal.

Finally, color may be added to the compound to produce the effect that the purchaser of the final part wants to achieve. This color comes in many forms and is produced by many different materials. *Dyes* are organic materials that dissolve in the polyvinyl chloride compound and give PVC and other plastics some of their brightest colors. Dyes, however, do not last as long as some other colorants. Generally, *pigments* (which are defined as colorants that do not dissolve in the compound in which they are used) have relatively more heat or light stability than dyes, and are used more often than dyes in PVC compounds. Pigments may be either organic or inorganic. The organic pigments, as a class, are relatively brighter and stronger than inorganic pigments but are usually more expensive, and they are usually less heat and light stable than inorganic pigments. No general statement can be made about what pigments are always best, since the specific use of the final part, and its other compounding ingredients must be known before a pigment can be selected.

For white polyvinyl chloride compounds the choice is almost always an inorganic pigment, and for black compounds the pigment of choice is almost universally carbon black. These pigments also have some other synergistic effects on the final part that leads to the decision to use them.

Many special effect colorants can be used to color PVC compounds, again depending upon the wants and needs of the purchaser of the compound or final product. Metallic pigments are finely ground flakes of metal used to make the plastic part simulate a metal part. A metallic effect may also be attained by the use of some coated mineral pigments manufactured specifically for this use. Fluorescent pigments allow the part to attain a surprising brightness by reflecting "brighter" light than it absorbs. Phosphorescent pigments allow parts to "glow in the dark," and pearlescent pigments allow parts to reflect a soft, silky appearance.

The majority of polyvinyl chloride compounds are solid (dry) materials and are provided to processors as powders or cubes. Some applications for PVC call for a liquid or similiquid form of the compound. To achieve this, a compound called a plastisol or organisol is made by adding a plasticizing liquid to the polyvinyl chloride resin in such quantities that the resulting material is a viscous liquid, sometimes so viscous that it resembles a paste. These plastisol and organosols are used in different processes and equipment than the solid (dry) compounds.

There are occasions when a processor of some polymer other than PVC wants to improve the physical properties of that polymer. There are many ways to change the physical properties of a polymer through the addition of many of the above-mentioned additives, but in some instances it is desirable to alloy that polymer with polyvinyl chloride. This addition of PVC to the other polymer will add many of the properties of polyvinyl chloride to that polymer, including resistance to ignition and burning. Indeed, the burning (or nonburning) characteristics of polyvinyl chloride make it a valuable alloying polymer to improve the overall combustibility properties of the other polymer.

Processing

Polyvinyl chloride, in its predominantly dry compound form, is processed mainly by extrusion, calendering, and injection molding. It can also be blow molded into bottles and other parts or made into film on a blown film line. The liquid (or paste) compounds are spread coated, slush molded, rotation cast, dip coated, or sprayed.

Applications

Adhesives
Anticorrosion sealants
Applicance housings and parts
Automotive parts
Banners
Binders and folders
Blister packaging
Book covers
Bottles
Box lids
Carpet backing
Change holders
Chemical storage tanks
Closures
Clothing
Computer disks
Computer parts
Conduit
Containers
Covers
Door frames and molding
Electrical insulation
Floor tile and other floor covering
Food wrap
Furniture parts
Garden hose
Gloves
Grips
Gutters and downspouts
Handbags
House siding
Housewares
Luggage
Meat wrapping
Mobile home skirting
Office machine parts
Outerwear
Packaging film and sheeting
Pipe and fittings
Phonograph records
Protective clothing
Protective coatings
Ribbons
Shoes
Shower curtains
Soffit and fascia
Sporting goods
Tapes

Applications (Continued)

- Tool handle covers
- Toys
- Tubing
- Upholstery
- Wall covering
- Wallets
- Window frames

Polyvinylidene Chloride

Polyvinylidene chloride (PVDC) is a chlorinated hydrocarbon thermoplastic better known by its trade name, Saran™. It has good chemical resistance and is an excellent barrier to liquids and gases. The molecular formula for vinylidene chloride, the monomer of polyvinylidene chloride, is shown in Figure 4.23.

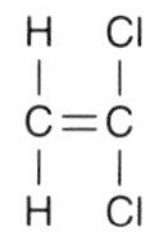

Figure 4.23 Structure of vinylidene chloride

Processing

Polyvinylidene chloride may be extruded, injection molded, melt blown, and film blown.

Applications

Polyvinylidene chloride's applications include food packaging, household film wrap, industrial packaging, pharmaceutical and cosmetic packages, and drum liners.

STYRENICS

This section includes all those thermoplastics that are commonly called styrenics, with the exception of the parent polymer, polystyrene, and ABS. Both these materials are treated separately because of the sheer volume and corresponding importance of the polymers.

Acrylic–Styrene–Acrylonitrile

Acrylic–styrene–acrylonitrile (ASA) is a terpolymer thermoplastic made from the three monomers that make up its name. Acrylic–styrene–acrylonitrile resins have physical properties similar to acrylonitrile–butadiene–styrene (ABS), but they have considerably better weathering properties. After outdoor weathering, ASA retains good gloss and impact resistance. Acrylic–styrene–acrylonitrile resins may be alloyed with many other polymers to improve their properties.

Acrylic–styrene–acrylonitrile may be processed by extrusion, injection molding, blow molding, or foam molding. Acrylic–styrene–acrylonitrile finds application where it can be coextruded over a less weatherable polymer to produce a finished product that will be weatherable. This includes products such as house siding, windows and doors, gutters and downspouts, pickup truck topper caps, pools and spas, boats, and recreational vehicles.

Acrylonitrile–Chlorinated Polyethylene–Styrene

Acrylonitrile–chlorinated polyethylene–styrene (ACS) is a terpolymer thermoplastic containing the three repeating units mentioned in its name. It is somewhat similar to acrylonitrile–butadiene–styrene (ABS) in some physical properties, but it has considerably better flame resistance, heat stability, and weatherability.

Acrylonitrile–chlorinated polyethylene–styrene can be processed on extruders and injection molding machines so long as certain processing temperatures are not exceeded. Its inherent antistatic properties lend it to uses where dust may be electrostatically attracted, for example, business machine housings, electronic equipment, video- and audiotape recorders, electrical connections, measuring equipment, and fire prevention equipment components.

Olefin-modified Styrene–Acylonitrile

Olefin-modified styrene–acrylonitrile (OSA) is a weatherable engineering polymer that is made by polymerizing styrene and acrylonitrile onto an olefinic elastomer. It can be coextruded over other polymers that degrade over time in outdoor applications, combining the properties of the substrate with its own, or it can be made into products itself.

It can be processed by extrusion, coextrusion, or injection molding. Its major applications are as protective layers over less weatherable polymers to make products such as house siding, windows and doors, gutters and downspouts, spa and swimming pool components, boat hulls, recreational vehicle components, pickup truck topper caps, and electronic equipment.

Styrene–Acrylonitrile

Styrene–acrylonitrile (SAN) is a thermoplastic copolymer that is transparent, has good gloss, good chemical resistance, high hardness and rigidity, and good dimensional stability. Some disadvantages to styrene–acrylonitrile are that it is hygroscopic (will absorb moisture from the air) and that is attacked by certain aromatic hydrocarbons, chlorinated hydrocarbons, and some esters and ketones.

Styrene–acrylonitrile's properties can be changed by varying either of the monomers, and specialty grades can be produced that are antistatic (will bleed off any static buildup) and that have superior gas transmission properties, have improved UV (ultraviolet rays) resistance, and good weatherability. Styrene–acrylonitrile resins can be extruded, injection molded, blow molded, and compression molded. Appplications include parts for the automotive, furniture, electronics, appliance, housewares, and medical industries.

Styrene–Maleic Anhydride

Styrene–maleic anhydride (SMA) is a styrenic thermoplastic made by the copolymerization of styrene and maleic anhydride. Styrene–maleic anhydride has higher heat resistance than either styrene or acrylonitrile–butadiene–styrene (ABS), good mechanical strength, good chemical resistance, and good clarity and gloss. Some styrene–maleic anhydride resins exhibit better than average plateability and plate adhesion (metal plating of the plastic part to make it resemble a metal part). Styrene–maleic anhydride resins may be processed by standard processing techniques. Applications include air conditioner grilles, automotive interior parts, food packaging, hospital food trays, microwave cookware, structural components (when glass reinforced), and dishwasher-safe tumblers.

Styrene–Butadiene

Styrene–butadiene (SB) is a thermoplastic copolymer that is useful because of its transparency, low haze, high gloss, and high impact strength. It may be extruded, injection molded, and blow molded, and film may be blown or cast. The material may also be used as an additive to other resins to enhance their properties. Applications include disposable trays, cups, lids, and other packaging items, toys, office supplies, medical equipment, shrink wrap, overwrap, and bags.

Thermoplastic Elastomers

Thermoplastic elastomers (TPE) are polymers that include a wide variety of types, depending upon the monomers used and the polymerization process. They include such diverse structures as ionic polymers, polyester/polyether copolymers, polyether/polyamide copolymers, styrene/ethylene–butylene/styrene copolymers, styrene/butadiene/styrene (SBS) copolymers, thermoplastic polyolefin elastomers (TPO), thermoplastic polyurethanes, and many other rubber/plastic alloys.

The properties of each of these materials can be built in during the polymerization process, and the hardness/softness and other properties may be modified by changing the proportions of the monomers, as well as the polymerization process itself. The methods include "grafting," and the resulting materials are referred to as "block" copolymers. Other thermoplastic elastomers are really compounds made by the blending together of different polymers. What the thermoplastic elastomers all have in common is that they combine the physical properties of a rubber (thermoset) with the processing economics of a thermoplastic.

Further alloying of these materials can produce yet more superior compounds. By altering the blending and polymerization methods even further, materials known as thermoplastic elastomer alloys are created. These materials have such technical-sounding names as melt-processable rubber (MPR) and thermoplastic vulcanizate (TPV).

Thermoplastic elastomers may be processed on all conventional processing equipment. Appplications for the thermoplastic elastomers include uses in the automotive, electrical, footwear, and mechanical goods industries.

THERMOSETS

Thermosetting plastics are polymers that have undergone a different chemical reaction than thermoplastics. In thermoplastics polymerized from one monomer, the chemical reaction has been between and among molecular fragments (free radicals) that have been produced by breaking a double bond in the simple molecule. All the material in the reactor is identical, and the reaction consists of the linking up of these simple molecular fragments *with themselves,* end to end. Where thermoplastic copolymers are concerned, the reaction consists of the linking up regular repeating units of the two monomers involved. These materials may be subjected to heat and pressure more than once (that is, they may be ground up and reprocessed, sometimes several times). Terpolymers are usually produced by the copolymerization of two monomers and the grafting of a third onto the chain.

On the other hand, the polymerization of thermosets involves not only the linking up of the various monomers to each other (end to end) and to other monomers (copolymerization), but also crosslinking (the formation of bonds between the atoms in several places along the chains). Instead of producing reasonably flexible materials characterized by long-chain molecules intertwining among themselves, a three-dimensional network is established. This tends to lead to the formation of much harder materials, and it also prevents the reapplication of heat or pressure to re-form the part. Heat and pressure subsequent to the original processing would cause those interchain bonds to break and the polymer to degrade (pyrolyze or break down), losing its properties.

Alkyds

Alkyd resins are thermoset polyester resins that are produced from polymerization reactions involving low amounts of unsaturated polyester resins. They are characterized by good electrical properties. Alkyds react rather rapidly and are processed by transfer and reciprocating screw injection molding. The highest volume alkyds are the phthalic anhydride type, whereas maleic anhyride types make up a small percentage of the total.

Applications for alkyds include aircraft windows, automotive ignition systems, appliance parts, electrical and electronic components, lenses, marine glazing, paint formulations, safety windows, vending machine windows, and watch crystals.

Allyls

Allyl resins are really a family of thermosetting plastics used as molding compounds. Parts made from allyl resins have good heat and chemical resistance, good mechanical strength and dimensional stability, and good electrical properties. The allyl esters most commonly used as monomers are diallyl phthalate and diallyl isophthalate. Applications include electrical and electronic components that are subjected to stressful environmental conditions. Allyls are also used as insulators, as television components, and as monomers themselves in the crosslinking of polyesters.

Aminos

The amino plastics include urea-formaldehyde and melamine–formaldehyde plastics, which are discussed individually.

Bismaleimide

Bismaleimide (BMI) is a thermoset polymer exhibiting high temperature resistance and good solvent resistance. It is usually processed as a prepreg and then used to make composite structures like printed circuit boards and aircraft parts.

Epoxies

Epoxy resins are a large family of thermosetting polymers that are valuable because of their extremely low shrinkage. Because of the reactants used to prepare them, they have such names as aliphatic epoxies, bisphenol A epoxies, brominated epoxies, multifunctional epoxies, novolac epoxies, and vinyl ester epoxies. They are all characterized by the unique oxirane structure in their molecules, which gives them their reactivity. They must all be cured by adding a hardener or crosslinking agent. Applications include adhesives, coatings of all types, electrical components, flooring, grouts, and laminations in a wide assortment of industries, including the aircraft, appliance, automotive, chemical, electrical/electronic, housing, mining, and oil industries. They are found literally everywhere.

Melamine–Formaldehyde

Melamine–formaldehyde, or more simply, melamine, resins are thermoset amino polymers made by the reaction of melamine with formaldehyde. They are closely related to the ureaformaldehyde resins, and both are sometimes referred to as amino molding compounds. They may be liquid or solid in form. Parts made of melamine–formaldehyde resins are extremely hard and tough, and they have good electrical properties. They are also very clear and have excellent laminating properties.

They may be processed by injection molding, compression molding, and transfer molding. Applications include adhesives and glues (by far the highest volume usage), ashtrays, buttons, cosmetic container closures, dinnerware, housings, table- and countertops, toilet seats, and wiring devices.

Phenolics

Phenolic resins are made from the reaction of phenol and formaldehyde and are heat cured thermosets, some of which have engineering properties. They are most often filled with mineral fillers and other reinforcements to give the strength properties required, which are high in strength, hardness, dimensional stability, and wear resistance, and have good electrical properties. Phenolics may be processed by injection molding, compression molding, and transfer molding. Applications include appliance connector plugs, automotive water pumps, brake pistons, buttons, electrical connectors, fuse blocks, handles, knobs, instrument panels, motor housings, pulleys, receptacles and switches, oil well valves, photographic development tanks, radio and television cabinets, and wheels.

Silicones

Silicones are a family of thermosetting polymers that are not usually classified as organic unless there are specific side groups. Organic (as opposed to inorganic) refers to the presence of carbon in the molecule, and instead of carbon in the chain as in every other type of polymer, there is silicon present in the material. The chain in the silicone polymer makeup is shown in Figure 4.24.

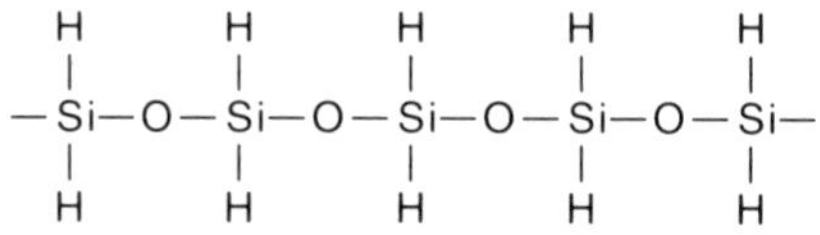

```
   H     H     H     H     H
   |     |     |     |     |
 —Si—O—Si—O—Si—O—Si—O—Si—
   |     |     |     |     |
   H     H     H     H     H
```

Figure 4.24 Structure of silicone

Si is the symbol for silicon, just as C is the symbol for carbon. Silicon is just below and in the same group as carbon on the periodic table of the elements. This means that silicon will have some of the same properties as carbon, including forming polymer chains. Whenever an organic group (that is, a molecular fragment containing carbon) is attached to one of the bonds, the polymer will be part organic and part inorganic.

Silicone polymers may be solid or liquid and even in between, as a gel. Silicones have a low coefficient of friction, which means they are very slippery and have good release properties. They have good electrical properties, excellent water resistance, good chemical resistance, and good weathering properties. Applications include gasket and sealing products, electrical insulation, caulking compounds, and as a base for some cosmetic products.

Thermoset Polyimides

Thermoset polyimide resins are used to make parts that are used in engineering applications because of their strength, their electrical and chemical resistance properties, and their usefulness at extremely high temperatures. They may be extruded, injection molded, compression molded, or transfer molded. Some thermoplastic polyimide resins must be processed in the same manner as metallic powders. Applications are in areas where the highest degree of quality is required, such as in aircraft and aerospace needs. Other uses include integrated chips, jet engine parts, printed circuit boards, gears, films, and coatings.

Unsaturated Polyesters

Unsaturated polyester resins are a family of thermosetting polymers that includes a large number of resins made from different raw materials. They are usually liquids and are processed by casting, spraying, spray/hand layup, pultrusion (a forming process whereby a material is pulled through a die as a solid material, rather than being heated and changed into a doughy mass and pushed through a die and cooled, as in extrusion), and high and low pressure molding.

Unsaturated polyesters may be cast as an unfilled resin, but they are almost always filled with a mineral filler or glass fibers. This is the material often referred to as "fiberglass." Applications determine the required properties of the resin, and these properties are built in during the compounding stage. They include low shrinkage, good electrical properties, water resistance, high gloss and other surface characteristics, mechanical strength, paintability, and outstanding dimensional stability. Applications include automotive and truck bodies, boat hulls, building facades, bulk molding compounds (BMC), flooring, highway and bridge repair, microwave cookware, sanitary ware (bathtubs, bathroom sinks, and shower wall-surrounds), and sheet molding compounds (SMC).

Urea-formaldehydes

Urea-formaldehyde resins are thermosetting amino resins made by the reaction of urea with formaldehyde and are often referred to as amino molding compounds. They are closely related to the melamine resins. They are available as liquids or solids. They are made into parts that very hard and tough and have good electrical properties. They may be processed by injection molding, compression molding, and transfer molding. Applications for urea-formaldehyde resins include adhesives and glues, bonding agents for particle board and plywood, coatings for abrasion paper, construction of boat hulls, and insulating foams.

5

Additives: Other Chemicals That Might be Found in Plastics

Although many plastics are used in their pure or "natural" form, many other plastics must be modified in some manner to make them acceptable to the ultimate consumer. When the modifiers, called additives, are added to a plastic, the resultant mixture is called a **plastic compound**. An **additive** is any substance that has been added to a polymer, resin, compound, blend, or alloy, to improve its processing characteristics in the manufacturing process; to impart specific physical, chemical, or appearance properties to it; or to modify its properties in some other way. The list of materials fitting this definition is very long, and new materials are being added to it constantly.

The many reasons for adding other materials to a plastic are described below. I have not tried to mention every additive within a specific class, or indeed, to list every class of additive. The descriptions in this chapter cover all the additives most likely to be found in a plastic compound, and subsequently in the finished part.

FILLERS

There are several reasons to add fillers (or extenders) to a plastic material. The most common is to extend its bulk and therefore lower its cost. Other reasons include the need to add chemical resistance or reinforcement properties; to raise or lower the gloss level of the surface or otherwise improve the surface of the final part; to improve processing properties; to improve flame retardant properties; to impart electrical conductivity or insulation properties; or to make the resulting compound uniform throughout in color or other visual properties. Some fillers serve more than one of these purposes and may perform a different function in one compound of one polymer than in a compound of another polymer. By volume produced, the fillers are the largest of the additives, and they are found in more compounds than any other additive.

Mineral Fillers

As indicated by the name, these materials are inorganic (ionic) in nature. They may be minerals that have been dug from the ground, purified, and reduced in size by grinding, or they may be exactly the same chemical compound as the naturally occurring mineral but produced chemically in an industrial operation. The type of filler and its purity, particle size and shape, and other properties are dictated by the required properties of the finished product as specified by the end user. Mineral fillers do not burn, and some of them function as flame retardants, smoke suppressants, or both. Therefore, mineral-filled plastic compounds, even those made from plastics that might burn readily in other formulations, will almost always burn much more slowly (if at all) than their unfilled counterparts, even if the filler was added for some other purpose than to reduce combustibility.

Mineral fillers include *calcium carbonate,* a water-insoluble material that may be produced from grinding limestone, marble, or the shells of sea and freshwater animals. It may also be precipitated (formed by a chemical reaction of other compounds, usually in a water solution in which it will not dissolve) in an industrial chemical process, in which case it is produced in a very fine particle size. It is used to impart cost reduction or dimensional stability, as a film antiblock agent (to keep plastic film from sticking together), or to impart impact strength and surface uniformity.

Other mineral fillers than calcium carbonate may also be precipitated, which means that the filler is a reaction product that is insoluble in water but is made from the chemical reaction of other chemicals that *are* soluble in water. The filler precipitates, or "falls out" of, the solution in a fine particle size, and it is then dried for bagging and use.

Clay (a form of aluminum silicate), may be used as a dried and ground material, or one that has been calcined (heated in a furnace to 1,800°F) and ground. It is not soluble in water and is used to lower the cost of the plastic compound and to add dimensional stability, electrical insulation, tensile strength, and thermal stability. Many clays have flat, platelike structures, and clays are the least expensive of materials with this structure. Clays are natural products that are found in various parts of the world and are mined for several uses. There are many types of clay, and not all are suitable for compounding into plastics. Many are suitable, however, and depending on the property required, clays may be the most suitable filler.

Mica is a water-insoluble mineral made of potassium–aluminum–magnesium silicate and is selected for its flat, platelike particles, which are ground down to a fine size. It used to impart chemical resistance, dimensional stability, tensile strength, low gas and water permeability, and electrical insulation properties to plastic compounds. Mica's flat particle shape gives it some of the same uses as clay, and which product is used may depend on other factors, such as cost, color, or other physical properties imparted.

Talc, which is mostly magnesium silicate, is a nonsoluble filler used to improve gas- and water-permeability barrier properties and to impart chemical resistance, dimensional stability, film antiblock properties, tensile strength, and water resistance to plastic compounds. Talc, like most other mineral fillers used in plastics, has a very fine particle size. This leads to the formation of very fine dust in the air, and all exposed persons must be protected from breathing dust of any composition.

Barytes is naturally occurring barium sulfate. It may be available as the ground mineral, bleached, or precipitated in an industrial chemical process. It is very high in specific gravity (that is, it is very heavy), and it is used to improve chemical resistance and dimensional stability, to increase the density of the filled part, to improve surface characteristics and thermal conduction, and to render a finished plastic article opaque to X rays.

Silica is the name given to at least three types of fillers: ground silica, precipitated silica, and diatomaceous earth (made up of the bodies of two-celled animals called diatoms). The different forms of silica may be used to lower the cost of the compound, to improve electrical properties, to impart good dimensional stability, as a film antiblocking agent, as a processing aid, and for controlling viscosity (resistance of a fluid to flow).

Antimony oxide is an insoluble mineral filler that has fairly good white hiding power, and doubles as a flame retardant agent. It is used to add opacity to a compound and as a pigment and opacifing agent.

Aluminum hydrate, also known as alumina hydrate, aluminum trihydrate, and ATH, is an insoluble filler used to improve mechanical properties and electrical insulation properties of plastic compounds. It also functions as a flame retardant and smoke suppressant.

Other mineral fillers include calcium sulfate, feldspar, molybdenum disulfide, nephiline syenite, and wollastonite (calcium metasilicate).

Other Fillers

Glass, in the shape of fibers, beads, or hollow or solid spheres, is often used as a filler in plastic compounds. It lowers the cost (in some cases), imparts chemical resistance and dimensional stability, improves mold flow, and adds mechanical strength and translucency (the ability of light to pass through a material). Glass fibers are used mainly to add strength as a reinforcement to a plastic compound.

Carbon is used either as a very fine powder or as a fiber. When used as a powder, its main function is that of a pigment, imparting black coloration throughout the compound. In higher concentrations such as in rubber tire compounds or plastics compounds, however, it is a reinforcement that adds tremendous strength and wear and abrasion resistance to the compound.

As a fiber, it is available in structures varying from the amorphous (noncrystalline) form of carbon to the crystalline form known as graphite. These fibers impart very good chemical resistance; corrosion resistance; electrical conductivity; and resistance to friction, mechanical, and thermal damage; and wear. Tremendous strength can be built into finished parts of light weight with the use of graphite fibers.

Metal fibers are used as fillers in plastics compounds mainly for the purpose of making them electrically conductive. These conductive compounds may be used to actually carry a current of electricity, to "bleed off" or dissipate an electrical charge or static electricity buildup, or to shield the surrounding environment from the electromagnetic interference (EMI) created by almost all electrical devices. Stainless steel fibers are the most commonly used, but nickel, aluminum, and silver powder and/or fibers may also be found in use. In some instances, the metal fibers may be coated by another material, usually to protect them during the compounding process.

Metal powders or flakes may also be found in many plastics compounds, but their presence there is mainly as a pigment to impart a metallic appearance to the finished part. They may also be added to impart many of the same properties added by metal fibers, but metal powders may not be as cost effective as the fibers. Additionally, metal powders are a serious safety hazard since the dusts will explode under certain conditions. Powdered aluminum, for example, is such a powerful explosive that it is often added to certain commercial explosives to increase their power. Any metallic powders used in the compounding of plastics must, for safety reasons, be made available dispersed in a liquid to prevent the formation of an explosive cloud.

Ceramic fibers are also used as reinforcements, and they offer many of the additional benefits provided by carbon/graphite fibers. These ceramic fibers include alumina fiber, boron fiber, refractory ceramic fiber, and silicon fiber. Ceramic fibers, as well as graphite and metallic fibers, must be added in a manner that does not break the fiber. As the mixed material passes through plastics processing equipment, care must be taken to protect these expensive fillers and to maintain the length of the fibers so that they will be able to perform the function for which they were added.

Organic fillers are used mainly to extend thermosetting plastics. Carbon black is considered by some to be an organic filler as well as a pigment. Carbon black may also be used as an ultraviolet (UV) ray absorber or as a reinforcing agent.

Other organic fillers include alpha cellulose, cellulosic fibers and flours, fly ash, paper, nut shell flours, synthetic fibers, wood flour, and woven fabrics.

COLORANTS

The next group of additives in volume (but probably highest in total cost) are the colorants. Colorants are those materials that impart the appearance property that we describe as color. They can be grouped into classes called pigments, dyes, and special effect colorants. The difference between a pigment and a dye is a technical one, but a dye is usually considered to be a colorant that will dissolve in the medium (in the context of this book, plastics) to which it is to impart color, whereas a pigment is considered to be insoluble in the medium in which it is dispersed (mixed). Dyes are usually organic chemicals, whereas pigments may be organic or inorganic. Special effect colorants may be metal particles, coated inorganic materials, or natural or synthetic minerals that impart some special visual effect (like pearlescence).

Inorganic pigments do not burn and do not add to the combustibility of the compound in which they are dispersed. In fact, in whatever amounts they are added, they will replace that percentage of combustible polymer, and the compound should be very slightly less combustible than uncolored compound. In some special cases, some flame retardant or smoke suppression properties might be added to the compound by the inorganic pigment used, but that is merely incidental to the purpose of this additive, which is to cause the finished article to have a desired color. However, if antimony oxide, aluminum trihydrate, or some other inorganic material that is also a flame retardant is chosen as the pigment, some flame retardant properties will be incorporated in the plastic compound.

Organic pigments and dyes do burn, but they are used in such small quantities that they should not affect the combustion properties of the uncolored compound in any appreciable way. The organic pigments and dyes are generally substituted hydrocarbons, which are mostly carbon and hydrogen, just like the polymer. These pigments and dyes do contain some elements that are not present in the original polymer, but the volume of these elements present is so low (because the organic pigment is present in small quantities) that combustion products of these elements are, for all practical purposes, undetectable in a real fire.

For example, very seldom does the amount of organic colorant in a compound that will be processed into a finished article exceed 1 percent of the material by weight, and in most instances the percentage of organic colorant is closer to .5 percent of the total weight. The greatest percentage of the weight of the organic pigment or dye will be material similar to the chemical composition of the polymer itself (that is, mostly carbon and hydrogen), so that there really is very little other material to change the final combustion products of the article itself.

Whether or not the special effect colorants burn depends upon their chemical makeup. If they are inorganic in nature, they won't burn, and they will, as already stated, have the same effect on the plastic part that other inorganic colorants have. If inorganic pigments are coated with another inorganic material, *they* won't burn. If there is an organic coating on the inorganic material, the amount of material is truly insignificant in its contribution to a fire. If the special effect pigment is a metal particle, that particle will burn. If the plastic part has a "speckled" appearance where large metal flakes have been incorporated into the plastic, the effect will hardly be noticeable in a fire, but if the plastic part has a metallic look imparted by large amounts of finely ground aluminum powder, the plastic part may burn fiercely. However, the plastic part that is burning must be looked at as part of all the other materials burning in the fire.

Chemically, inorganic pigments include, but are not limited to, the chemical compounds listed in Table 5.1.

Organic pigments and dyes are often very complex chemicals, but they are almost always hydrocarbon derivatives (that is, the "backbone" of the molecule is carbon and hydrogen, and the rest of the molecule is the chromophore, or color-producing portion). Some organic pigments contain metal ions, which might sound confusing in light of some of the previous discussion about ionic and covalent chemical compounds. Be advised that the chemistry of these organic colorants is very complex, and far beyond the scope of this text. Remember, the goal here is to recognize what will burn and how, not to become polymer or pigment chemists.

TABLE 5.1 Typical Inorganic Pigments

antimony oxide
barium sulfate
barium titanate
cadmium sulfide
cadmium sulfoselenide
chromium oxide
cobalt aluminate
iron oxide
lead carbonate
lead chromate
lead molybdate
lithopone
manganese titanate
nickle titanate
titanium dioxide
ultramarine blue, pink, and violet
zinc chromate
zinc ferrite
zinc oxide
zinc phosphate
zinc sulfide
and complex structures containing chromium, cobalt, iron, manganese, molybdenum, tin, and other metals. Many inorganic naturally occurring materials are also useful as pigments.

Anything organic will burn, so organic pigments and dyes will burn. However, as stated above, their presence in the final plastic part is so small (typical compounds colored only with organic colorants will usually have a maximum of 0.5% colorant, and many don't have that much) that it will have only a negligible influence on the burning characteristics of the compound. The presence of toxic combustion products attributed solely to the organic coloration of a plastic part in a real fire would be in quantities so small as to be undetectable for all practical purposes. Not all organic pigments are complex chemical compounds. Carbon black is a very strong pigment (as far as coloring power is concerned), and it is basically finely divided elemental carbon.

Organic pigments and dyes are chosen over inorganic pigments because the organic colorants are stronger, brighter, and will not add opacity to the compound. If the end use of the plastic part requires a bright color, or if the part must be transparent, the choice will almost always be an organic pigment or dye. However, these organic colorants are not as heat- or light-stable as the inorganic pigments. They may fade during manufacture or in use if they must face the ultraviolet rays of the sun without proper protection built into the compound. Also, organic colorants, as a class, are more expensive than their inorganic colorants, not to mention the cost of ultraviolet protection in the form of chemical light stazbilizers. If the end user can put up with the duller, more drab earth tones and other colors produced by the less expensive inorganic colorants, these will then be the choice.

Special effect colorants pose even less contribution to the fire load, unless the part is designed to have a metallic appearance as described above. Even then, the relationship of the burning plastic part containing powdered aluminum to the rest of the fire load must be considered. Other metallic particles such as copper and bronze do not burn in the same manner as aluminum powder, and would burn much more slowly.

Some colorants are mistaken for pigments or dyes because of their physical form. A form of colorant called "dry color" is really a pigment or dye that is mixed with something else to help it flow or disperse. That something else might be calcium carbonate (or some other mineral usually used as an extender in plastics compounds), a plastic resin, or a wax of some sort. Burning characteristics of the pigments will be changed by the medium in which they are dispersed. Obviously, inorganic pigments dispersed in calcium carbonate (or other mineral filler) will not burn, but not so obviously, inorganic pigments will not burn dispersed in any material. If the dispersing medium is a resin or a wax, the resin or wax may burn, but more slowly because of the presense of a noncombustible material such as the inorganic pigment. Organic pigments and dyes will burn in any medium, unless that medium is a flame retardant substance.

In summary, colorants (excluding aluminum powder) have little or no detectable effect on the burning characteristics of plastic compounds when used in normal concentrations. Whenever combustible or noncombustible colorants are present in "abnormal" concentrations, as in products called color concentrates, the burning properties of the base polymer or compound will be changed. They will either be increased or decreased, depending on the burning (or nonburning) characteristics of the colorant and the polymer involved.

STABILIZERS

Stabilizers are chemical compounds that modify a polymer or plastic compound so that it will withstand a particular hostile environment. That hostile environment might be the ultraviolet rays of the sun, the varying temperatures and moisture conditions in which the part is used, or it might be the heat of processing during the creation of the plastic compound or during the forming of the finished part. Processing (sometimes referred to as conversion) a polymer or plastics compound means to take it from its powder, pellet, cube, bead, liquid, slab, or other form and put it through special equipment to form a desired part. Processing usually involves mixing action, heat, and pressure.

Polyvinyl chloride (PVC) in its pure form is practically impossible to process into a usable article, since the pure polymer is so sensitive to high heat. Even trying to process it in a low temperature technique produces frictional heat within the polymer mass that will cause it to begin to decompose, rendering the entire "charge" in the equipment useless, and perhaps even damaging the equipment.

To allow polyvinyl chloride to be converted (processed) into useful parts, a heat stabilizer must be added before processing (in the compounding stage) to resist the degradation (decomposition) of the polymer in the processing equipment. These stabilizers are organic chemical compounds, usually liquids but sometimes solids. In any event, being organic in composition, they will burn. Since other materials are usually added to polyvinyl chloride, the amount of stabilizer added will not change PVC's burning characteristic materially. Stabilizers may be added at a rate of one to two parts per one hundred parts of vinyl resin. With the addition of other materials, the proportion of stabilizer in the final part is often considerably less than 2 percent.

Polyvinyl chloride and several other polymers are also degraded by ultraviolet rays from the sun, so light stabilizers (also called UV absorbers) are added to the polymer in small amounts to give the compound the protection it needs to withstand outdoor exposure. These chemical compounds are organic in nature and they will burn. Carbon black and titanium dioxide, a strong white pigment, will function as efficient UV absorbers in certain polymers such as PVC and the polyolefins (the polyolefins are polyethylene, polypropylene, polybutylene, and polymethylpentene).

PLASTICIZERS

Plasticizers are chemical compounds (usually liquids) that, when added to a polymer, reduce its rigidity and increase its impact strength. In other words, a plasticizer will make a stiff, rigid, brittle plastic softer and less brittle, depending on the amount and type added.

The classic example is the addition of plasticizers to polyvinyl chloride (PVC). Polyvinyl chloride resin is very hard and brittle. In fact, it is so hard and brittle that there are no uses for the pure resin, and all polyvinyl chloride resin must be modified (usually by the addition of a heat stabilizer, as described above) to some extent before it can be processed. Other polymers are naturally less stiff and brittle and require no plasticizer for processing into a useful product.

The amount of plasticizer needed to change the hardness of rigid polyvinyl chloride (RPVC) to the softness of a shower curtain is sufficient to radically alter its burning characteristics. Rigid polyvinyl chloride is very difficult to ignite. It burns so slowly and with such a low rate of heat release that parts made from rigid polyvinyl chloride will not support their own combustion. However, most of the plasticizers commonly used in polyvinyl chloride compounds are combustible liquids, and the addition of thirty to forty parts of these liquids will allow flexible polyvinyl chloride to burn very rapidly. Varying amounts of plasticizer are added to polyvinyl chloride resin to reach the particular degree of softness and/or flexibility specified by the customer, so the burning characteristics of polyvinyl chloride will follow a fairly straight-line relationship with the amount of plasticizer used. Rigid polyvinyl chloride (no plasticizer) will burn only if there is another material burning and producing a supporting flame (or some other source of very high energy), whereas very soft polyvinyl chloride will burn relatively rapidly.

Some plasticizers for polyvinyl chloride are actually flame retardants. The use of these materials to produce flexible vinyl compounds will maintain the difficult-to-burn characteristics of rigid polyvinyl chloride. The selection of the particular plasticizer to be used always depends upon the end-use application of the finished article and the performance specifications demanded by the customer.

The plasticizer used does not have to be liquid. Most plasticizers being used today, however, are liquid esters, combined with a few solid ester compounds. Rigid polyvinyl chloride, as well as other rigid polymers, may also be plasticized by the addition of an solid elastomer or another compatible, flexible polymer. The resulting compound is usually known as a polymer blend or an alloy.

As with the use of stabilizers, polyvinyl chloride is not the only resin to which a plasticizer may be added. Cellulosic plastics, nylons, and polyvinyl chloride–

acetate are among those rigid polymers that will accept liquid plasticizers. Other rigid polymers may be plasticized by the above-mentioned method of incorporating a softer polymer into the rigid one.

Plasticizers may also function within the compound as an extender, impact modifier, lubricant, or flame retardant.

LUBRICANTS

Lubricants are organic chemical compounds that are added to a plastics compound to help the material move against itself and through equipment during processing or finishing. They are classified as internal or external lubricants, respectively, and they are selected on the basis of one or more of several properties of the plastic compound, its processing characteristics, and the end use of the finished part. They range in chemistry from hydrocarbon waxes, esters, fatty acids, fatty alcohols, and fatty amides to metallic stearates. They are used in relatively small amounts and should not materially affect the burning characteristics of the polymer. In some cases, such as mold-release lubricants to allow the easy release of a part from the mold of an injection molding machine, the lubricant may not be added to the plastic compound but may be sprayed on the surface of the mold itself.

FLAME RETARDANTS

Flame retardants are covered in greater detail in Chapter 7. The following paragraphs will serve as an introduction to the topic.

Flame retardants are chemical substances that are added to reduce the combustibility (ease of ignition and rate of burning) of a polymer or one of its compounds. They may be organic or inorganic in composition, and they may double as a filler, pigment, or plasticizer. Ideally, flame retardants operate by decomposing when exposed to heat levels below the level sufficient to ignite the compound. The decomposition of the flame retardant reduces the ignitability of the compound and lowers its rate of burning. When the igniting flame is removed, the fire either goes out or burns at a controlled rate, well below that of the nonretarded compound. Typical flame retardants include aluminum trihydrate (ATH), antimony oxide, halogenated compounds, and phosphate esters.

The chemistry behind the operation of the different flame retardant chemicals is as different as the chemicals themselves. Once again, it is not the purpose of this text to describe the different (and sometimes complex) chemical reactions that occur when the flame retardant material operates. However, it is only common sense to realize that the very purpose of the addition of a flame retardant is to alter the manner in which a combustible material burns. In doing this, depending on the type of flame retardant used, the combustion products of the altered compound may also be altered. If flame retardants are used that have significantly different chemical compositions than the plastic compound to which it is being added (and if they are added in significant quantities), the makeup of the final combustion products will be changed. That is, if a halogenated hydrocarbon (a compound containing a hydrocarbon backbone and a halogen such as chlorine or bromine) is used to alter the burning characteristics of a hydrocarbon polymer such as polystyrene, the final combustion products will include carbon dioxide, carbon monoxide, carbon, and water (all the final combustion products of pure polystyrene) plus hydrogen chloride or hydrogen bromide, depending on which halogen is used in the compound.

The amount of hydrogen chloride or hydrogen bromide released in the pyrolysis or combustion of the finished part depends on many factors, including how much of the flame retardant is present in the compound, and how much of the flame-retarded compound has burned, decomposed, or otherwise degraded when exposed to heat, fire, or both. The amount of these irritant gases found in the total fire gases liberated by the room in which the fire takes place may be insignificant, depending on the entire fire load under consideration.

MISCELLANEOUS ADDITIVES

Other additives, whose class names are mostly self-explanatory, include antioxidants, antistatic agents, coagulants, coupling agents, crosslinking agents, curing agents, emulsifiers, foaming and defoaming agents, fragrances, fungicides, impact modifiers, mildew preventatives, mold releases, preservatives, smoke depressants, solution modifiers, ultraviolet brighteners, and viscosity depressants. Again, these are mostly organic chemicals but are used in such small quantities that they should not materially change the burning characteristics of the polymer or the content of the fire gases liberated.

SUMMARY

- Additives for plastics fall into many categories; the choice of additive depends entirely on the property desired for the plastic compound under consideration.
- Additives may be organic, inorganic, or, in special effect instances, metals or coated fibers.
- Additives may be solid or liquid, and the solids may differ in size, from slabs of material down to fine, dust cloud–forming dusts.
- The only additive that will materially affect the burning characteristics of the

compounds in which they are incorporated are large amounts of powdered aluminum; the non–flame retardant plasticizers when used in large amounts of polyvinyl chloride and other polymers (which will increase combustibility); and flame retardants themselves, which, of course, will decrease the combustibility of the resulting plastic compound.

- Large amounts of impact modifiers added to polymers with lower combustibility than the compounds themselves will also raise the combustible properties of the whole compound by acting as a flame retardant.

6

The Combustion of Plastics

The way any substance burns depends on many parameters and conditions. Among those parameters and conditions (and this list is not all-inclusive) are

- The chemical makeup of the substance (the elements within its molecules and the way they are arranged)
- The state of matter in which it exists (solid, liquid, or gas)
- The shape and size in which it exists when it is ignited
- Its configuration during ignition and combustion (horizontal, vertical, or in-between)
- The source of ignition (how "hot" and how big is the ignition source)
- The substance's ease of ignition
- The amount of oxygen needed for its combustion
- Whether the substance is free-standing or laminated against another material
- The rate of heat released as the material burns
- The material's thermal conductivity
- The rate of flame spread in the burning substance
- Ventilation and/or environmental conditions
- Whether or not other materials are burning (the total fire load)
- The moisture content of the substance
- The material's net heat of combustion
- The total mass loss or gas generation of the material
- Whether or not it forms a char when it burns; char is the carbonaceous (or mostly carbon) material formed by pyrolysis or incomplete combustion
- The manner in which it pyrolyzes
- Its melting temperature
- Its ignition temperature

All of these parameters and conditions, and probably a few more, make up a product's combustion or burning characteristics. These are the major reasons why the combustibility of plastics is not a simple subject with a simple explanation.

Some of these parameters are more important for one class of materials than another, and to study them independently might lead one to believe that one material is more dangerous than another where fires or the threat of fires is involved. Although it may be of some interest to a curious investigator to determine the burning characteristics of a pure substance in a particular form, this information might be of questionable value if the material is never (or seldom ever) used in that shape or form in the real world. For example, wood dust will explode if mixed in air in the proper proportions, as will plastics dusts. Metal dusts, however, particularly aluminum and magnesium, will detonate with much greater force than either wood or plastics dust. Yet to try to label metal as an explosive killer would be meaningless, except in the rarest of instances where metallic powders are created or used in a finishing operation (or where fragmentation bombs are used to kill people deliberately). In such instances the metal items *have* exploded and killed or injured unwary users and victims, and their use *must* be limited to only those circumstances where the explosion hazard has been eliminated by the use of strict safety measures.

Wood is the most common building material used in the United States today, particularly in residential dwellings. Its combustibility is well known, and firefighters have a tremendous amount of experience fighting wood fires. The combustion products of wood are deadly, but those firefighters who have learned that they must use respiratory protection in a wood fire have learned to protect themselves from such hazards. Also, firefighters have had almost all their training and experience fighting wood fires, and they have become very familiar with its combustion characteristics, even under all the variables listed at the beginning of this chapter. Wood and the structures made from it, as anyone who is familiar with the fire problem in the United States knows, are very combustible, and pose many toxicity and collapse hazards.

However, wood can be made safer in its use, and in some cases it has been. Firefighters have also learned how to protect themselves from the deadly combustion products of wood. Plastics are less well known, and since their uses are rapidly growing in all conceivable applications in the United States and around the world, concerns are being voiced over a fire problem that is perceived by some to be different. Since plastics are synthetic, most people feel they will behave in a different manner than so-called "natural" or "traditional" materials to which everyone has become accustomed. Their fears are fed when researchers test individual materials and proclaim plastics (or certain plastics) to be many, many times more dangerous in their toxicity properties than wood when the plastic or articles made from the plastic burn. The toxicity of combustion products is discussed in Chapter 9, but it must be understood that this issue is very highly charged emotionally, and very little scientific fact has been presented in relation to real-world (accidental and/or unexpected) fires.

How plastics burn must be described on two levels, since this is what happens in the real world. The two levels or types of real-world fires are (1) contrived test fires in a laboratory, and (2) accidental, unplanned, and/or unexpected fires that occur in the real world. Of course, the laboratory is part of the real world, and in that sense a test fire in the lab is a real-world fire. However, the test fire may consist of a simple burning of one material under one set of controlled conditions—or it may be a so-called large-scale test, involving a much greater and more varied fuel load and a much larger fire, also under controlled conditions. Either situation qualifies as a laboratory test. The test may also consist of an "actual" burn of a real house with real furnishings. This is still a test with all the variables being controlled instead of being truly variable. In this case the laboratory is the test house itself, but the situation is far from that of a real fire. For purposes of all following discussions in this book:

- A *real-world fire* is an accidental, unplanned, unexpected, noncontrived fire of any size that is not deliberately set as a test
- A *test fire* is a fire of any size deliberately contrived, controlled, and carried out *for test purposes* only: The laboratory may be (a) the stereotyped idea of a scientific place for carrying out experiments; (b) a real laboratory; (c) the fire scene at a test burn of a house, some other structure, or vehicle; or (d) anything in between.

In one circumstance, plastics samples are burned in the laboratory (or outside a "traditional" laboratory in a very large setting) as described in the several small-scale (or large-scale) combustibility tests as described in Chapter 8. On the other hand, plastics burn in almost every accidental or unexpected fire in the real world. The fact is that the way these plastics burn in these two situations is different enough to suggest that the laboratory tests are nothing more than the satisfaction of someone's curiosity—which is, of course, what is really going on in any experiment. That is not to say that the tests are worthless or meaningless. But the fact remains that seldom ever do plastics burn in a real-world fire as they did in a test fire, regardless of scale or location. The test may still be valuable to experimenters so long as it is recognized for what it really is: *an artificial setting and a contrived real fire observed under controlled conditions, and absolutely nothing more*. Even the importance that seems to be lent to a given test or group of tests by a renowned organization such as a government agency, a standards-setting organization, or a prestigious university, should not blind one to its limitations. As a contrived, controlled experiment, no matter how well managed, it has little or no relevance to real-world, accidental fires.

It is important to remember that this dichotomy is true for all materials, not just plastics. Most researchers try to find a small-scale test that will accurately predict everything that will happen in a full-scale situation. Nevertheless, even a full-scale test fire is still very different from a real-world fire occurring accidentally in someone's house or office. It is true that computers have helped reseachers to create models, but these only begin an attempt to handle the myriad details that must be considered in a real-world fire. Progress toward simulating a real-world fire has been almost negligible. Elsewhere in this book, I argue that the inherent fallacy of these tests is that the computer model is necessarily based on certain assumptions, and these assumptions are simplifications that just do not exist in the real-world, accidental-fire situation. Certainly, firefighters will talk about the similarities of certain fires, and firefighting training must necessarily make assumptions that fires are somewhat similar so that the firefighter can get *some* training. True fire experts all readily admit that there have never been two fires exactly alike: There are more possible combinations and permutations of materials, ignition sources, configurations, shapes, times, temperatures, ventilation possibilities, combustion characteristics, and other conditions than the total number of fires that have existed or ever will exist worldwide.

Knowing the complexity of real-world fires, the impossibility of creating all possible combinations, and the shortage of money needed to conduct tests, why are people seeking very simple solutions to one of the world's most complex problems? *It just doesn't work!* Should we then give up testing? Should we quit seeking solutions? Should we quit trying to improve the materials we use in our everyday lives? Should we quit worrying about fires and the lives they take or destroy and the property and environmental damage they do? Should we stop searching for the answers to how all sorts of materials burn, and how to provide a safe environment for everyone? Should we stop fighting fires? Should we stop trying to prevent fires? *No!* But we *do* need to remove the *emotion* from the search for answers. We need to approach real-world problems by searching for real-world solutions. We need to study real-world accidental fires to learn how to prevent them from occurring in the first place; to keep them from spreading in the second place; and last, but certainly most important, to minimize the danger to people, the environment, societal systems, and property if and when they do occur. We must get our priorities straight and leave economic and competitive hidden agendas out of our science.

HOW PLASTICS BURN: TESTS VERSUS THE REAL WORLD

When one keeps in mind all the listed parameters and conditions that control the way a material (or an article made from that material) burns, it must then become intuitively obvious even to the untrained observer that one cannot describe in simple terms how each different plastic burns. However, the question "How does X polymer burn?" is asked so often, particularly by firefighters, that a way must be devised to answer it simply. Therefore, acknowledging that there is no way to describe the process simply, let us set up the mechanism to do just that. But first, to repeat: Knowing how this particular article burns *by itself in test conditions tells us almost nothing* about how it burns in a real-world accidental fire.

What should interest people most is not how a particular plastic burns, but how it burns when fabricated into a particular familiar article and put into the use for which it was originally designed or a use that can be reasonably foreseen. Therefore, to see how a plastic, and an article made from that plastic, burns, a common article of known composition should be selected. The burning should always be done outside in the open air or in a vented laboratory hood—not because of some special hidden danger in burning a plastic article but rather for general safety purposes. It is just common sense to experiment under the safest conditions. No one, much less a firefighter or some other member of a community's safety forces, wants to start an accidental, unexpected, damaging fire while performing an experiment to see how something burns. Needless to say, whenever any material is burned, proper ventilation is required to prevent a buildup of toxic combustion gases, principally carbon monoxide. This, of course, includes the precaution of having a fire extinguisher handy. The extinguishment of burning plastics is discussed at the end of this chapter.

The following descriptions of how individual plastics burn imply that either a common fabricated article or a test strip of the particular polymer is used. Using plastic articles is the better practice. It will identify some difficulties that certain materials in certain forms, such as films and foams, experience in burning. It will also demonstrate how some plastics will melt during the heating process.

CLASSIFYING PLASTIC ARTICLES BY TYPE

It may be difficult to identify the actual plastic used in making the article selected for burning, because many companies producing identical products may use different plastics, or different formulations of compounds of the same plastics, to make them. The use of flame retardants will inhibit burning, but valuable information can be acquired by the experimenter following up with the manufacturer to learn as much as possible about the composition of the flame retarded article's material. Again, I must warn against drawing incorrect conclusions about the burning characteristics of *all* articles made from the same plastic as the sample article. They may be of different thicknesses, shapes, configurations, and compound compositions, to name only a few differing properties. Experimenters can be sure only that the particular article they test in a particular manner on a particular day in an experiment run in a particular, scientific (or nonscientific) way will behave in a particular way.

The following descriptions are given, in general, for articles made of a particular plastic in its solid form. That is, liquid resins or compounds are not discussed, but cellular or foamed, products may be. Foamed products will burn differently than noncellular (solid or nonfoamed) articles, with the degree of difference depending upon the density of the product (weight per unit of volume), which may may be translated as the degree of foaming (cell formation) per unit of weight. That is, resins or compounds may be "blown" or expanded (cells created to produce a solid foamed product) during processing to form articles of differing density, depending upon the size of the cells formed. If, in the description of the way an individual plastic burns, no mention is made of the article being foamed (or cellular in nature), the noncellular or solid form is to be inferred. Whenever a cellular product is discussed, it will be described as such.

The description will also be of a rigid product, rather than a flexible one, unless defined to the contrary. In the case of thicknesses, differentiation will have to be made between sheeting and film, since films behave very differently in a fire than will thicker sheets. The assumption will always be that the composition of the article being burned is essentially the pure polymer under observation, with only small additions of processing materials and/or color. These additives should not materially affect the burning characteristics of the plastic being discussed. However, in at least one volume thermoplastic, polyvinyl chloride, the addition of a plasticizer to impart the flexibility *does* materially affect combustion properties. These differences will be noted as they arise.

GROUPING OF POLYMERS BY CHEMICAL COMPOSITION

Generally, one may create different groups of plastics based on the elements present in their molecules. Doing this will give a general indication of how the article made from this group of plastics will burn. An added benefit of this grouping is that the combustion products of those groups of plastics will be known. It is possible to take almost all (99%) of the volume of plastics made in the world and apply this grouping. The groupings are:

- Polymers that contain only carbon and hydrogen in their molecules
- Polymers that contain only carbon, hydrogen, and oxygen
- Polymers containing carbon, hydrogen, and nitrogen (and occasionally oxygen)
- Polymers containing only carbon, hydrogen, and a halogen (fluorine or chlorine)

There are other plastics that contain other elements, such as sulfur or a metal, but they are made in such small volume and constitute such a small and specialized use that they need not be considered.

Polymers Containing Only Carbon and Hydrogen (see Table 6.1)

polybutadiene	polypropylene
polybutylene	polystyrene
polyethylene	

Predictably, these polymers are known as the hydrocarbon plastics, since the monomers from which they are polymerized contain only carbon and hydrogen. As a group, they will burn hotter than the other groups of plastics since the polymer chain is all fuel. This can be verified by examining Table 6.1. For comparison of the heats of combustion of other materials, please refer to Appendix E.

Since these polymers contain only carbon and hydrogen in their molecules, the final combustion products will be carbon, carbon dioxide, carbon monoxide, and water. There is always the possibility that longer-chain intermediate combustion products will form if combustion is not complete or pyrolysis products are carried away by the thermal column before they burn. However, all intermediate combustion products of these materials are combustible, so the vast majority of these newly formed compounds will be burned in turn, forming the final combustion products already mentioned.

Polymers Containing Carbon, Hydrogen, and Oxygen

acetals	polycarbonate
acrylics (any polymer with "acrylate" in its name)	thermoplastic polyesters (polyethylene terephthalate and polybutylene terephthalate)
cellulosics	unsaturated polyesters
epoxies	ureaformaldehyde resins
phenolics	

With the exception of acetals and cellulose nitrate, the polymers containing oxygen in addition to carbon and hydrogen burn more slowly than the hydrocarbon plastics. This is because the presence of oxygen in the molecule replaces potentially available fuel so that it is not available to support combustion.

TABLE 6.1 Heats of Combustion of Selected Polymers

Polymer	*Heat of Combusion, BTU/lb*
acetal	7,300
acrylonitrile–butadiene–styrene (ABS)	15,500
cellulose acetate	10,100
epoxy	14,175 to 16,750
melamine	8,175
natural rubber	19,450
nylon 6	14,175 to 16,750
nylon 66	13,750
phenolics	11,175 to 12,025
polyacrylonitrile	13,750 to 15,475
polybutadiene	19,850
polybutylene	20,175
polycarbonate	12,900 to 18,900
polyester (thermoset, glass-filled)	7,725
polyethylene	20,000
polyethylene oxide	11,475
polyethylene terephthalate	9,250 to 11,600
polymethyl methacrylate	11,175
polyphenylene oxide	9,875 to 14,175
polyphenylene sulfide	12,350
polypropyene	14,625 to 19,775
polysulfone	7,725 to 11,175
polystyrene	18,000
polytetrafluoroethylene	125 to 1,950
polyurethane	10,275 to 12,025
rayon (viscose)	8,175
rigid polyvinyl chloride	8,175 to 8,600
polyvinyl fluoride	9,025
polyvinylidene chloride	4,300
polyvinylidene fluoride	3,875
silicone rubber	5,600 to 6,650
styrene–butadiene rubber	18,925
ureaformaldehyde	7,650

Oxygen, of course, does not burn, and therefore the entire polymer molecule burns more slowly than if it were pure fuel.

The exceptions, acetal and cellulose nitrate (which contains oxygen combined in a different manner than other plastics containing oxygen), exist because of the availability of the oxygen in the polymer molecule to support oxidation. In these exception polymers, the bond linkage allows the oxygen to break free of the molecule during pyrolysis, and it then becomes available to support combustion. Both acetal and cellulose nitrate burn hot and "clean" (that is, very little smoke and carbon monoxide generation). As a matter of fact, pure, unstabilized cellulose nitrate is a very dangerous material when subjected to heat. Once ignited, it is almost impossible to extinguish, since the molecule carries its own oxygen. Cellulose nitrate was first used in movie film, and its flammability caused many deaths in movie houses when the structure was ignited by burning cellulose nitrate film. Today's cellulose nitrate is chemically stabilized against rapid burning and/or decomposition, so the danger of the past is gone. However, it does burn faster than the other polymers containing carbon, hydrogen and oxygen.

The heats of combustion of these polymers are lower than those of the hydrocarbon plastics, again because oxygen in the molecule replaces fuel that could burn to produce more heat energy.

Combustion products of the polymers in this group are carbon, carbon dioxide, carbon monoxide, and water. These are the same combustion products as the hydrocarbon plastics, since they both contain the same fuels: carbon and hydrogen. Other intermediate combustion products are possible, including the short-chain aldehydes. These combustion products will most closely approximate those of wood, since the major ingredient in wood and wood products is the naturally occurring polymer cellulose.

Polymers Containing Carbon, Hydrogen, and Nitrogen

acrylonitrile–butadiene–styrene (ABS)
melamine
nylons
polyacrylonitrile
polyurethanes (both thermoplastic and thermosetting)
urea-formaldehyde resins
any copolymer, blend, alloy, or composite containing them

The heats of combustion of this group fall in about the middle of the range of all the polymers. They burn at about the same rate as the plastics that contain carbon, hydrogen, and oxygen for the same reason: The nitrogen, which occupies space on the polymer chain, does not burn, but its presence replaces fuel. The rate of burning also is about the same for both groups. The polyurethanes gained a reputation for rapid burning mainly from the combustion of unprotected polyurethane foam. Today's polyurethane foam is flame retarded, and the solid polyurethane never did burn with the same speed as the foam.

Combustion products of the polymers in this group are carbon, carbon dioxide, carbon monoxide, water, and hydrogen cyanide. The hydrogen cyanide is formed because of the presence and subsequent release of nitrogen from the polymer molecule. The hydrogen cyanide, like carbon monoxide and any intermediate combustion products, is flammable, and the vast majority of hydrogen cyanide produced during combustion is burned in the fire.

As stated in the foregoing section, cellulose nitrate contains nitrogen (in addition to carbon, hydrogen, and oxygen), but it is held in this compound differently than it is in other nitrogen-containing plastics. The nitrogen is bonded in a different manner than in the amino, polyacrylonitrile, nylon, and polyurethane molecules. In those molecules, the possibility of formation of the cyanide (—CN) radical usually occurs with oxidation, and this does not happen with the nitrogen in cellulose nitrate. In cellulose nitrate the nitrogen will not combine with carbon to form the cyanide radical (—CN), which otherwise would combine with hydrogen to form hydrogen cyanide (HCN). However the cyanide radical can, under certain conditions, be liberated in a group of gases known as the nitrogen oxides, most of which are toxic and can produced serious, often delayed, results.

Polymers Containing Carbon, Hydrogen, and a Halogen

Polymers Containing Chlorine:

chlorinated polyethylene (CPE)
chlorinated polyvinyl chloride (CPVC)
polycarbonate (PC)
polyvinylidene chloride (PVDC)
polyvinyl chloride

Polymers Containing Fluorine:

chlorotrifluoroethylene (CTFE)
ethylene–chlorotrifluoroethylene (ECTFE)
ethylene–tetrafluoroethylene (ETFE)
fluorinated ethylene propylene (FEP)
polychlorotrifluoroethylene (PCTFE)
perfluoroalkoxy (PFA)
polytetrafluroethylene (PTFE)
polyvinylidene fluoride (PVDF)
polyvinyl fluoride (PVF)

The halogens that appear in polymer chains are fluorine and chlorine. The other two halogens, bromine and iodine, are much too large in size to fit on a polymer chain, and the bonds that would be formed if they did fit would be too weak too hold them on.

When these polymers burn, the final combustion products are carbon, carbon dioxide, carbon monoxide, water, and hydrogen chloride or hydrogen fluoride, depending on whether the polymer contains chlorine or fluorine. These products have the lowest heats of combustion and the lowest rate of flame spread of all the polymers. With one possible exception, they are very difficult to ignite, and they will all stop burning when the supporting flame is removed. The possible exception is chlorinated polyethylene.

Rigid polyvinyl chloride, the largest volume plastic in this group and the second largest volume plastic overall, is much maligned on the question of combustion. When plasticizer is added, and the plasticizer is a combustible liquid, the Limiting Oxygen Index of polyvinyl chloride will be lowered as the amount of this plasticizer is added. **The Limiting Oxygen Index (LOI)** is the limiting concentration of oxygen in the atmosphere necessary for sustained combustion. A material with an LOI of more than 21 should not burn in air at room temperature. Flexible polyvinyl chloride will burn and support its own combustion if enough combustible plasticizer is added. Rigid polyvinyl chloride will *not* support its own combustion, and its reaction in fire is typical of the halogen-containing plastics. For the Limiting Oxygen Index of various polymers, please refer to Appendix E.

In a fire, the halogen is held on the polymer chain until enough heat energy is input to break its covalent bond with carbon. The halogen then comes off the chain (in an "unzipping" fashion) as a free radical, not as a halogen molecule. This free radical then bonds with a hydrogen radical coming off the same chain in the same manner, and removes it from possible combustion. Because of this sequence, polyvinyl chloride and similar halogen-containing plastics will not burn once the supporting flame is removed. The combustion product formed is either hydrogen chloride or hydrogen fluoride.

These above four groups represent 99% of the volume of plastics made and used. Detailed below are the individual plastics that make up these groups.

HOW INDIVIDUAL PLASTICS BURN

Throughout this discussion on how various individual plastics burn, I continually refer to the performance of a simple burn test to see how the particular plastic in question burns. *It is extremely important to realize* that I am *not* referring to *strict scientific tests run by accredited laboratories operated professionally by trained technicians or to any other scientific test run by qualified experimenters at a university or business enterprise*. I am addressing the people for whom this book was written: architects, building code officials, engineers, firefighters and other emergency responders, insurance personnel, interior designers, government regulators, safety managers, and others who are just interested in how individual plastics burn. If you want to have some idea how these materials respond to fire, simple, *totally nonscientific* method of demonstration is available to you. Simply take an article made from the particular plastic in question and burn it in a safe manner, usually just outdoors, or in a vented laboratory hood available to you at a local high school, college campus, or industry.

All one must do to realize the truth of these statements is to study the tests described by standards-setting organizations such as the American Society for

Testing and Materials (ASTM), the International Standards Organization (ISO), and Underwriters Laboratories (UL). Once you have familiarized yourself with some of these procedures, you will see clearly that testing for fire properties, as well as other properties, is complicated and therefore must be precisely described and controlled. Anything less than these rigid procedures produces meaningless results. Moreover, even if one carries out the tests correctly and exactly in accordance with the procedures established by the above-mentioned organizations, the only thing one can be sure of is that the material will behave the way it did under *only* the circumstances of the test performed.

Let me stress again that this simple burning of an article is absolutely not a scientific test, and the results and observations you obtain can *not* be generalized to apply to a real, unplanned fire involving the material in question. Anyone who suggests otherwise—anyone who tries to portray the simple burning of a piece of plastic as having scientific significance—either misunderstands the scientific method or is pursuing some hidden agenda. You are burning a piece of plastic for two purposes: (1) simply to see how it burns or how it reacts in a flame and (2) to help identify the plastic, if you do not know the composition of an article. These considerations apply to burning an article made out of any material, not only a plastic.

Nor should sinister meanings be read into the admonishment—repeated several times—that the simple burn test should be run outdoors or in a vented laboratory hood. It is just simple common sense to want to burn something in the safest manner possible. Anyone who deliberately sets anything on fire indoors without confining the fire and evacuating the products of combustion is not safe to be around! (Of course, this does not apply to the lighting of decorative candles or the ignition of a fuel for heating or cooking, although there are some people who should not even be allowed to do that!)

To understand why the article burned the way it did, it is very important to identify the polymer from which each article test-burned is made. This includes knowing what additives are compounded into the polymer (see Chapter 5), since those additives may have some influence on how the article burns. For example, a flame retardant will reduce or materially change the combustion characteristics of the plastic in that article. Other additives may increase the combustibility of a polymer, for example, certain liquid plasticizers added to rigid polyvinyl chloride. One cannot simply take a piece of plasticized polyvinyl chloride, burn it, and assume that all other articles made of plasticized polyvinyl chloride will burn the same way. Indeed, the results may be 180 degrees apart, since the plasticizer used in another sample of flexible polyvinyl chloride may have flame retardant properties. Many additives, because of their chemical makeup or amount added to a particular polymer, will have no perceptible influence on how the article behaves in a fire, whereas other additives will influence the burning characteristics one way or another.

These are just a few cautions that must be heeded before doing a simple burn. The most important thing to remember is that the simple burning of an article as described below will not predict any behavior of the material in a real, unplanned fire.

INDIVIDUAL PLASTICS

Thermoplastics

Acetal

Acetal, also known as polyoxymethylene or polyformaldehyde, burns with a hot, blue flame. Very little smoke is generated by the combustion of acetal plastics, since the molecule contains oxygen that makes itself available to support the burning process. With this added oxygen available during combustion, the reaction will be more complete. The presence of smoke is indicative of incomplete combustion, with the release of unburned carbon particles producing the black portion of the smoke. Not all plastics that contain oxygen burn in this manner, since in the molecules of other oxygen-containing plastics the oxygen is arranged in the structure in a way that does not allow it to be freed and subsequently available to support combustion. The drops of molten acetal may or may not be burning as they drip.

Since acetal is really polyformaldehyde, the decomposition of its molecule will produce formaldehyde gas. Formaldehyde is flammable and will burn if its ignition temperature of 806°F is reached by whatever heat source is producing the decomposition of the polymer. This process of decomposition of thermoplastic polymers below their ignition temperatures is called **depolymerization,** and as its name implies, the process is the reverse of polymerization.

Final combustion products of acetal are carbon monoxide, carbon dioxide, and water. Formaldehyde will be present as well as minute amounts of carbon, but they will also burn.

Acrylonitrile–Butadiene–Styrene (ABS)

Acrylonitrile–butadiene–styrene (ABS) plastics represent an extremely large volume for a material that was originally classified as an engineering plastic. It ranks in the top five or six thermoplastics in volume produced and is available in many grades, depending upon the end-use requirements and the composition of the final compound. Burning characteristics somewhat resemble the behavior of each of the three polymers that make up ABS and will be determined by the relative amounts of each of them used in making the particular ABS part under examination. Therefore, since there may be a great difference in the amounts of each of these monomers used from article to article, a description of the burning process must of necessity be general.

The ignition temperature of ABS is in the range of 780°F to 915°F, depending on its composition. It will begin to melt (soften) in the range of 190°F to 257°F.

The total heat (heat of combustion) released by burning ABS places it at about the middle of the scale of different plastics, between polyester near the bottom and polyethylene at the top. It burns with a smoky flame, producing large amounts of soot (charcteristic of polystyrene) and may smell of burnt rubber (owing to the presence of polybutadiene). ABS will burn with a luminous yellow flame and will support its own combustion (that is, it will continue to burn after the igniting flame has been removed). In some cases, the molten drops from burning ABS will burn as they fall.

It is not easy to identify an article made from ABS, since manufacturers may switch back and forth between and among different plastics from which to produce each product. Certain lines of luggage are made from ABS, as are many rigid automobile parts. Most door liners in refrigerators are made from ABS, but some are made from polystyrene. The best way to select an article for testing is to check with the manufacturer of that article to determine if ABS was used.

It is also important to check with the manufacturer to determine whether the ABS compound contains any flame retardant material, since this will certainly affect the way the article burns. For example, many plastic television cabinets are made from ABS, and for many appliance applications a flame retardant must be used to reduce the product's combustibility. An ABS part made from a compound containing flame retardant will burn only with great difficulty, and only if the flame is very hot and in contact with the part throughout the burning process.

Assuming a non–flame retarded compound was used to make the article being tested, and if the article burns "normally" (that is, as described above), the final combustion products of ABS will include carbon, carbon monoxide, carbon dioxide, water, and hydrogen cyanide (because of the presence of polyacrylonitrile in the polymer). Intermediate products of combustion may be liberated, which might include the short-chain aldehydes. Hydrogen cyanide, as well as carbon monoxide, is flammable, and the vast majority of each gas liberated is consumed in the burning process.

For safety (and housecleaning) reasons, the burning of plastics for demonstration purposes should be done outside or in a laboratory hood that will safely evacuate the combustion products. Once again, the size, thickness, and configuration of the part will play an important part in how it burns.

Acrylics

The term acrylic refers to an entire family of plastics, as explained in Chapter 3. Many other plastics described in that chapter would be classified differently in other reference books, but as explained there, this book focuses on the combustibility of plastics. All polymers containing the acrylate group are therefore grouped with the acrylics for this discussion.

There are a great number of acrylic polymers, and many variations of compounds and copolymers are possible among them. But, since they burn in a somewhat similar manner, polymethyl methacrylate (PMMA), by far the most commonly used acrylic polymer, is used as the model whenever the term "acrylic" is used. Be aware, nonetheless, that certain additives greatly modify the thermal properties and burning characteristics of any given polymer, let alone a family of polymers as large as the acrylics.

Acrylics have a very wide ignition temperature range, depending on the particular polymer. That range extends from 806°F to 1040°F, which is the reported ignition temperature range for polymethyl methacrylate as listed by various references.

Acrylics burn with a slightly crackling, luminous yellow-topped blue flame and may emit a sweet fruitlike odor. Smoke production may be very low. At temperatures below the ignition temperature (from 527°F to 752°F), the polymer depolymerizes, producing a gaseous monomer.

The different acrylics depolymerize to different extents, but polymethyl methacrylate depolymerizes the most, often to the extent of 90 percent of its weight. Therefore, the final combustion products will be carbon (in small amounts), carbon dioxide, carbon monoxide, and water, with few intermediate combustion products or none. The pyrolysis of polymethyl methacrylate also produce few aldehydes or other products, with the main pyrolysis product being methyl methacrylate monomer. At the other extreme is polymethyl acrylate, which liberates almost no monomer (does not depolymerize). Its combustion products include carbon, carbon dioxide, carbon monoxide, and water, with some longer-chain compounds as intermediate combustion or pyrolysis products.

Acrylics rank near the middle of the range in heat of combustion.

Cellophane

Cellophane decomposes in the range from 349°F to 399°F, does not melt, and burns readily with a bright flame. It has an odor of burned paper. Its final combustion products are carbon, carbon dioxide, carbon monoxide, and water. Attempts to make it flame retardant will also make it opaque. It may burn too rapidly to be used as a demonstration of how plastics burn.

Celluloid

Celluloid is a trade name for a type of cellulose nitrate also known as pyroxylin. It is made by combining cellulose nitrate and a plasticizer plus a flame retardant. Many older references referred to celluloid as "fiercely burning" and used the rate of burning of celluloid as a benchmark, although its rate of burning had not been measured or further defined. Current references list celluloid as flammable or very flammable, and the U.S. Department of Transportation (DOT) classifies it as a Flammable Solid. Celluloid scrap is classified by the Department of Transportation as a Spontaneously Combustible material. The indications are that current versions also burn rapidly, although the rate of burning has been reduced by the presence of the flame retardant.

Since it is still basically cellulose nitrate, celluloid burns with a bright flame, producing reddish-brown gases. Its odor will be similar to camphor. Its final combustion products are carbon, carbon dioxide, carbon monoxide, and water, in addition to intermediate longer-chain combustion products plus the nitrogen oxides. Celluloid is probably too flammable to be used in a test to demonstrate how plastics burn.

Cellulose Acetate

Cellulose acetate has an ignition temperature of 887°F and burns with a yellow flame. Sparks may be present as it burns, and as the polymer melts, the drops may be burning as they fall. The odor of burning cellulose acetate resembles that of acetic acid (vinegar smell), and final combustion products are carbon, carbon dioxide, carbon monoxide, and water, with some of the longer-chain aldehydes and other derivatives as intermediate combustion or pyrolysis products.

Cellulose Acetate Butyrate

Cellulose acetate butyrate burns with a yellow-topped blue flame, and as the plastic melts the drops may be burning as they fall. A moderate amount of soot is produced as it burns, and the flame may produce some sparks. The odor of burning cellulose acetate butyrate resembles that of rancid butter or burned paper. Its final combustion products are carbon, carbon dioxide, carbon monoxide, and water, and there may be some other intermediate longer-chain combustion products.

Cellulose Acetate Propionate

Cellulose acetate propionate is combustible, as most of the cellulosics are, and its combustion products are similar (carbon, carbon dioxide, carbon monoxide, and water, with some of the short-chain aldehydes and other compounds as intermediate combustion or pyrolysis products). Cellulose acetate propionate burns with a dark yellow sparking flame and generates a moderate amount of soot. As the cellulose propionate burns it melts and drips, and the drops burn as they fall. The odor of burning cellulose acetate propionate is similar to that of cellulose propionate, which is the odor of burned paper.

Cellulose Nitrate

Cellulose nitrate is much more flammable than the rest of the cellulosic plastics because of the presence of the nitrate group in its molecule. This nitrate radical will release its oxygen if the article is heated, making it available to support combustion. Under some circumstances, when cellulose nitrate burns—and *always* when it decomposes slowly—a group of toxic and dangerous gases known as the nitrogen oxides are liberated.

Cellulose nitrate has a long and unpleasant history of fire-related tragedies, and the manufacturers of this useful polymer have worked long and hard to make it safe. Today's cellulose nitrate is compounded with many materials to make it as safe for use as most other plastics. However, cellulose nitrate is still flammable. It will ignite easily at the relatively low temperature of 286°F, burning rapidly (assuming the product is not flame retarded) with a bright flame. In addition to its final combustion products of carbon, carbon dioxide, carbon monoxide, and water, there will be intermediate longer-chain combustion products plus the nitrogen oxides. The presence of the nitrogen oxides (as reddish-brown vapors) makes the combustion of cellulose nitrate very hazardous, because exposure to these gases may produce serious, delayed damage to the lungs.

Cellulose nitrate's uses include the production of lacquers, carriers for colorants for various plastics, printing inks, high explosives, smokeless powder, bookbinder coatings, and the manufacture of celluloid.

For safety reasons, cellulose nitrate probably should not be used for a demonstration of how plastics burn, nor should any article be burned that might be made from or contain cellulose nitrate. If articles made from cellulose nitrate are burned, the only place this combustion should take place is in a vented laboratory hood, and only under strict safety procedures.

Cellulose Propionate

Cellulose propionate burns was a dark yellow sparking flame and generates a moderate amount of soot. As the cellulose propionate burns it melts and drips, and the drops burn as they fall. The burning cellulose propionate has an odor of burned paper. Its final combustion products are carbon, carbon dioxide, carbon monoxide, and water, and there may be intermediate longer-chain combustion products.

Cellulose Triacetate

Cellulose triacetate has an ignition temperature of 1004°F. It is relatively easy to ignite, burns with a *dark* yellow flame, and produces a considerable amount of soot. The burning product has an odor of vinegar. Its final combustion products are carbon, carbon dioxide, carbon monoxide, and water, and there may be intermediate longer-chain combustion products.

Ethyl Cellulose

Ethyl cellulose has an ignition temperature of 565°F. Its final combustion products are carbon, carbon dioxide, carbon monoxide, and water, and there may be intermediate longer-chain combustion products.

Methyl Cellulose

Methyl cellulose burns with a green-yellowish flame and produces a "sweet" odor of burnt paper. The portion that is not burned melts and decompose to form a char. The final combustion products of methyl cellulose are carbon, carbon dioxide, carbon monoxide, and water, and there may be intermediate short-chain combustion and pyrolysis products.

Miscellaneous Cellulosic Plastics

Carboxymethyl cellulose, cellulose acetate phthalate, cellulose xanthate, and hydroxyalkyl cellulose are four small-volume cellulosic polymers with specialized uses. There is not a great deal of information published about their burning characteristics, probably because they are not used a great deal. Their combustion characteristics should be similar to those of all the other cellulosic polymers, with the exception of cellulose nitrate. What they have in common with the other cellulosics is their chemical composition, which means their molecules contain only carbon, hydrogen, and oxygen. This means their final combustion products will consist of carbon, carbon dioxide, carbon monoxide, and water, and there may be intermediate short-chain combustion and pyrolysis products.

Nylon

Nylons are really polyamides, but since the consuming public knows them as nylons, that is how they are listed here. The most important nylons commercially are nylon 6, nylon 6/6, nylon 6/10, nylon 11, and nylon 12. Nylon 6 is the most common. It will be used as the model for all nylons, since they will all burn in a similar fashion. Nylons have an ignition temperature in the range from 795°F to 990°F, with nylon 6 at 842°F. Nylons burn with yellowish orange flames with blue edges. Their final combustion products are carbon, carbon dioxide, carbon monoxide, and water; in addition, amines, ammonia, cyclic ketones, esters, and the nitrogen oxides may form because nitrogen is present in the polymer. At least one reference indicates hydrogen cyanide may evolve from burning nylon. Hydrogen cyanide formation may depend on the type of nylon burning and all the other parameters related to the combustion characteristics of the polymer. Sometimes the flame crackles as it burns, and the nylon melts, bubbles, chars, and gets thick and "stringy."

Polybutadiene

Polybutadiene is the polymer of 1,3-butadiene. Polybutadiene is the second highest volume synthetic rubber, and its blends are used heavily in the manufacture of tires. It is considered to be a rubber rather than a plastic since it is crosslinked or "cured" to be useful in its final form.

It burns very much like other rubber compounds, producing large quantities of carbon (and thus, heavy black smoke) and carbon monoxide. Since it is a hydrocarbon polymer, its final combustion products are carbon, carbon dioxide, carbon monoxide, and water. There is always the possibility that other short-chain hydrocarbons and hydrocarbon derivatives will be formed, but they are all combustible.

Polybutylene

Polybutylene is the third compound in the polyolefin series, the first two being polyethylene and polypropylene. The monomers for these three polymers are ethylene, propylene, and butylene (or ethene, propene, and butene), the first three alkenes (olefins). Polybutylene is considerably tougher than polyethylene and polypropylene but is relatively rare. In many jurisdictions, however, polybutylene pipe has been approved for use in fast-acting sprinkler systems. This would make samples of polybutylene pipe available from suppliers for burn tests. Polybutylene (pipe grade) has been rated slow burning (UL 94-HB).

Polybutylene has a specific gravity ranging from 0.909 to 0.925. It melts in the range of 244°F to 259°F. It will burn in the same manner as polyethylene and polypropylene, with a yellow-tipped blue flame, emitting burning drops and the odor of candle wax. Polybutylene is used to make film, but it is relatively rare and may be positively identified only by the manufacturer.

The final combustion products of polybutylene are carbon, carbon dioxide, carbon monoxide, and water. Polybutylene, like other polyolefins, does not depolymerize, so some long-chain hydrocarbons and short-chain aldehydes will be present as intermediate combustion products and pyrolysis products. Polybutylene has probably the highest of the heats of combustion of all common polymers.

Polybutylene Terephthalate

Polybutylene terephthalate is a thermoplastic polyester polymer related very closely to polyethylene terephthalate, and therefore it has very similar burning characteristics. Polybutylene terephthalate burns with a yellowish orange, sooty flame. As the polybutylene terephthalate burns it melts and drips, and the drops may be burning as they fall. Its final combustion products are carbon, carbon dioxide, carbon monoxide, and water, but short-chain hydrocarbon derivatives may be produced as intermediate combustion products.

It's much more difficult to obtain an article made from polybutylene terephthalate than one made from polyethylene terephthalate because the volume of polybutylene terephthalate used is so much smaller. However, there probably will be no significant difference in the way it burns, so trying to get samples made from polybutylene terephthalate to test may not be worthwhile.

Polycarbonate

Polycarbonate ignites at 968°F and burns with a bright to dark yellow, sooty flame that "spits" occasionally. It melts as it burns, decomposes, foams, and leaves a char. As it burns, it produces an odor of phenol (disinfectant smell). Its final combustion products are carbon, carbon dioxide, carbon monoxide, and water, but intermediate products will also include short-chain aldehydes and ketone.

Polycarbonate ranks in the upper half of the heats of combustion of various thermoplastics, liberating heat in the amount about halfway between polyethylene terephthalate and polystyrene.

Polyethylene

Polyethylene is the highest volume plastic in the world and is used in many common articles. The use of polyethylene most familiar to the general public (and the one most easily idenfitied as polyethylene) is the one-gallon plastic milk jug.

If you intend to determine how polyethylene burns, the one-gallon milk jug is as good an article to use as there exists. Remove the cap (it may also be made of polyethylene, but one cannot be absolutely sure) and discard it. Decide whether you will burn the entire bottle or just a long strip cut from it.

All samples should be burned in the same manner. That is, a decision should be made concerning where the flame will be applied. This is important because many of the fire tests mentioned elsewhere in this book specify where the flame shall be applied, and some of them direct that the flame be applied at the top of the test strip to produce "candlelike" burning. The most severe test would be to use a mechanical device to hold the sample in a vertical position and then apply the flame at the bottom of the sample. Since you are not conducting a scientific test but are simply observing how a sample or particular article burns, you should probably repeat this procedure on any sample of any plastic article selected.

Polyethylene ignites at a relatively low temperature, usually around 660°F. This means it can be ignited with any common ignition source, so long as that ignition source is held in place long enough to raise the polyethylene to its ignition temperature. The ease with which it ignites depends on the many parameters mentioned elsewhere.

Polyethylene burns with a smoky blue, yellow-tipped flame. The polyethylene will melt (its melting point is in the range from 224°F to 284°F, depending on its molecular weight and other factors), and the drops produced may sometimes burn as they fall. The specific gravity of polyethylene homopolymer ranges from 0.917 to 0.940 for low density polyethylene (LDPE) and medium density polyethylene (MDPE), and from 0.952 to 0.965 for high density polyethylene (HDPE). The specific gravity goes over 1.0 with some copolymers, even those unfilled.

The odor emitted by burning polyethylene is distinctly similar to candle wax (paraffin), since polyethylene and other polyolefins are chemically similar to it. These fumes are neutral, neither acidic nor caustic. The drops will either solidify when they hit a cooler surface or, if the radiated heat from the fire is high enough, the drops will stay liquid and form a pool of molten polyethylene. This liquid may continue to produce vapors that burn, and the resulting fire might resemble a burning combustible liquid.

If a polyethylene film (a trash bag or a positively identified plastic food wrapping film) is selected to be burned, you may find it difficult to ignite the sample. Like most plastics when reduced to a thin film, polyethylene will shrink away from the source of heat very rapidly. In many instances, the film will either shrink away or melt so rapidly that ignition of the sample may be impossible. Rolling the film into a wand or folding it over many times to form a thick wad may allow ignition, but then the sample might burn like a rigid sheet, rather than the film under observation.

If an article of cellular (foamed) polyethylene is selected to be burned, difficulty in ignition may once again occur. However, there is a somewhat better chance of igniting the foamed article than a thin film, especially if the cell size of the foam is very small. Generally, the smaller the cell size (that is, the higher the density of the material), the more easily the ignition will occur. Foamed sheet or other cellular articles made from large cell polyethylene will melt very rapidly, and the desired ignition will be difficult to accomplish.

The final combustion products of polyethylene include carbon, carbon monoxide, carbon dioxide, and water. The pyrolysis of polyethylene may be difficult to observe, but smoldering polyethylene may be a good example. Here, a whitish smoke is seen, and the pyrolysis products may include the short-chain aldehydes and some longer-chain hydrocarbons. Decomposition of polyethylene at temperatures below its ignition temperature does not result in depolymerization as it does the decomposition of some other polymers. In other words, when the combustion gases of burning polyethylene are analyzed, very little or no ethylene monomer is detected.

Polyethylenes rank at the top of the list in generation of heat as they burn. Their heats of combustion are the highest or second highest, depending on the reference.

Polyethylene Terephthalate

Polyethylene terephthalate ignites at 896°F and burns with a yellowish orange flame, producing a moderate amount of soot. As the polyethylene terephthalate burns it melts and drips, and the drops may be burning as they fall. Its final combustion products are carbon, carbon dioxide, carbon monoxide, and water, but short-chain hydrocarbon derivatives may be produced as intermediate combustion products.

It is relatively easy to find articles made from polyethylene terephthalate, since this polymer is growing rapidly in popularity and use. Plastic soft drink bottles are all made from polyethylene terephthalate. To test-burn a bottle (outside or in a vented laboratory hood), first remove the metal cap and the paper label, then the polyethylene base cup, if present. The bottle may be difficult to ignite, since its side wall is relatively thin and may melt before it ignites.

Polyethylene terephthalate ranks about in the middle of the rankings of heats of combustion for the thermoplastics, yielding about half the heat that polystyrene yields when it burns.

Polyimide

Polyimide resins include polyamide–imide as well as polyether–imide and polyester–imides. These materials exhibit highly unusual thermal stability and have service temperatures up to 932°F for very short periods. They are listed by some references as nonburning, while other references have an ignition temperature of greater than 1000°F. Another reference states that polyimides cannot be ignited by an external flame. In any event, they are very fire resistant. Thermal decomposition products include carbon, carbon dioxide, carbon monoxide, water, hydrogen, aniline, phenol, and benzene.

Polypropylene

Polypropylene, like polyethylene, is a polyolefin, so-called because their monomers, ethylene and propylene, are members of the alkene or olefin family. And since their monomers are similar, the polymers will be similar. Although polypropylene is generally tougher than polyethylene in comparable articles, it burns in almost exactly the same manner. To test the way polypropylene burns, one must be sure of the composition of the article to be burned. Since the volume of polypropylene is considerably less than polyethylene, it may be more difficult to find an article that you can be sure is made from polypropylene.

Once the product has been identified as polypropylene and the burning begins, however, you will notice that it burns in almost exactly the same manner as polyethylene. The burning rate may be different because the configuration of the article may be different. Since polypropylene is so tough, it has many uses that require strength, and the parts made from it may be thicker and more massive than articles made of polyethylene. Polypropylene's ignition temperature is from 806°F to 824°F. The polypropylene will burn with a yellow-tipped blue flame and will produce burning drops as it burns and melts. There will be the distinctive odor of candle wax, since polypropylene is related to paraffin. Polypropylene melts in the range of 302°F to 347°F, depending on whether it is a homopolymer or a copolymer and whether or not it is filled. The specific gravity of polypropylene ranges from 0.90 for unfilled homopolymer and 1.27 for

40-percent talc filled, to 0.88 for unfilled copolymer and 1.24 for filled. Polypropylene ignites in the range from 660°F to 700°F.

Cellular (foamed) polypropylene and polypropylene films are not as common as injection molded solid polypropylene parts, so one must make positive identification of these articles for the burning experiment to be educational. Polypropylene films, if found and burned, react in the same manner as other films exposed to radiated or conducted heat by shrinking away from the heat. It may not be possible to burn such a sample. Polypropylene foams, since their cell walls will tend to be more rigid than those of polyethylene, will probably ignite more easily than a polyethylene foam. However, there is still a chance that the foamed article will melt too rapidly to ignite.

The final products of combustion of burning polypropylene are carbon, carbon dioxide, carbon monoxide, and water. During pyrolysis or thermal decomposition below its ignition temperature, polypropylene may generate the short-chain aldehydes and some longer-chained hydrocarbons, but no depolymerization is detected. This conclusion is based on the absence of propylene in the analysis of combustion gases sampled during the burning of polypropylene. Polypropylene, like other polyolefins, ranks among the highest generators of heat when heats of combustion of polymeric materials are measured.

Polystyrene

Polystyrene (often incorrectly called styrene, which is the name of its monomer) ignites in the range of from 910°F to 925°F (some references claim that the ignition point is really over 1000°F). It burns with a bright yellow flame and produces great quantities of soot. The burning polystyrene emits a sweet odor, and its final combustion products are carbon, carbon dioxide, carbon monoxide, and water. Polystyrene will depolymerize at elevated temperatures below the ignition temperature, producing styrene monomer. Styrene monomer will ignite and burn at 914°F.

Polystyrene has a relatively high heat of combustion, ranking third below polyethylene and polypropylene.

High impact grades of polystyrene will produce odors of burnt rubber when they burn, since rubber is the material added to the polystyrene to give it impact resistance.

As with any other material, tests of the burning characteristics of polystyrene should be carried out outdoors or in a vented laboratory hood. In the case of polystyrene, the production of soot is so great that the inside of any enclosure will be difficult to clean for a long time.

Hot drink cups made of foamed material are invariably polystyrene, so this is one article of which one can be sure of the makeup. However, it is very difficult to ignite a foamed polystyrene cup, since the material is mostly air, and the heat from any flame will cause the material to melt so rapidly that it will not have a chance to ignite (the process of melting acts as a heat sink and prevents the thin cell walls of polystyrene from igniting). To assure enough material to ignite, one should stack several cups together before applying a flame.

Polystyrene is also used in many applications in which it is formed into a sheet. If a sheet of plastic can be positively identified as polystyrene, a better example will then be available to demonstrate how this plastic burns. The characteristic flame color and great production of soot will be present whether a solid piece of polystyrene or foamed polystyrene is burned. Many appliance housings may be made of polystyrene, but they must be flame retarded. It might be interesting to compare how flame-retarded styrene products react in the same flame and conditions as non–flame retarded styrene products (the base polystyrene should be the same), assuming that properly and correctly identified articles made from both materials are available.

Polytetrafluoroethylene (PTFE)

Polytetrafluoroethylene (Teflon™) is considered by many as a plastic that does not burn, even though some references show ignition temperatures of from 986°F to 1076°F. This is due to the presence of four fluorine atoms in the monomer, and therefore in the repeating unit of the polymer. An examination of the drawing of its molecular structure in Figure 4.8 shows that all the hydrogen atoms that were present in ethylene have been replaced by fluorine atoms. If and when polytetrafluoroenthylele does burn, its total heat of combustion is so low (less than one-tenth that of polyethylene) that polytetrafluoroethylene will not support its own combustion.

There have been instances where firefighters have reported that pots and pans coated with polytetrafluoroethylene and involved in kitchen fires have been found with the polytetrafluoroethylene totally gone from the metal it had previously covered, and the bare metal exposed. For this to have occurred, the heat from the burning materials in the room must have produced temperatures so high that the polytetrafluoroethylene actually evaporated off the red-hot metal. At those temperatures, the hydrogen fluoride gas (an irritant like hydrogen chloride) formed as a decomposition product poses no threat to anyone exposed to it, since the extreme temperatures themselves will cause death to anyone in the vicinity.

Anyone wishing to test the burning characteristics of polytetrafluoroethylene must use an ignition source with a considerably higher flame temperature than wood or paper matches. It may be necessary to try to burn polytetrafluoroethylene with a propane torch. Once again, any such burning test should be performed outdoors or in a vented laboratory hood. Experimenters must decide for themselves whether the polytetrafluoroethylene actually ignites and burns.

Most of the other fluorocarbons mentioned in Chapter 3 will react in the same manner of polytetrafluoroethylene; that is, they will not burn. In general, the fluorocarbons cannot be ignited by normal energy sources such as other "natural" materials burning. A very high energy source must be present to force the decomposition of these materials. Although hydrogen fluoride is one of the decomposition products, the volume of fluorocarbons existing in any one place (with the exception of where the polymer is made or stored) is too low to cause any appreciable production of this halogen acid gas.

Polyurethane (PUR)

Polyurethanes are so versatile in their chemistry that thermoplastic, thermosetting, and elastomeric foams may be made in rigid, semirigid, and flexible forms. Thermosetting flexible polyurethane sheet and film is also possible. However, polyisocyanates are always reacted with diols and/or polyols (compounds with two or more —OH groups attached) when polyurethanes are produced, and if they burn, their final combustion products are carbon, carbon dioxide, carbon monoxide, water, and hydrogen cyanide or nitriles.

They ignite at around 780°F and burn with a light yellow to orange flame, melting and dripping as they burn. There may be considerable frothing, and the polyurethane continues to burn after the flame is removed. All polyurethanes will decompose at temperatures below their ignition temperature. Decomposition in some polyurethanes will begin at 428°F, while no polyurethane is stable above 482°F. Decomposition is by depolymerization, liberating the polyol and the isocyanate. Domestically produced polyurethane foams are flame retarded, but decomposition is still likely in high heat situations.

Manufacturers have tried to produce a more fire resistant type of polyurethane, and a new material, polyisocyanurate (PIR), is the result. Polyisocyanurates are more brittle and more thermally stable than polyurethanes, having anywhere from 36°F to 90°F more stability than the corresponding polyurethanes. Polyisocyanurates do not depolymerize, so thermal decomposition almost always occurs by burning. Since nitrogen is present in the molecule connected to carbon, hydrogen cyanide and nitriles are possible combustion products in addition to carbon, carbon dioxide, carbon monoxide, and water.

Polyvinyl Chloride (PVC)

Polyvinyl chloride is one of the most controversial plastics whenever the discussion of fire properties takes place. This is so because even though polyvinyl chloride is the second highest thermoplastic in terms of volume produced, it is number one in the number of uses, and so it is more visible to the consumer (and has more competitors in metals and other "natural" materials). The range of compounds available with polyvinyl chloride (and therefore the range of properties possessed by those compounds) is virtually limitless, however, so the combustion behavior of a particular polyvinyl chloride compound depends directly upon its composition (and other factors controlling ignitability and combustibility). In this way it is like any other material, natural or synthetic.

No commercial article directly available to consumers is made of pure polyvinyl chloride, because pure polyvinyl chloride cannot be heat-formed into any useful shape. If pure polyvinyl chloride is subjected to the temperatures needed to form a part, it degrades and decomposes, turning brittle and charring. Special additives (heat stabilizers) must be compounded into polyvinyl chloride to allow it to pass through the heat of the processing equipment without pyrolyzing. Mineral fillers such as calcium carbonate may be added to it to lower its cost, and very small amounts of organic pigments or larger amounts of inorganic pigments may be used to add color. Lubricants and impact modifiers may also be added to help it pass through the processing equipment and resist breaking under sudden stress. The final compound may only be 50 to 75 percent polyvinyl chloride, but since no plasticizer has been added, it will be called a rigid polyvinyl chloride compound.

When polyvinyl chloride burns, it burns with a yellowish orange flame that may have some green on the edge. Its final combustion products are carbon, carbon dioxide, carbon monoxide, and water, plus the short-chain aldehydes. The additional combustion product formed when polyvinyl chloride burns (as opposed to a polymer containing only carbon and hydrogen or only carbon, hydrogen, and oxygen) is hydrogen chloride, an irritant gas that forms hydrochloric acid when it dissolves in water. This evolution of hydrogen chloride has prompted the accusation by makers of competing "natural" materials that the combustion of polyvinyl chloride is more dangerous than the combustion of other materials.

At temperature below its ignition temperature, polyvinyl chloride will pyrolyze by yielding some hydrocarbons, hydrocarbon derivatives, and hydrogen chloride. Although this decomposition procedure is often referred to as "unzipping," the reaction occurs slowly enough to allow a processor of polyvinyl chloride to lower the temperature on the equipment to stop the degradation. As polyvinyl chloride degrades, it turns yellow, then amber, and finally brown and black as it forms a char. Burning rigid polyvinyl chloride will also form a char during combustion.

To discuss how polyvinyl chloride burns, one must first separate the myriad of possible polyvinyl chloride compounds into rigid and nonrigid or flexible compounds. Rigid polyvinyl chloride is very difficult to ignite (its autoignition temperature is 1035°F). Once ignited, it does not produce enough heat during combustion to maintain the combustion process. Its heat of combustion is a little more than 40 percent of that of polyethylene, a little lower than halfway down the list of thermoplastics listed in order of their decreasing heats of combustion (see Table 6.1). In other words, rigid polyvinyl chloride will *not* support its own combustion, and for rigid polyvinyl chloride to burn and continue burning, some other material must be burning to provide the required heat energy.

Rigid polyvinyl chloride cannot be used in every application for which polyvinyl chloride is the material of choice, because of its stiffness and rigidity. To reduce its rigidity and increase its flexibility, a substance called a plasticizer is added. This material is usually a hydrocarbon derivative called an ester, and it is usually in liquid form. The plasticizer, being an organic compound containing carbon, hydrogen, and oxygen, is usually a combustible liquid with a flash point greater than 400°F. The addition of this combustible liquid to rigid polyvinyl chloride will change not only the rigidity of the polymer but also its burning characteristics. The ignition temperature of flexible polyvinyl chloride may be as low as 815°F, depending on the amount of plasticizer used. Changes in the burning characteristics of polyvinyl chloride can be measured by experimenting with the compounding of polyvinyl chloride. As more plasticizer such as di-2-ethyl hexyl phthalate (DEHP), the most common PVC plasticizer, is added, the combustibility of the polyvinyl chloride compound increases. If the compound is very soft and pliable, that is, the more plasticizer is added to it, the amount of actual polyvinyl chloride resin in the compound may drop to as low as 35 percent.

Although rigid polyvinyl chloride will not burn unless there is a supporting flame (i.e., when the flame from other burning material is removed, the burning polyvinyl chloride ceases to burn), plasticized or flexible polyvinyl chloride will burn, depending upon the amount of plasticizer added. Very soft shower curtains will burn readily, especially in the vertical orientation in which they are used, whereas less plasticized polyvinyl chloride compounds (and articles made from them) may also quit burning when the supporting flame is removed.

Because hydrogen chloride gas is very irritating (it will readily dissolve in the upper respiratory system, seldom reaching the lungs), any test burning of polyvinyl chloride should be done outdoors or in a vented laboratory hood. It is important to positively identify a test sample as being made from rigid polyvinyl chloride, since any test burns on an unknown might be misleading. Rigid polyvinyl chloride pipe or house siding are good materials for a test burn to demonstrate the combustion characteristics of this material.

Although polyvinyl chloride is nearly 57 percent chlorine, free chlorine is not released by its burning or decomposition. Instead, the chlorine comes off the polymer chain as a free radical. The chlorine free radical suppresses the combustion process by combining with hydrogen free radicals, "unzipping" the hydrogen off the polymer chain. This rapid formation of hydrogen chloride effectively removes the hydrogen as a source of fuel for the fire, and the flame goes out. The noncombustible hydrogen chloride gas also dilutes the oxygen available to support combustion and forms a layer of gas over the remaining fuel. These are some of the reasons why rigid polyvinyl chloride stops burning once the flame from some other burning material is removed.

Most of the hydrogen chloride released by burning or degrading polyvinyl chloride never reaches the upper respiratory system of anyone exposed to the fire, because hydrogen chloride is so soluble in water. Most of it dissolves in the moisture in the air and condenses onto cool surfaces instead of remaining dispersed in the air as a gas.

Polyvinyl chloride may be alloyed or blended with other thermoplastics to achieve certain properties not possible in other polymers, alloys, or blends. When an alloy or blend of this type is encountered, the burning characteristics of both (or all) polymers will be present but modified by the mixture. Although polyvinyl chloride can be made more combustible by adding other materials to it, when used as an alloying polymer with another plastic it actually makes the other polymer *less* combustible. The resulting alloy or blend will have final combustion products representative of all the polymers in it. This means that hydrogen chloride will be a final combustion product of such alloys and blends.

Polyvinylidene Chloride

Polyvinylidene chloride (Saran™) is usually used as a film, and in that configuration it is very difficult to ignite. Most films will shrink away from the heat or will melt before they have the chance to ignite, and polyvinylidene chloride is no exception. Even if found as a solid article and tested for combustion properties, polyvinylidene chloride is very difficult to ignite. And since polyvinylidene chloride contains two atoms of chlorine compared to the one atom of chlorine in polyvinyl chloride's repeating unit, it will be even more difficult to burn than PVC.

Polyvinylidene chloride will ignite at temperatures higher than 986°F, and if the article can be made to burn, it will burn with a yellowish orange flame with a green tinge on its edges. It will behave much like polyvinyl chloride in that it will change colors as it degrades at temperatures below its ignition temperature, first turning yellow, then amber, then dark brown as it chars. Its final combustion products are carbon, carbon monoxide, carbon dioxide, water, and hydrogen chloride. Short-chain intermediate combustion or decomposition products may also be formed.

The heat of combustion of polyvinylidene chloride is about one-half that of rigid polyvinyl chloride, which places it near the bottom of the ranking of how much heat is produced during the complete combustion of polymers. The reason for the low heat of combustion is that the very large chlorine atom displaces the very small hydrogen (fuel) atom on the polymer chain.

Rayon

Rayon, or viscose rayon as it is sometimes called (or regenerated cellulose), ignites at 788°F and burns rapidly to completion. It burns so rapidly that for it to be used commercially, it must be flame retarded. Its final combustion products are carbon, carbon dioxide, carbon monoxide, and water, plus the short-chain aldehydes.

Thermosets

Thermosetting plastics and their compounds differ from thermoplastics and their compounds in the way the polymer chains line up. Thermoplastic polymer chains are more or less free to move among themselves as the material melts, forming a hot liquid. The thermosetting polymer chains are not free to move among themselves because there are extra covalent bonds that connect the chains to each other. This formation of bonding between the chains is referred to as **crosslinking.** Therefore, when thermosetting materials are subjected to heat (after they have been cured or "set up," they will not absorb heat and melt as do the thermoplastics. Instead of acting as a heat sink, the polymer chains will begin to break up (pyrolyze). Eventually the thermosets will burn much like wood and other materials containing the same elements in their molecules. However, the thermosets are as different from one another as the thermoplastics are from each other.

Alkyds

Alkyd resins burn with a brilliant flame, producing a harsh smell of acrolein. They continue to burn when the flame is removed, and ease of ignition ranges from easy to difficult, depending on the particular alkyd resin tested. It will be difficult to determine whether a particular article is composed of alkyd resins. You will have to ask the manufacturer of the article if you plan a test.

Allyls

Allyl resins burn with a yellow flame having blue edges. The odor emitted is sweet, and the article will burn without melting. It is difficult to determine if a particular article is composed of allyl resin, so you will have to contact the manufacturer of the article to be tested for positive identification of the polymer.

Epoxies

The thermal stability of epoxy resins depends on the monomers used in their polymerization. Some epoxies begin to decompose thermally at temperatures ranging from 464°F to 662°F, whereas other have ranges of 482°F to 842°F. Their decomposition products include the short-chain aldehydes and ketones, whereas their final combustion products include carbon, carbon dioxide, carbon monoxide, and water. Epoxies are more flammable than most other thermosetting plastics, but incorporating mineral fillers such as calcium carbonate and glass will reduce the burning characteristics.

Epoxies more often used as films and coatings than as solid parts.

Melamines

Melamine resins are really melamine–formaldehyde resins. The melamine–formaldehyde resins will ignite somewhere in the range above 1112°F and burn with little smoke evolution. The flame will be yellow, and the solid material will swell, crack, char, and turn white on the edge of the burning article. Melamine contains a great deal of nitrogen, and when the nitrogen breaks off the polymer structure it combines with hydrogen to form ammonia, rather than other decomposition or combustion products. Final combustion products include carbon, carbon dioxide, carbon monoxide, water, and ammonia. When ammonia burns, several nitrogen oxides will be formed. At temperatures below the ignition temperature, decomposition products may include ammonia and melamine. This decomposition temperature is determined by the type and amount of filler used in the compound. Melamine is sometimes used as a flame retardant additive in other thermosets.

At least one reference indicates that hydrogen cyanide may evolve from burning melamines. Whether cyanide is released may depend on the type of melamine burning and all the other parameters of the combustion characteristics of the polymer.

Phenolics

Phenolic resins (or phenol–formaldehyde resins) are resistant to temperatures of from 248°F to 346°F, and thermal decomposition begins around 582°F, depending upon whether any fillers were used (and they usually are). Phenols ignite at about 1060°F. They will burn with a bright sooty flame, with little smoke being liberated. When the igniting or supporting flame is removed, the material usually stops burning. Charring occurs readily, which further reduces the chance of burning. Final combustion products include carbon, carbon dioxide, carbon monoxide, and water. Thermal decomposition products at conditions below ignition temperatures include acetone, carbon monoxide, formaldehyde, methane, and phenol.

As with other plastics, the burning characteristics of the phenols depends largely upon additives used in the compound. Combustibility of the compound increases if the fillers consist of material like wood flour, wood pulp, sisal fibers, or chopped polymer fibers, or of metal flakes or powders such as aluminum or bronze. Conversely, if glass fibers, talc, minerals, clay, or diatomaceous earth are used as fillers, they increase the bulk of the compound with materials that do not burn, so the resulting compound has lower combustibility characteristics.

Phenolic materials will not depolymerize, even though some of the thermal decomposition products include the monomers used to make the thermosetting plastic.

Unsaturated Polyesters

Unsaturated polyesters start to decompose at temperatures above 284°F and ignite at from 900°F to 910°F. They burn with a bright yellow flame, producing soot and a sweet odor. Unsaturated polyesters char as they burn and do not melt.

The thermal stability of unsaturated polyesters can be improved by adding mineral fillers such as glass fibers. Final combustion products include carbon, carbon dioxide, carbon monoxide, and water. Unsaturated polyester resins are usually liquids, some of which are flammable and some of which are combustible. Unsaturated polyester resin that is burning may escalate in reaction speed and cause a runaway polymerization explosion.

Urea-formaldehyde

Urea-formaldehyde resins ignite somewhere in the range of 1100°F and burn with little smoke evolution. The flame will be yellow, and the solid material will swell, crack, char, and turn white at the edge of the burning article. Final combustion products include carbon, carbon dioxide, carbon monoxide, water, and ammonia. At temperatures below the ignition temperature, decomposition products may include ammonia and melamine. At least one reference indicates hydrogen cyanide may evolve from burning ureaformaldehyde resins. It may depend on the type of ureaformaldehyde resins burning, as well as all the other parameters of the combustion characteristics of the polymer.

THE EXTINGUISHMENT OF BURNING PLASTICS

Having spent so much time and energy discussing combustion characteristics of these materials called plastics, I will devote very little space to extinguishing them. The process is so simple and straightforward that it does not require a tremendous amount of information.

Plastics are Class A materials, and when they are burning, the fire may be extinguished by using Class A extinguishing agents. The one exception to this is cellulose nitrate, and it will be discussed later.

A scenario in which plastics and only plastics are burning is extremely rare. Plastics are everywhere, but in occupancy fires, with perhaps one exception (a warehouse where only plastics are stored), plastics articles are interspersed with articles made from other materials. In a "normal" residence, plastics may be found as all or part of wall, ceiling, and floor coverings, draperies, furniture and furniture upholstery, appliances, window glazing, lighting fixtures, housewares, toys, hobby, and recreational articles, clothing, decorations, finishings on articles made of other materials, bathroom fixtures and sanitary ware, tools, packaging materials, pipes, wire insulation, machine housings, exterior siding, gutters and downspouts, hoses, and a myriad of other uses. Even with this seemingly endless list of products and articles, plastics will still make up the smaller portion of combustible materials in the total fire load in the occupancy, when compared to wood and wood products (including paper and paper products) and other articles made out of "natural" material.

Water is the extinguishing agent of choice in all Class A fires, and it is the extinguishing agent of choice where plastics are involved in fires alone or when burning with other materials. An occupancy fire where plastics are present should be fought as would any other fire in the same circumstances. The same care should be exercised if charged electrical equipment becomes involved, or if any flammable or combustible liquids are present in significant quantities. In other words, the training that firefighters have been receiving for years in extinguishing occupancy fires does not have to be modified as more and more plastics are used in those occupancies.

Here is where arguments are raised concerning fires where more and more plastics are involved. The question is asked: "Aren't fires different today than they were thirty or forty years ago?" The answer certainly has to be yes, but not for reasons most people assume. Fires are different today than in any distinct and identifiable time period in the past, not because plastics are or will be present in different quantities, but because the total fire load is different. **Fire load** is defined not only as the total amount of materials present that will burn (total fuel), but more importantly as the total amount of energy released by the material (total heat of combustion), its ignitability (how easily it becomes involved in the fire), and how fast the energy is released (rate of heat release).

It is true that certain plastics (the so-called hydrocarbon plastics—polyethylene, polypropylene, polybutylene, and polystyrene) have high heats of combustion, but the amounts and forms in which they are used do not contribute greatly to the total heat of combustion of all the materials in the total fire load. Other plastics, like polyvinyl chloride and other halogen-containing plastics, are very difficult to ignite. Once ignited, they have a low heat of combustion and a rate of heat release so low that they usually will not support their own combustion. Those plastics in between (those containing oxygen or nitrogen in addition to carbon and hydrogen) have heats of combustion and rates of heat release between the hydrocarbon plastics and the halogen-containing plastics. Additionally, many of these plastics compounds are flame retarded, lowering their rate of heat release even more.

This could mean that today's fires *are* different than those of thirty or forty years ago—in that they may not be as hot burning. Many people feel that this is not true, and many of those people are firefighters. One reason for this apparent contradiction is that many of today's fires are in houses and other buildings that are more tightly constructed than is the past, so that more of the heat of the fire is held in. Tight construction might also account for what appear to be an increasing number of flashover fires and an increased need for forced ventilation of fire buildings. If occupancy fires actually are hotter today than in the past, the increased presence of plastics may have little or nothing to do with it.

Firefighters also feel that today's fires are smokier than those of the past, and this might be true, again because of the changed fire load. Some plastics do evolve more smoke than other materials, but not all plastics fall into this category. Also, flame retarded materials, both plastics and wood, may evolve more smoke than nontreated materials. Smokiness is sometimes attributed to the difficulty with which a material burns. Certainly this is true for most carbon-containing materials.

Finally, many have expressed the opinion that today's fires are deadlier than those of the past because of the combustion products of plastics. If this were true, one would think that the number of fire deaths would climb each year in direct proportion to the amount of plastics used by the consuming public. This, of course, has not happened. Many of the same people feel that the death rate has dropped because of the widespread use of smoke detectors and fast-acting sprinkler systems. This is very likely, and the material of choice in the construction of these devices and systems coincidentally is plastic.

This is a long, roundabout way of saying that plastics should be treated in an occupancy fire in the same manner as other Class A materials and, of course, *all* firefighters in *any* fire should be equipped with (and should use at *all* times at *all* fires) a positive-pressure, self-contained breathing apparatus (SCBA).

Cellulose nitrate is the exception in the way burning plastics can be extinguished. Because of its molecular makeup, cellulose nitrate is not only a combustible material but also an oxidizing agent. That is, it carries its own oxygen in its structure, and this oxygen can be released in a form that supports combustion. This makes cellulose nitrate difficult if not impossible to extinguish. Water is still the choice of extinguishing agent, but firefighters should know that water may just cool down the fire, without extinguishing it. Under certain circumstances, cellulose nitrate might burn under water. Other plastics containing oxygen in their molecules do not have this hazard because the manner in which oxygen is bound in those molecules will not allow it to be released in a form in which it will support combustion.

The only location in which a fire involving cellulose nitrate should be a problem is where it is either manufactured or stored in large quantities, and not in a "normal" occupancy. This is true for all plastics. Polyethylene may be used in many applications in a normal occupancy, but not enough is present to be the biggest part of the fuel. However, in a manufacturing operation where the plastics are polymerized or stored in very large quantities, the majority of the fuel will be polyethylene (or any other polymer made or used there).

In this situation, the emergency responder might be faced with a Class A material converting to a Class B material as the plastic melts and begins to flow. In this situation, water is still the extinguishing agent of choice, but it should be applied as one would apply water to a Class B fire, as a fog or high pressure spray. This technique will rapidly cool the material down, extinguishing the fire as the plastic solidifies. Where halogen-containing plastics are involved in a fire, the additional hazard of the generation of hydrogen chloride or hydrogen fluoride liberation must be considered. If the firefighter is wearing a SCBA and full protective gear, there will be no need for concern for respiratory problems. Of course, the SCBA will be necessary to protect all emergency responders from the deadly effects of carbon monoxide and lowered oxygen levels.

If polyvinyl chloride resin is burning, some other material must be burning and supplying the energy to burn to the PVC, which cannot support its own combustion. If the burning material is attacked and extinguished, the polyvinyl chloride resin will stop burning.

Care should be taken not to use straight streams of water wherever powdered plastics resins and/or compounds may be located. Plastics are organic materials, and any powdered or dusty organic material can explode if dispersed in the air in the "right" amounts (within the explosive range) and the proper ignition energy is provided.

On small fires, wherever they may occur, carbon dioxide and dry chemical extinguishers may also be used.

7

Flame Retardants and Flame Retarded Plastics

Since all carbon-containing polymers are organic, they will all burn under one or more sets of conditions. The demand for plastics, particularly thermoplastics, for use in construction and other applications that require strict code enforcement involving combustion characteristics is constantly increasing. A group of special additives have been developed to allow these plastic compounds to pass safety specifications. Today, nearly all thermoplastics are available in one or more flame retarded formulations. The materials used to impart this resistance to fire are chemical compounds called **flame retardants.**

CONSIDERATIONS IN SELECTION

Flame retardants are chemical compounds or mixtures of chemical compounds that are added to plastics resins, compounds, alloys, and blends to reduce the combustibility (or otherwise alter the combustion characteristics) of the finished article. They are defined as substances that, when added to another substance, material, or product, will suppress, reduce, or delay the propagation of flame through the mass and/or the surface of that substance, material, or product. Flame retardants are added to plastics resins, compounds, alloys, and blends, for two purposes: (1) to change the combustion characteristics of the plastic material so that it will be more difficult to ignite or (2) once ignition has occurred, to cause the fire to die out or cause the material to burn more slowly so that the flame spread, rate of heat release, or both will be substantially lowered.

It would be ideal if the addition of flame retardants to plastics would totally prevent ignition of the article, or, in the alternative, cause the fire to go out once ignition had occurred. This usually does not happen, because of the chemical and physical nature of the flame retardants. Although they might favorably influence the combustion characteristics of the plastic, they usually cause other properties of the plastic to which they are added to deteriorate to such an extent that the flame retarded plastic compound is no longer suitable for its originally intended use.

For example, plastics used in home appliances generally must pass certain tests concerning their behavior under fire conditions. High impact polystyrene and acrylonitrile–butadiene–styrene (ABS), as cases in point, are sometimes used to manufacture housings for television sets. Whether the housing will be processed by being injection molded in one piece or fabricated by thermoforming or some other process will have a bearing on how the plastic compound is supposed to behave under these processing conditions. To perform properly in the equipment during processing the compound must meet certain specifications for fluxing temperature, melt flow, and other characteristics if it is to be shaped properly into the final finished product.

The finished article (the TV housing) must also meet the television manufacturer's specifications for such properties as impact resistance (resistance to breakage when dropped or otherwise impacted), strength in all dimensions, and machinability, as well as any other properties specific to the end use of the product. And as if processing and finishing specifications weren't difficult enough, certain appearance properties such as gloss, color, and the lack of surface imperfections must be met. Television manufacturers want their housings to arrive at the assembly plant ready for the installation of the working parts. They do not want to have to perform additional finishing steps.

All of the foregoing specifications are severely affected by the addition of a flame retardant chemical. To satisfy all requirements for the finished article, the processor must choose a flame retardant carefully.

First and foremost, anything that is added to any polymer will materially affect the properties of that polymer. All of the requirements just mentioned (which represent less than 10% of the required properties of the finished article) will be changed by any additive, including those added to improve the characteristics of the polymer! Therefore, processing conditions, behavior of the plastic compound in the processing equipment, the physical properties of the finished article, its finishing characteristics, its color, and its overall strength will be affected by the flame retardant added.

Plastics compounders or engineers are therefore constantly seeking the proper loading level of flame retardants to provide the proper balance to flame retardation and all the other properties required by the end user. These decision-makers must, for reasons of economy, seek the exact level of flame retardants that will allow efficient processing in the compounding operation as well as in the processing equipment and the factory where the final product will be assembled. It is no easy task. The compounder in the processing plant or the chemical engineer in the polymerization plant (the flame retardant may be added in either place) must constantly search for the ideal flame retardant for the particular plastic compound and end application of the finished product.

The ideal flame retardant material will, at a minimum:

- Be efficient at a low loading level in the compound
- Provide maximum flame retardation safely
- Be totally compatible with the compound
- Process easily in all processing equipment normally used for the particular thermoplastic in question
- Operate at a temperature high enough so it will not degrade during processing but will release its flame retardant efficiency early in a fire
- Not volatilize out of (or otherwise leave) the compound
- Not produce hazardous byproducts either during processing or when functioning as a flame retardant
- Not detract from the performance of the plastic material
- Allow the final product to have proper surface characteristics
- Be easily colorable to the color desired by the end user
- Be low in cost

Should the ideal flame retardant ever be developed, it must be understood that no organic material, no matter how well protected by any additive, can resist ignition, degradation, or decomposition for very long when exposed to a large, intense, long-lasting fire or the radiated heat from this fire or another high-heat source.

TYPES OF FLAME RETARDANTS

Flame retardants, for the purpose of this book, are broken into additive and reactive types and, within this classification, are broken down by chemical type. This necessitates the use of some very long chemical names, since most organic flame retardants are complicated chemicals. Don't worry about these chemical names. You are not expected to remember, let alone pronounce, some of the names of these chemicals. They are being presented for illustrative purposes only. However, if you encounter a flame retarded plastic, and if you can determine (from the manufacturer or compounder) the name of the particular flame retardant used, you can deduce some valuable information, for example, possible combustion products of the flame retarded plastic article, or degradation products of the flame retardant chemical.

Flame retarded plastics ignite with more difficulty than do the same plastics when not protected by a flame retardant. However, the decomposition products of the flame retarded plastic may be different from those of the unprotected material. This will be especially true if new elements are introduced into the plastic by the flame retardant, and these new elements are present in a chemical compound that will allow them, upon decomposition, to react with elements present in the polymer molecule. With this information, the curious reader will be able to determine the nature of some of those decomposition products. For instance, unprotected polystyrene will liberate only hydrocarbons and some hydrocarbon derivatives containing carbon, hydrogen, and oxygen. Polystyrene protected with a brominated or chlorinated hydrocarbon will contain, in addition, hydrocarbon derivatives containing bromine and chlorine (depending on whether a brominated or chlorinated flame retardant was used), and its final decomposition products will include hydrogen bromide or hydrogen chloride (again depending on which type was used).

Again it must be stressed that the objective of this book is to provide basic information about how plastics do or do not burn, and complicated chemistry will not be included. In this light, it is not really important to the reader whether a particular flame retardant is an additive or a reactive type, I will define these materials just to make the chapter complete.

Additive Flame Retardants

Typical additive flame retardants:

ammonium phosphate
alumina trihydrate
brominated paraffin
chlorinated paraffin
decabromodiphenyl oxide
octabromodiphenyl oxide
tricresyl phosphate
trioctyl phosphate
triphenyl phosphate
tris(2,3-dibromopropyl)phosphate
tris(trichloropropyl)phosphate
zinc borate

Polymeric additive flame retardants:

brominated polycarbonate
brominated polystyrene
chlorinated polyethylene
polyvinyl chloride

The *additive flame retardants* are materials that will change the combustion characteristics of a particular polymer, compound, resin, or alloy when compounded (mixed into) or added to the plastic resin or compound. This can be done in the compounding stage when other materials such as stabilizers, fillers, colorants, plasticizers, or fillers are added. The flame retardant is incorporated into the plastic compound by dispersing the additive uniformly throughout the mass, usually in a simple mixing process. No chemical reaction between the flame retardant and the plastics compound occurs during this compounding stage: The resulting mass, whether it has been fused (softened) and re-formed into pellets or cubes ground into a powder, or allowed to stay in the powdered phase, is a simple physical mixture and not a new chemical compound. This is an important difference between the additive-type flame retardants and the reactive-type flame retardants (see next section).

Some polymers, particularly those that contain halogens and burn with difficulty, may be added to other polymers, compounds, and blends to produce alloys. When used in this manner, these halogenated polymers act as flame retardants.

Reactive Flame Retardants

Typical reactive retardants:

bis(2-hydroxyethylamino)octachlorobiphenyl
diethyl *N,N*-bis(2-hydroxyethyl)aminomethyl phosphonate
chlorendic anhydride
tetrabromophthalic anhydride
tetrachlorophthalic anhydride

The *reactive flame retardants* are those materials that are added during the polymerization process and therefore become an integral part of the polymer. Since polymerization is a chemical reaction, the reactive flame retardant can be considered a monomer. This means that a plastics compound made with the use of a reactive-type flame retardant is really a new chemical compound. In this situation, it is more likely that decomposition of the plastic article at temperatures below the ignition temperature of the plastic will include elements from the flame retardant chemical. This means that decomposition products from a flame retarded plastic of this type will be different from what would be expected if the unprotected (no flame retardant added) plastic material decomposed at the same temperatures.

Chemical Types

Flame retardant chemicals may be organic or inorganic in chemical makeup, but the vast majority of them are in the organic class. Halogens are the most efficient flame retardant discovered thus far, because of the manner in which they interrupt the combustion process. Iodine is the most efficient of the halogens, followed by bromine, chlorine, and fluorine, in order of decreasing efficiency.

Iodine compounds used as flame retardants are not common, in part for reasons of safety, cost, or availability. The most important reason, however, is that organic iodine compounds have such weak covalent bonding that the flame retardants break down too quickly in advance of when they are needed. Many times, these iodine compounds will break down in the compounding process, and the final product will not be protected at all. Fluorine compounds are also not used as flame retardants because their bonding is so strong that the activation temperatures of the fluorine compounds are too high, and the breakdown of the flame retardant compound comes much too late in the fire scenario. Of the two remaining halogens, bromine compounds are the most common class of organic flame retardant material, followed by chlorine compounds. As a flame retardant, bromine is the more efficient of the two.

The third class of organic flame retardant chemicals in volume is the group known as esters of phosphoric acid. The differences in the mode of operation of these flame retardants are discussed in the next section.

The most common *inorganic* flame retardant compound in use is aluminum hydroxide, also known as alumina trihydrate, ATH, or hydrated aluminum oxide. Antimony oxide, another inorganic compound, is used in conjunction with the halogen flame retardants for its synergistic effect. Other inorganic flame retardants include barium metaborate, calcium borate, magnesium carbonate, magnesium hydroxide, and zinc borate.

HOW FLAME RETARDANTS WORK

Thermoplastics (as well as thermosets), being organic, will pyrolyze and burn in a manner reasonably similar to that of other organic materials. The basic difference in how they behave is that the pyrolysis of thermoplastics first proceeds by absorbing some of the energy when they melt (or degrade, as with thermosetting materials, which do not melt). The individual, precise chemical mechanisms that are involved in the combustion process have been identified, but such detail is not necessary in a book of this type. Suffice it to say that the combustion process of plastics involves the breaking of covalent bonds, the subsequent forming of very reactive free radicals, and finally the re-formation of those free radicals to constitute the intermediate combustion products of the burning plastics.. Those intermediate products then go through the same chain fracture (breaking of covalent bonds) as the original polymer molecules, the creation of more free radicals, and new end product formation. This time, the final combustion products of the burning plastic are formed. All of these reactions are continuous and very fast (reaction times being measured in thousandths and millionths of a second). Key to this process is the formation and subsequent reaction of a particular free radical, without which the combustion process would not exist or would cease. The inclusion of flame retardants in a plastic compound acts to interrupt the above-mentioned combustion processes in one of three different ways.

Halogenated Flame Retardants

First, the halogenated (bromine and chlorine) flame retardants behave as any other organic material do when subjected to the heat of a fire. They break down, forming free radicals, which then seek to re-form into stable chemical compounds. These free radicals generated by the flame retardants, the bromine (—Br) free radical and the chlorine (—Cl) free radical (depending, of course, on whether a brominated or chlorinated flame retardant was used), then interfere with (or actually replace) the hydroxyl (—OH) radical in the pyrolysis reaction. Hydroxyl radical formation is the key to the combustion of any hydrocarbon or hydrocarbon derivative. If the production of the hydroxyl radical is interfered with, or if it the hydroxyl radical is prevented from being formed, the fire will go out. This process is exactly the same as the method by which Halon™ fire extinguishing agents operate. The ideal flame retardant operates at temperatures just below the decomposition temperature of the plastic, so that when the polymer molecule begins to degrade a "free radical quenchant" or free radical "trap" will be present to capture the hydroxyl radical or prevent its creation.

The efficiency of halogenated flame retardants is usually increased significantly by a particular material called a **synergist.** A synergist does not always function by itself as a flame retardant, nor does it always function as a flame retardant in the compound into which it is added with the halogenated compounds. *Synergism* is defined as a phenomenon in which the chemical effect of two materials acting together is greater than the effect that would be produced if each material acted individually and the effects were added together. In other words, a particular flame retardant chemical, when used alone in a plastics compound, will exhibit a certain degree of efficiency. However, when another particular chemical, whether it is a flame retardant on its own or not, is added to the same plastic compound, the combined flame retardant efficiency is greater in this compound than the efficiency of either material alone.

The chemistry of synergism is very complicated, but the presence of a material like *antimony oxide* (also called antimony trioxide) greatly increases the formation of the halogen free radical (—X), which then is available in larger quantities to extinguish the fire. Antimony oxide, in addition to a few other synergists, therefore, is a very common ingredient in many flame retarded compounds, where it greatly increases the efficiency of some other flame retardant.

Phosphate Ester Flame Retardants

The second mechanism by which flame retardants retard the burning process in a plastic material is represented by the way *phosphate esters* operate. The phosphorus compounds operate initially by speeding up the decomposition of the polymer molecule by means of their own chemical decomposition. This chemical breakdown of the phosphorus-type flame retardant causes the formation of phosphoric acid, which then reacts with the decomposing polymer to form a char. This char, forming a barrier between the source of heat and the rest of the surface of the product, protects the polymer from continued heat degradation. Again, during the decomposition of the polymer molecule, the hydroxyl radical is formed. If the flame retardant is effective, the layer of char acts as an insulating blanket between the polymer and the source of the heat, which then prevents the further formation of the hydroxyl radical by preventing thermal decomposition of the polymer.

Also formed in the process of decomposition of the phosphorus-type flame retarding agents are water and some nonflammable gases, which together probably make some minor (some may say major) contribution to the flame prevention and/or flame extinguishment. The water will act as a coolant and the nonflammable gases will act as a diluent as described below.

Another manner in which phosphorus compounds may be thought to operate as flame retardants is the way they form the char, rather than allowing the carbon to be oxidized to carbon monoxide and/or carbon dioxide. This is an indication that, in one way or another, the phosphorus compound is interfering with the oxidation process, the type of chemical reaction that includes combustion. If oxidation is hindered in *any* way, the combustion process will be hindered. Interference with the oxidation process would also reduce the total amount of energy (heat) available to the entire system. This would have the effect of interfering with the pyrolysis reaction, and this would also operate to hinder the combustion process by limiting the amount of fuel available for burning.

Inorganic Compounds

The third manner in which flame retardants interrupts the combustion process in plastics is represented by a group of inorganic compounds, including alumina trihydrate, barium metaborate, calcium borate, magnesium carbonate, magnesium hydroxide, and zinc borate. Some of these inorganic flame retardants act in three separate ways: They may (1) liberate water, (2) liberate nonflammable gases, and (3) the reactions in which they participate are endothermic (heat absorbing). If each of these individual reactions were to occur alone, it is likely that none of them would be efficient enough to prevent combustion by itself. In any of the reactions, the water vapor liberated is insufficient to cool the fire below its ignition temperature, the nonflammable gases liberated are insufficient to "smother" the fire (remove the oxygen), and the heat absorption is insufficient to lower the temperature below the ignition temperature of the polymer. If all three reactions occur simultaneously, however, their combined reactions will smother the fire. In addition, each one of these actions can prevent formation of the hydroxyl radical, although none as effectively as the mode of operation of the halogenated hydrocarbons.

Alumina trihydrate contains a significant amount of water (34.5% of its weight), and at the onset of heat absorption it begins to release it (ATH begins to decompose at about 400°F). This endothermic reaction preferentially absorbs the heat, thereby protecting the polymer chain from fracture. The specific heat of water and its latent heat of vaporization together act as the second heat sink, and finally the water released dilutes any combustible gases present. *Magnesium hydroxide* operates in the same manner, producing 31 percent of its weight in water, beginning at 608°F. On the other hand, *magnesium carbonate* releases carbon dioxide (36.1% of its weight) in addition to water (20.2%), starting at 445°F. Note that a compounder or chemical engineer chooses a particular flame retardant not only on the basis of efficiency but also because of its activation temperature.

Flame retardant compounds containing *boron* may operate in a completely different fashion. Some of these compounds are thought to work by forming a glassy surface over the char that is formed, thereby protecting the surface of the polymer from further degradation. In forming a protective layer over the unburned polymer, the char prevents the hydroxyl radical from forming by preventing the breaking of covalent bonds in the polymer molecule caused by heat.

Silicone-based flame retardants function by raising the oxygen index of the polymer. The **oxygen index** represents the amount of oxygen that must be present at room temperature for the polymer to burn. Any increase in this index indicates a degree of flame retardant action. The **Limiting Oxygen Index** for common polymers is presented in Appendix D. Silicones also promote the formation of a layer of char to further reduce the flammability of certain polymers by separating the unburned polymer surface from the oxygen in the atmosphere.

DECOMPOSITION PRODUCTS OF FLAME RETARDED PLASTICS AND FLAME RETARDANTS

For a flame retardant to be effective in altering the ignition or combustion characteristics of the polymer, compound, alloy, or blend, it must, of necessity, break down chemically into the very reactive materials that actually prevent fires or cause them to die out. Some flame retardant chemicals do, in a manner of speaking, break down, though not necessarily by a chemical reaction. If a hydrated material is heated, the reaction sought is to release the water of hydration to the compound, thereby altering its combustion characteristics. Water is not a hazardous decomposition product in this mechanism. Whether this thermal decomposition of the flame retardant occurs at temperatures below the ignition temperature of the plastic compound or during combustion of the plastic compound, it is the final decomposition products of the total reaction that interest those concerned with human exposure to these decomposition products.

The thermal decomposition of any organic material will produce some products that can cause harm to those exposed to them. The final combustion products of all organic materials will include carbon monoxide, the combustion gas that is responsible for almost all deaths caused by exposure to combustion gases. The intermediate combustion products of those same organic materials will include hydrocarbons, aldehydes, and other hydrocarbon derivatives formed as the organic material breaks down under heat. Both the final combustion products and the intermediate products will be superheated and may cause harm to exposed persons from their energy content alone. Products of pyrolysis include very active chemicals, usually heated to temperatures close to or at their ignition temperature.

Many critics claim that when the decomposition products of flame retardant chemicals are added to this mixture of flammable, irritating, and toxic gases, the result is very hazardous. It is claimed that when experiments are carried out on isolated samples of plastics—whether protected or unprotected by the addition of flame retardant chemicals—the resultant gases are deadly to any animal or human life exposed to them.

The problem is that these experiments *are* isolated. When halogenated plastics are decomposed, halogen chlorides (such as hydrogen chloride) and halogenated hydrocarbons are indeed liberated, as are other products of pyrolysis such as those already mentioned. Indeed, animals exposed to these gases over time may die. In conducting these tests, however, researchers expose the animals only to the specific materials being examined. What the researchers fail to do is to run tests on all the materials likely to be found in the situation they should be trying to duplicate: the home environment of humans. In *any* burning of *any* organic material; in *any* situation in which organic materials are pyrolyzing; in *any* situation in which the oxygen level is reduced and the temperature increased—unwanted compounds *will* displace air and be breathed by some living organism. And *any* time that this occurs, some harm to the organism will occur.

Plastics are used in the experiments cited by their critics precisely because it is plastics that are under attack. Critics claim that adding flame retardants to plastics makes the decomposition products worse than if unprotected polymers were decomposing. This claim makes sense only in the isolated atmosphere of the laboratory, and not the real world where fires take place. Flame retarded plastics are seldom if ever, the source of ignition in real-world fires, simply because they *are* flame retarded. The first materials to be ignited are other materials, such as wood and wood products—including paper and paper products. And if pyrolysis occurs in any additional materials, it is because of the heat generated and radiated or conducted to those other materials by the burning wood or paper products. If additional materials present do begin to burn, again the energy is being supplied by those other "natural" products that are already burning. Moreover, these "natural" materials produce combustion products that could be deadly to anyone exposed to them. If researchers investigating the toxic combustion products of plastics were to take into account the materials that are already burning, the lab animals might be dead long before the products of plastics combustion got to them.

The problem is not the decomposition products of flame retardant chemicals or the combustion products of protected or unprotected plastics, but *the fire itself*. If the focus were shifted to the prevention of fire, the early detection of fire, and the rapid extinguishment of fire, there would be no reason to look for scapegoats among economically competitive materials. It is ironic, indeed, that the use of plastics in smoke detectors and fast-acting sprinkler systems has placed these lifesaving devices within the financial reach of everyone. And it is also ironic that the large loss-of-life fires that receive so much publicity occur in buildings whose owners have steadfastly refused to install these lifesaving devices because they "were not required by the building code when the buiding was built." To save a few dollars, such owners put many lives at risk. In the late 1970s and early 1980s, a number of widely publicized fires led to attempts to blame plastics for the many deaths that occurred. These fires all were related to serious code violations, a lack of safety systems and devices, or both. Had these safety systems and devices been present, people would have been warned of the fire hazards and their lives might have been saved.

The addition of flame retardants to plastics is a safe, efficient way of allowing plastics to be used in the ways that people can best use them—in systems and products that meet all safety requirements for those products. To save lives, one must identify the real enemy. That enemy is fire, and it is aided and abetted by owners who do not make their occupancies safer by strict adherence to building codes (no cheating on construction methods and materials), the inclusion of fire detection systems, and proper and efficient fire extinguishment systems.

HOW FLAME RETARDANTS ARE USED IN DIFFERENT PLASTICS

Like any other additive, a flame retardant will be selected for the particular properties it imparts to make it satisfy the specifications for the final compound established by the customer. Different flame retardants may be chosen to give different levels of protection, depending upon the flame retardant specification the finished article must meet. New products are constantly being introduced, and the specifications of finished products are also constantly being changed, so that what may be most commonly used in a particular polymer today may not be used tomorrow. A more complete list of flame retardant chemicals is found in Tables 7.1 and 7.2 at the end of this chapter.

Very different flame retardants may be used in polymers that burn in a similar manner but may behave differently during processing. Since the temperature at which a polymer must be processed will influence the decision on which flame retardant will be used, and different chemicals will produce different properties in different polymers, the same flame retardant chemicals may or may not be used in similar polymers, compounds, blends and alloys.

Common polymers and the materials most often used in making them flame retardant are discussed below in alphabetical order. No attempt is made to list them in order of volume or in order of frequency with which any flame retardant chemical is used in them. Neither has any attempt been made to present a comprehensive listing. The materials discussed do represent the vast majority of flame retarded plastics in terms of volume of use.

Acrylics

Acrylics, such as polymethyl methacrylate (PMMA), are flame retarded by the use of halogenated materials or phosphorus-containing flame retardants. Another effective way of increasing the polymer's favorable combustion characteristics is to alloy it with polyvinyl chloride.

Acrylonitrile–Butadiene–Styrene (ABS)

Acrylonitrile–butadiene–styrene (ABS) is a higher impact material than high impact polystyrene and has greater strength. Among the most commonly used flame retardants for ABS are tris(tribromophenoxy)ethane, octabromodiphenyl oxide, and tetrabromobisphenol A. Small amounts of antimony oxide may be used for synergism, and polyvinyl chloride (PVC) is often used as a flame retardant added to acrylonitrile–butadiene–styrene to form an ABS/PVC alloy.

Cellulose Nitrate

Some references state that the best way to eliminate the fire hazards of cellulose nitrate is to replace it with another polymer, such as cellulose acetate. For those compounders or end users who specify cellulose nitrate, it may be flame retarded by the addition of phosphate plasticizers, particularly halogenated phosphate plasticizers.

Cellulosics

Cellulosics such as cellulose acetate, cellulose acetate butyrate, and cellulose acetate propionate, but excluding cellulose nitrate, are all combustible, but none to the degree of cellulose nitrate. They may be modified for better combustion properties by the use of phosphate plasticizers (particularly halogenated phosphate plasticizers), other phosphorus- and halogen-containing compounds, or both.

Epoxies

Epoxies may be flame retarded by the introduction of certain halogenated compounds into part of the system. Those materials include tetrabromobisphenols and tetrochlorobisphenols. Epoxies may also be filled with inorganic fillers, as in the case of the unsaturated polyesters.

Nylon

Nylons have been flame retarded with several different chemicals, including alicyclic chlorine compounds used with antimony oxide or ferric oxide as a synergist, brominated epoxy, brominated polystyrene, or poly(dibromophenylene oxide). Some of these flame retardants are recommended for nylon 6 and nylon 6/6, but manufacturers of flame retardants should be consulted for their specific recommendations for specific nylons.

Polycarbonate

Polycarbonate is an inherently flame retarded polymer. Its resistance to flame can be enhanced by the use of other flame retardants such as a modified version of tetrabromobisphenol A. Antimony oxide is not used in polycarbonate because it causes the polycarbonate to depolymerize, a highly undesirable reaction. Sodium antimonate may be used instead.

Polycarbonate may be alloyed with other polymers to achieve very specific properties. Popular among these alloys are polycarbonate/acrylonitrile–butadiene–styrene (PC/ABS) and polycarbonate/high impact polystyrene (PC/HIPS). These alloys may be flame retarded by use of tricresyl phosphate and triphenyl phosphate.

Polyesters

Polyesters, particularly unsaturated (thermosetting) polyesters, burn readily, but a very wide range of materials may be used to add flame retardance to the compounds. Halogenated paraffins and phosphate flame retardants may be used, or, quite simply, large amounts of inorganic (mineral) fillers may be used. These inorganic fillers, in addition to flame retardant properties, will add strength to the final part and may lower the total cost considerably if the right filler is chosen. Materials such as chlorendic anhydride, tetrabromophthalic anhydride, and tetrachlorphthalic anhydride may also be used.

Polyethylene

The uses of polyethylene in which resistance to flame is required are considerably fewer than those for polypropylene and are usually limited to films, electrical wire and cable insulation, and a few other applications.

Polyethylene burns very much like polypropylene, so many of the flame retardant chemicals will be similar. However, other differences in the polymers, such as the way polyethylene melts and flows, may contribute to the differing amounts of flame retardant needed. These peculiar properties are "so what's!" for firefighters and other emergency responders and safety managers, but they are important to the polymer chemist. Different amounts of these materials may be used to compensate for the physical properties and processing requirements of the polymer.

Among the most common flame retardants used in polyethylene are chlorinated cyclic compounds, decabromodiphenyl oxide and other aromatic bromine compounds, chlorinated paraffin, antimony oxide, dithiopyrophosphate, and magnesium hydroxide. The amounts used may be significantly different between high and low density polyethylene, and it is known that antimony oxide is much more efficient in polyethylene than polypropylene.

Another method to flame retard polyethylene is to chlorinate the polymer to create chlorinated polyethylene, a polymer that has its own applications.

Polypropylene

Polypropylene is usually flame retarded commercially by using a chlorinated cyclic compound with antimony oxide added as a synergist. The most common chlorinated cyclic compound used contains 65 percent chlorine, and the antimony oxide is added at a 13 percent level. In some cases, some of the antimony oxide is replaced by zinc borate.

When a brominated flame retardant is desired, decabromodiphenyl oxide is used. This material contains 83 percent bromine by weight, and it is usually used in conjunction with talc. Talc is usually used in polypropylene as a filler to reduce the overall cost of the compound, but in this case it also reduces the amount of dripping that usually occurs when polypropylene burns.

Other substances added to produce some degree of flame retardant action in polypropylene (and to produce other desired effects as well) include fumed silica to decrease the amount of dripping, special grades of chlorinated paraffin to improve thermal stability, and alumina trihydrate or magnesium hydroxide to lower smoke production.

Silicone-based flame retardants will raise the oxygen index of polypropylene and help produce a hard char, especially when used with magnesium stearate. Magnesium stearate is a different type of synergist, since other materials discussed are usually inorganic materials that will not themselves burn. Magnesium stearate is an organic compound that burns.

Intumescent coatings are another method of flame retarding polypropylene. These surface coatings swell and form a hard char on the surface to discourage further burning. Nitrogen and phosphorus-containing materials, including ammonium pyrophosphate, work well in this application.

Polystyrene

Polystyrene may be either general purpose or impact modified. The general purpose polystyrenes that must meet specifications calling for flame retardant properties are usually the foamed materials used for insulation and other purposes.

Among the most common flame retardants used in extruded foam or expandable polystyrene (EPS) bead board are dibromoethyldibromocyclohexane, pentabromochlorocyclohexane, and hexabromocyclododecane. The use of antimony oxide in general purpose polystyrene is not common: It appears not only is there no synergism, but the addition of antimony oxide actually seems to lower the oxygen index. This of course, is undesirable.

On the other hand, antimony oxide is used in impact grade polystyrene compounds. Impact grade polystyrene, used in the manufacture of television cabinets and other home appliance housings, is more difficult than general purpose polystyrene to flame retard with retardands that work in general purpose polystyrene. Perhaps the presence of rubber, the ingredient that makes the polystyrene impact resistant, also helps the synergistic effect to be present. The antimony oxide is used with the most popular flame retardant for impact polystyrene, decabromodiphenyl oxide.

Polyurethane

Polyurethanes are among the most widely used foams, and fire protection requirements must be met in buildings, furniture, refrigeration, and transportation. A very wide range of flame retardants and synergists are used in polyurethanes, including antimony oxide, chlorinated paraffins, polyvinyl chloride, hexachloroendomethylenetetrahydrophthalic acid (HET acid), dibromoethyldibromocyclohexane, dibromoneopentyl glycol, red phosphorus (encapsulated), and aluminum hydroxide.

Polyvinyl Chloride (PVC)

Rigid polyvinyl chloride is itself inherently flame retardant since its molecules contains 57% chlorine by weight. In its rigid state (no plasticizer added), polyvinyl chloride will not support its own combustion. When a supporting flame is hot enough and big enough, however, the rigid polyvinyl chloride will ignite. As soon as the supporting flame is removed or extinguished, the rigid polyvinyl chloride will stop burning. However, once plasticizer is added to rigid polyvinyl chloride to make it flexible, the resulting polyvinyl chloride compound burns, assuming enough plasticizer has been added and the plasticizer added is a combustible liquid. The popularity of polyvinyl chloride means it is used to make many products that must meet certain fire resistant specifications.

Many flame retardant chemicals have been added to polyvinyl chloride, including alumina trihydrate, antimony oxide, barium borate, chlorinated paraffins, phosphate esters, and zinc borate. Tricresyl phosphate is a plasticizer that, when added to rigid polyvinyl chloride in high enough quantities, will make the resulting compound flexible. But since tricresyl phosphate is a flame retarding plasticizer, the flexible polyvinyl chloride compound will not burn like those compounds in which phthalate plasticizers are used.

Styrene–Acrylonitrile (SAN)

Styrene-acrylonitrile copolymer resins burn with some of the characteristics of polystyrene. The addition of antimony oxide actually lowers the oxygen index, as it does in polystyrene, so its use must be carefully considered. The flame retardants of choice here are alicyclic bromide compounds.

Thermosetting Plastics

Some thermosetting plastics do not require the addition of flame retardants simply because their combustion characteristics are good enough to meet many

TABLE 7.1 Commonly Used Flame Retardants

ammonium bromide
ammonium polyphosphates
alumina trihydrate (aluminum hydroxide or hydrated aluminum oxide)
antimony fluoborate
antimony pentoxide
antimony oxide (antimony trioxide)
barium metaborate
bis(2,3-dibromyl) fumarate
bis(tetrabromophthalimido)ethane
bis(tribromophenoxy) ethane
bis(tribromophenoxy) ethylene
bis(pentabromophenoxy) ethane
brominated epoxy resin
brominated polycarbonate
brominated polystyrene
p-bromobenzaldehyde
bromochlorinated paraffin
n-butyl bis(hydroxypropyl) phosphine oxide
t-butyl phenyl diphenyl phosphate
t-butylphenyl phosphate
calcium borate
calcium phosphate
calcium sulfate
calcium sulfate dihydrate
chlorendic anhydride
chlorinated paraffin
chlorinated polystyrene
cresyl diphenyl phosphate
cyclooctyl hydroxypropyl phosphine oxide
decabromodiphenyl ether
decabromodiphenyl oxide
diallyl tetrabromophthalate
dibromoethyldibromocyclohexane
dibromoneopentyl glycol
dibromophenol
O,O-diethyl-1-N,N-bis(2-hydroxyethyl) aminoethylphosphonate
O,O-diethyl-1-N,N-bis(2-hydroxyethyl) aminomethylphosphonate
diisopropylphenyl phenyl phosphate
di-(polyoxyethylene)hydromethyl phosphonate
dodecachlorododecahydrodimethanodibenzo-(a,e) cyclooctene diphenyl phosphate
ethylene bis(dibromonorborane) dicarboximide
ethylene bis(tetrabromophthalimide)
ethylene bis(tribromophenyl ether)
hexabromobenzene
hexabromocyclododecane
hexachlorocyclopentadiene
hexachloroendomethylenetetrahydrophthalic acid (HET acid)
isodecyl diphenyl phosphate
isopropylphenyl diphenyl phosphate
isopropylphenyl phosphate
magnesium carbonate
magnesium hydroxide
molybdic oxide (molybdenum trioxide)
neoalkoxy tri(dioctyl phosphate) titanate
neoalkoxy tri(dioctyl pyrophosphate) titanate
octabromodiphenyl oxide
octyl diphenyl phosphate
(pentabromobenzyl)acrylate
pentabromochlorocyclohexane
pentabromodiphenyl ether
pentabromodiphenyl oxide
pentabromodiphenyl oxide/triaryl phosphate blend
pentabromoethylbenzene
pentabromomethylbenzene
pentabromophenol
pentabromophenyl benzoate
perchloropentacyclodecane
phosphorus, red, encapsulated
poly(dibromophenylene oxide)
poly(pentabromobenzyl)acrylate
poly(tribromostyrene)
polyvinyl chloride
sodium antimonate
tetrabromobisphenol A
tetrabromobisphenol A-bis(allyl ether)
tetrabromobisphenol A bis(2,3-dibromopropyl ether)
tetrabromobisphenol A bis(2-hydroxyethyl ether)
tetrabromobisphenol A carbonate oligomer
tetrabromobisphenol A ethoxylate
tetrabromoxylene
tetrabromophthalic anhydride
tetrachlorobisphenol A
terrachlorophthalic anhydride
tetradecabromodiphenoxybenzene
tetrakis(hydroxymethyl)phosphonium chloride

TABLE 7.1 *(Continued)*

tribromoneopentyl alcohol	tris(betachlorethyl) phosphate [tris(2-chloro-ethyl)phosphate]
tribromophenol	tris-butoxyethyl phosphate
tribromophenyl allyl ether	tris (chloropropyl) phosphate
tribromophenylmaleimide	tris(2,3-dibromopropyl) phosphate
tributoxyethyl phosphate	tris (dichloropropyl) phosphate
tributyl phosphate	tris(dichloroisopropyl) phosphate
trichlorethylene	tris(hydroxypropyl) phosphine
trichloromethyltetrabromobenzene	tris(tribromophenoxy)ethane
tricresyl phosphate	trixylene phosphate
triethyl phosphate	vinyl bromide
triisopropylphenyl phosphate	zinc borate
trioctyl phosphate	zinc borophosphate
triphenyl phosphate	

specifications. Phenolics, ureas, and melamines fall into this class, and on some occasions, melamines are added to other plastics as a flame retardant.

However, polyesters and epoxies do need protection, since their burning characteristics will not allow them to meet most fire specifications. Thermosetting polyesters are mentioned above, under polyesters, but there are a few more specific flame retardants used only in the thermosetting resins. These include hexachloroendomethylenetetrahydrophthalic acid (HET acid) and its anhydride, and bromoneopentyl glycol.

When phenolics must be flame retarded, often tetrabromobisphenol A and *p*-bromobenzaldehyde are used, in addition to phosphorus oxychloride and boric acid. Chlorinated paraffins and some bromine compounds are also used, usually with antimony trioxide added as a synergist.

FLAME RETARDANTS

Table 7.1 lists chemical compounds that have been or are being used as flame retardants in one or more plastics compounds. They are presented in alphabetical order and are listed only for the reader who is curious about what chemicals might be in various plastics in some quantity. The list is not complete.

PLASTICS AND THE FLAME RETARDANTS USED IN THEM

The list in Table 7.2 is presented by type of plastic, followed by the particular chemical compound that might be used as a flame retardant in a compound of that polymer. The list is not complete, since new materials are being developed almost every day. Some materials appearing on the list may no longer be used, for any number of reasons.

TABLE 7.2 Flame Retardants Listed by the Plastic in Which They Are Used

Acrylic
- alumina trihydrate
- ammonium polyphosphate
- antimony oxide
- barium metaborate
- brominated polystyrene
- t-butyl phenyl diphenyl phosphate
- chlorinated paraffin
- cresyl diphenyl phosphate
- decabromodiphenyl oxide
- dibromoneopentyl glycol
- hexabromocyclododecane
- magnesium hydroxide
- octabromodiphenyl oxide
- octyl diphenyl phosphate
- pentabromodiphenyl oxide
- phosphorus, red, encapsulated
- sodium antimonate
- tribromoneopentyl alcohol
- tributoxyethyl phosphate
- tricresyl phosphate
- triisopropylphenyl phosphate
- tris(betachlorethyl) phosphate
- trixylenyl phosphate
- vinyl bromide
- zinc bromide

Acrylonitrile–Butadiene–Styrene
- alumina trihydrate
- ammonium polyphosphate
- antimony oxide
- barium metaborate
- brominated polystyrene
- chlorinated paraffin
- decabromodiphenyl oxide
- halogenated hydrocarbons
- magnesium hydroxide
- octabromodiphenyl oxide
- octyl diphenyl phosphate
- pentabromodiphenyl oxide
- pentabromomethylbenzene
- pentabromophenyl benzoate
- phosphorus, red, encapsulated
- polyvinyl chloride
- sodium antimonate
- tetrabromobisphenol A
- trichloromethyltetrabromobenzene
- tris(tribromophenoxy)ethane
- triphenyl phosphate
- zinc borate

Cellulose Acetate
- alumina trihydrate
- ammonium bromide
- ammonium polyphosphate
- antimony oxide
- t-butyl phenyl diphenyl phosphate
- chlorinated paraffin
- cresyl diphenyl phosphate
- decabromodiphenyl oxide
- dibromoneopentyl glycol
- hexachlorocyclopentadiene
- octyl diphenyl phosphate
- pentabromodiphenyl oxide
- phosphorus, red, encapsulated
- tribromoneopentyl alcohol
- tributoxyethyl phosphate
- tributyl phosphate
- tricresyl phosphate
- triethyl phosphate
- triisopropylphenyl phosphate
- tris(betachlorethyl) phosphate
- triphenyl phosphate

Cellulose Acetate Butyrate
- alumina trihydrate
- ammonium polyphosphate
- antimony oxide
- t-butyl phenyl diphenyl phosphate
- chlorinated paraffin
- cresyl diphenyl phosphate
- hexachlorocyclopentadiene
- magnesium hydroxide
- octyl diphenyl phosphate
- pentabromodiphenyl oxide
- phosphorus, red, encapsulated
- tributoxyethyl phosphate
- tributyl phosphate
- tricresyl phosphate
- triethyl phosphate
- triisopropylphenyl phosphate
- triphenyl phosphate
- tris(betachlorethyl) phosphate
- tris(dichloropropyl) phosphate

TABLE 7.2 *(Continued)*

Cellulose Nitrate
alumina trihydrate
ammonium polyphosphate
antimony oxide
t-butyl phenyl diphenyl phosphate
chlorinated paraffin
cresyl diphenyl phosphate
di-(polyoxyethylene)hydromethyl phosphonate
hexachlorocyclopentadiene
magnesium hydroxide
octyl diphenyl phosphate
pentabromodiphenyl oxide
phosphorus, red, encapsulated
sodium antimonate
tribromoneopentyl alcohol
tributoxyethyl phosphate
tributyl phosphate
tricresyl phosphate
triethyl phosphate
triisopropylphenyl phosphate
triphenyl phosphate
tris(betachlorethyl) phosphate
trixylenyl phosphate

Chlorinated Polyethylene
antimony oxide

Epoxy
alumina trihydrate
ammonium polyphosphate
antimony oxide
barium metaborate
t-butyl phenyl diphenyl phosphate
brominated polystyrene
chlorinated paraffin
chlorendic anhydride
cresyl diphenyl phosphate
decabromodiphenyl oxide
dibromoneopentyl glycol
magnesium hydroxide
neoalkoxy tri(dioctyl phosphate) titanate
octyl diphenyl phosphate
pentabromophenol
pentabromodiphenyl oxide
phosphorus, red, encapsulated
sodium antimonate
tetrabromobisphenol A
tetrachlorobisphenol A
tetrachlorophthalic anhydride
tribromophenol
tributoxyethyl phosphate
tributyl phosphate
tricresyl phosphate
triisopropylphenyl phosphate
tris(betachlorethyl) phosphate
tris(dichloropropyl) phosphate
trixylenyl phosphate
zinc borate

Ethyl Cellulose
alumina trihydrate
ammonium polyphosphate
antimony oxide
t-butyl phenyl diphenyl phosphate
chlorinated paraffin
cresyl diphenyl phosphate
hexachlorocyclopentadiene
magnesium hydroxide
octyl diphenyl phosphate
pentabromodiphenyl oxide
phosphorus, red, encapsulated
tributoxyethyl phosphate
tributyl phosphate
tricresyl phosphate
triethyl phosphate
triisopropylphenyl phosphate
tris(betachlorethyl) phosphate
tris(dichlorophropyl) phosphate
trixylenyl phosphate

Ethylene Propylene Diene Modified
brominated polystyrene
decabromodiphenyl oxide

Ethylene Propylene Rubber
magnesium carbonate
magnesium hydroxide

Ethylene Vinyl Acetate
Alumina trihydrate
Antimony oxide
magnesium carbonate
magnesium hydroxide

Phenolic
alumina trihydrate

TABLE 7.2 *(Continued)*

ammonium polyphosphate
antimony oxide
barium metaborate
p-bromobenzaldehyde
t-butyl phenyl diphenyl phosphate
calcium phosphate
chlorinated paraffin
cresyl diphenyl phosphate
decabromodiphenyl oxide
dibromophenol
di-(polyoxyethylene)hydromethyl phosphonate
magnesium hydroxide
octyl diphenyl phosphate
pentabromodiphenyl oxide
pentabromophenol
phosphorus oxychloride
phosphorus, red, encapsulated
tetrabromobisphenol A
tetrachlorophthalic anhydride
triisophropylphenyl phosphate
tris(betachlorethyl) phosphate
tris(dichloropropyl) phosphate
zinc borate

Polyacrylonitrile
alumina trihydrate
ammonium polyphosphate
antimony oxide
barium metaborate
t-butyl phenyl diphenyl phosphate
chlorinated paraffin
decabromodiphenyl oxide
magnesium carbonate
octyl diphenyl phosphate
pentabromodiphenyl oxide
phosphorus, red, encapsulated
sodium antimonate
triisopropylphenyl phosphate
tributoxyethyl phosphate
tris(dichloropropyl) phosphate
zinc borate

Polyamide (Nylons)
antimony oxide
brominated epoxy
brominated polystyrene
ferric oxide
hexabromobenzene
octabromodiphenyl oxide
phosphorus, red, encapsulated
poly(dibromophenylene oxide)

Polybutylene
chlorinated paraffin

Polybutylene Terephthalate
antimony oxide
brominated polycarbonate
brominated polystyrene
pentabromomethylbenzene
pentabromophenyl benzoate
phosphorus, red, encapsulated
trichloromethyltetrabromobenzene

Polycarbonate
alumina trihydrate
ammonium polyphosphate
barium metaborate
bisphenol A
calcium sulfate
magnesium hydroxide
phosphorus, red, encapsulated
sodium antimonate
tetrabromobisphenol A
tetrachlorobisphenol A
tris(betachlorethyl) phosphate
zinc borate

Polyethylene
alumina trihydrate
ammonium polyphosphate
antimony oxide
barium metaborate
brominated paraffin
brominated polystyrene
chlorinated paraffin
decabromodiphenyl oxide
dithiopyrophosphate
hexabromocyclododecane
magnesium hydroxide
neoalkoxy tri(dioctyl phosphate) titanate
octabromodiphenyl oxide
pentabromodiphenyl oxide
phosphorus, red, encapsulated
sodium antimonate
zinc borate

Polyethylene Terephthalate
brominated polystyrene
pentabromomethylbenzene

TABLE 7.2 *(Continued)*

penabromophenyl benzoate
phosphorus, red, encapsulated
trichloromethyltetrabromobenzene

Polyphenylene Oxide
isopropylphenyl diphenyl phosphate
tricresyl phosphate
triphenyl phosphate

Polypropylene
alumina trihydrate
ammonium bromide
ammonium polyphosphate
antimony oxide
barium metaborate
brominated paraffin
brominated polystyrene
chlorinated paraffin
decabromodiphenyl oxide
hexabromocyclododecane
magnesium hydroxide
neoalkoxy tri(dioctyl phosphate) titanate
pentabromodiphenyl oxide
phosphorus, red, encapsulated
sodium antimonate
triisopropylphenyl phosphate
zinc borate

Polystyrene
alumina trihydrate
ammonium polyphosphate
antimony oxide
barium metaborate
brominated polystyrene
chlorinated paraffin
chlorinated polystyrene
cresyl diphenyl phosphate
decabromodiphenyl ether
decabromodiphenyl oxide
dibromoethyldibromocyclohexane
hexabromocyclododecane
magnesium hydroxide
neoalkoxy tri(dioctyl phosphate) titanate
octabromodiphenyl oxide
octyl diphenyl phosphate
pentabromochlorocyclohexane
pentabromodiphenyl oxide
pentabromomethylbenzene
pentabromophenyl benzoate
phosphorus, red, encapsulated
sodium antimonate
tributoxyethyl phosphate
tributyl phosphate
trichloromethyltetrabromobenzene
tricresyl phosphate
triisopropylphenyl phosphate
tris(betachlorethyl) phosphate
tris (dichloropropyl) phosphate
trixylenyl phosphate
zinc borate

Polyurethane
alumina trihydrate
aluminium hydroxide
ammonium bromide
ammonium polyphosphate
antimony oxide
barium metaborate
brominated paraffin
t-butyl phenyl diphenyl phosphate
chlorendic anhydride
chlorinated paraffin
cresyl diphenyl phosphate
dibromoethyldibromocyclohexane
decabromodiphenyl oxide
dibromoneopentyl glycol
O,O-diethyl-1-N,N-bis(2-hydroxyethyl) aminomethylphosphonate
di-(polyoxyethylene)hydromethyl phosphonate
hexabromocyclododecane
hexachloroendomethylenetetrahydrophthalic acid (HET acid)
magnesium hydroxide
molybdic oxide
neoalkoxy tri(dioctyl phosphate) titanate
octyl diphenyl phosphate
pentabromodiphenyl oxide
pentabromophenol
phosphorus, red, encapsulated
sodium antimonate
tetrabromophthalic anhydride
tetrachlorophthalic anhydride
tribromoneopentyl alcohol
tricresyl phosphate
triethyl phosphate

TABLE 7.2 *(Continued)*

triisopropylphenyl phosphate
triphenyl phosphate
tris(betachlorethyl) phosphate
tris(chloropropyl) phosphate
tris(dichloropropyl) phosphate
trixylenyl phosphate
zinc borate

Polyvinyl Acetate
alumina trihydrate
ammonium polyphosphate
antimony oxide
barium metaborate
t-butyl phenyl diphenyl phosphate
chlorinated paraffin
cresyl diphenyl phosphate
decabromodiphenyl oxide
magnesium hydroxide
octyl diphenyl phosphate
pentabromodiphenyl oxide
phosphorus, red, encapsulated
tributoxyethyl phosphate
tributyl phosphate
tricresyl phosphate
triisopropylphenyl phosphate
triphenyl phosphate
tris(betachlorethyl) phosphate
trixylenyl phosphate
vinyl bromide
zinc borate

Polyvinyl Chloride
alumina trihydrate
ammonium polyphosphate
antimony oxide
barium metaborate
t-butyl phenyl diphenyl phosphate
brominated paraffin
chlorinated paraffin
cresyl diphenyl phosphate
decabromodiphenyl oxide
diisopropylphenyl phenyl phosphate
isodecyl diphenyl phosphate
isopropylphenyl diphenyl phosphate
magnesium carbonate
magnesium hydroxide
molybdic oxide
neoalkoxy tri(dioctyl phosphate) titanate
octyl diphenyl phosphate
pentabromodiphenyl oxide
phosphorus, red, encapsulated
sodium antimonate
tributoxyethyl phosphate
tributyl phosphate
tricresyl phosphate
triethyl phosphate
triisopropylphenyl phosphate
trioctyl phosphate
triphenyl phosphate
tris(betachlorethyl) phosphate
tris(dichloropropyl) phosphate
trixylenyl phosphate
vinyl bromide
zinc borate

Styrene–Acrylonitrile
antimony oxide
phosphorus, red, encapsulated

Thermoplastic Elastomer
magnesium carbonate
magnesium hydroxide
phosphorus, red, encapsulated

Thermoplastic Rubber
antimony oxide

Unsaturated Polyesters
alumina trihydrate
ammonium polyphosphate
antimony oxide
barium metaborate
calcium sulfate
chlorendic anhydride
chlorinated paraffin
decabromodiphenyl oxide
dibromoneopentyl glycol
di-(polyoxyethylene)hydromethyl phosphonate
hexabromocyclododecane
magnesium carbonate
magnesium hydroxide
molybdic oxide
neoalkoxy tri(dioctyl phosphate) titanate
(pentabromobenzyl)acrylate
pentabromodiphenyl oxide
phosphorus, red, encapsulated
sodium antimonate

TABLE 7.2 *(Continued)*

tetrabromobisphenol A	triphenyl phosphate
tetrabromophthalic anhydride	tris(betachlorethyl) phosphate
tetrachlorophthalic anhydride	tris(dichloropropyl) phosphate
tribromoneopentyl alcohol	trixylenyl phosphate
triethyl phosphate	zinc borate

SUMMARY

- It is important to know why a flame retardant is compounded into plastics, and what the consequences of that use are.
- Flame retardants are used to improve the combustion characteristics of the plastic in which it has been compounded; this use will alter the way the plastic behaves in a fire and may change the composition of gases contained in the final combustion or decomposition products of the plastic.
- The end use of articles made from the plastic profoundly affects the choice of flame retardant as well as the government-regulated requirements of the finished article.
- It is important to know whether a particular plastic contains a flame retardant or not.
- Each polymer is different, and its flame retardancy requirements are different.
- Materials are constantly being tested to determine their efficiency as flame retardants; no list can be considered complete at any given time, since some flame retardants are becoming obsolete while others are being developed.

8

Fire Tests

The subject of testing plastics for their performance in a real-world fire situation is very complex. The manufacturer of any material, including any plastic resin, compound, alloy, or finished part, wants to see the material tested fairly and impartially. The problem, which on the surface appears relatively simple, quickly becomes emotionally charged when manufacturers or other interested parties see that an economic disadvantage is being assigned to their material or part while an economic advantage is being granted to a competitive material or part on the basis of certain tests.

The manufacturers of plastics and plastics products and others with an economic interest in the plastics industry are no different in this respect than anyone else with an economic interest in a particular product or service. If anything, those in the plastics industry feel they have been singled out for unfair economic punishment quite apart from the everyday competitive battle to satisfy the marketplace. The battle has become heated (no pun intended) with the entry of segments of the public that claim to have no economic interest in the rise or fall of the plastics industry. These groups vociferously claim that plastics have some adverse effect on society, either from a safety aspect (plastics involved in fires), as an allocation of nonrenewable resources (the diversion of oil into the petrochemical segment of industry as a raw material for plastics), or as a solid waste (plastics will "last forever" in the environment).

The last two issues will not be discussed here, since the focus of this work is on combustibility. Here the plastics industry feels the attack is most unwarranted. This feeling comes from the use of specific tests that show plastics to be more dangerous in a fire than "traditional" materials (wood and/or wood products). So far no small-scale test will accurately and completely correlate the behavior of materials in a test fire with what happens in a real-world fire. With a few notable exceptions, the tests described in this chapter are tests of one material, standing alone, subjected to different stimuli in an attempt to approximate the conditions of fire. Some of the tests are very simple procedures that take a specified piece of material, hold it in a specified manner, ignite it with a specified flame or other ignition source, and record what happens. These results are then gathered and compared with the way other materials behaved under these identical specified conditions. Despite warnings that the results of those tests are not to be used to select one class of materials over another, important decisions are made regarding the final use of that material in one application or another.

I will not comment on the tests except to present shortened descriptions of the procedures involved. Not all the tests listed in Appendix D are described here. Only fire tests for plastics are described below in any length, although some other tests of interest (such as smoke tests) may be mentioned. *These descriptions should not be used to actually run the tests.* The full description of each test has been prepared by its sponsoring organization, and only *its* description of the correct procedure should be used in conducting the specific test. Full descriptions of the latest version of each of the tests may be obtained by contacting the sponsoring organization at the addresses given in the appendices.

When this work is published, all tests that are clearly obsolete at the time of publication will have been removed. Arguments will certainly arise to the effect that one or more of these tests are indeed obsolete or have no relevance to the testing of plastics (or other materials). Many tests have been changed to handle plastics, which were not originally designed to do so. Again, I offer no judgment about the relevancy, accuracy, or credibility of any test. Its inclusion in this list means only that it has been used to test plastics in one or more combustion methods.

The field of fire testing is also undergoing some very rapid changes. Tests that were not in use when this material was compiled will obviously be omitted.

DESCRIPTIONS OF INDIVIDUAL FIRE TESTS

Following are short descriptions of selected guides, standards, and tests as set forth by the sponsoring organization whose name appears in parentheses. These guides, standards, and tests are "living" documents, which means they will be modified from time to time. The changes may be small or large, and the entire guide, standard, or test may be made obsolete, replaced, or adopted by another sponsoring organization. *No attempt should be made to run a test or otherwise use these guides, standards, and tests to determine the acceptance or rejection of a product, material, or assembly without contacting the sponsoring organization for the latest copy of the document describing the guides, standards, and tests.*

As you survey the procedures described below, and then read the full description as set out by the sponsoring organization, it should become abundantly clear why I have continually stressed that the simple burning of a plastic article or a sample of the plastic resin, compound, blend, or alloy from which it was made does not constitute a scientific test, nor does it even come close. Certified, professional testing laboratories have invested a great deal of time, effort, and money in personnel and equipment to run these tests, and sometimes *they* have problems.

Once again, these tests are being presented for information only. The descriptions do not comment about their accuracy, credibility, or relevancy to any particular use of the material being tested. Nevertheless, I believe that there is still no small-, medium-, or large-scale test that will duplicate all the reactions and byproducts that occur in a real-world, unplanned, unexpected, uncontrolled fire.

Berkeley Test (University of California)

The Berkeley Test, named for the University of California at Berkeley, is an 8 foot corner test for flame spread, using a modification of the ASTM E 108 test to determine flame spread up a vertical exterior wall.

D–229 (ASTM)
Standard Method of Testing Rigid Sheet and Plate Materials Used for Electrical Insulation

This test is under the jurisdiction of ASTM Committee D-9 on Electrical Insulating Materials. It stated purpose is to develop a method to "cover procedures for testing rigid electrical insulation normally manufactured in sheet or plate form."

Although this standard does not mention burning rate and flame resistance, thirteen sections of the standard are dedicated to describing two methods of testing materials for these properties, since electrical insulating materials are often subjected to temperatures that might be above the ignition temperature of the materials themselves. This is possible when the electrical apparatus of which the insulation is a part malfunctions, if other equipment in the same system malfunctions or fails, if the insulating material fails to resist ignition in normal usage when it is exposed to electrical arcing, or any combination of these occurrences.

In Method I, a relatively simple way of establishing relative flame resistance

Source: Test descriptions printed with permission from the Annual Book of ASTM Standards, copyright American Society for Testing and Materials, 1916 Race Street, Philadelphia, PA 19103.

among several materials is described. It provides for the measurement of resistance to ignition and resistance to continued burning of a single material tested alone in a controlled laboratory environment. It is meant to be a simple screening test to determine quickly whether or not a material is relatively flame resistant. It does not provide for any quantitative measurement of how the material will behave in actual use. Neither Method I or Method II can predict the flammability of any material under actual use conditions, nor does either test allow for a valid comparison of burning characteristics of different materials or for determination of a material's tendency to contribute to the spread of a fire.

In Method I, twenty test specimens 127 mm (± 1.6 mm) long by 12.7 mm (± 0.51 mm) wide by the normal thickness of the sheet as usually found in use are prepared and conditioned for use. The sample is clamped within 6.3 mm of the top within a draft-free flame cabinet and held in a vertical position allowing for a 9.5 mm space between the bottom of the sample strip and the top of the burner tube. A methane or natural gas–supplied Tirrill burner will be ignited and adjusted to produce a 19.1 mm blue flame with no yellow top. The burner is then placed beneath the sample for 10 seconds and removed, and flaming time of the sample is measured and recorded. Immediately upon cessation of the flaming of the sample, the burner is placed beneath the sample for another 10 seconds and removed. Flaming plus glowing time of the sample is measured and recorded. A table of required conditions is provided within the method, and instructions are given for samples not meeting these conditions.

The test described in Method II is considerably more complex than in Method I. A more complex apparatus is used for testing, a specified heater coil is used as the ignition source rather than a Tirrill burner, difference size samples than Method I are used, and much more precise measurement equipment is used. Drawings of the flame cabinet and an electrical diagram of the control cabinet are provided, in addition to a drawing of the mandrel for the heater coil and coil spacing gage. Calibration of the equipment is necessary before the test can be carried out on conditioned samples. Methods of calculation of burning time and average ignition time are provided.

D–470 (ASTM)
Standard Methods of Testing Crosslinked Insulations and Jackets for Wire and Cable

The test is under the jurisdiction of ASTM Committee D-9 on Electrical Insulating Materials and is the direct responsibility of Subcommittee D 09.18 on Elastomeric Insulation on Cables. The stated purpose of this test is to "cover procedures for testing crosslinked insulations and jackets for wire and cable. . . . These methods do not apply to the class of products known as flexible cords."

ASTM D 470 contains a horizontal flame test described by four sections of the standard. The apparatus used is a test chamber made of sheet metal measuring 305 mm wide, 355 mm deep, and 610 mm high, open at the top and front. A support must be used to hold the sample in a horizontal position. A Tirrill burner using "ordinary illuminating gas" at normal pressure is used to burn the samples of insulation over untreated surgical cotton, using a watch or clock "having a hand that makes one complete revolution per minute." A ventilation hood may be used for the test if a room free of drafts is not available.

The flame is to be adjusted and brought into contact with the specimen in a particular manner for 30 seconds, then removed. During contact of the flame with the sample and after, it must be observed and reported whether or not flame extended on the sample to an area outside the area of flame exposure. The behavior and duration of the flaming of the specimen after the application of the test flame must also be reported. Finally, it must be reported whether or not sparks or flaming drops from the burning test specimen of insulation ignited the cotton lining on the floor of the test chamber.

D 568 (ASTM)
Standard Test Method for Rate of Burning and/or Extent and Time of Burning of Flexible Plastics in a Vertical Position

It is under the jurisdiction of ASTM Committee D-20 on Plastics and is the direct responsibility of Subcommittee D 20.30 on Thermal Properties (Section D 20.30.03). The stated purpose of this test is to develop a procedure "for comparing the relative rate of burning and/or extent and time of burning of plastics in the form of flexible, thin sheets or films, tested in the vertical position."

The test is to be run within a shielded area, the dimensions of which are provided in the description of the test. The sample (10 specimens are to be prepared and conditioned), 25 mm wide by 45 cm long (the thickness is to be determined by Method B of Test Methods D 374 or equivalent), is to be marked 70 mm from one end, so that a 38 cm length is established over which the burning rate may be measured. The sample will hang from the clamp so that 43 cm is exposed below the hood, and the clamp and shield are placed inside a ventilated hood (with no ventilation during the test). A stopwatch is used as a timer, and it is started as soon as the flame is brought into contact with the sample. A bunsen burner is used with a 25 mm flame, and the flame is brought into contact with the bottom end of the sample until it is ignited, but for no longer than 15 seconds. The flame is removed once ignition occurs but may be moved back if the sample does not ignite. If the sample melts without burning and/or the sample continuously shrinks away from the flame, the test is not applicable.

If the sample ignites, the shield is closed and the time is recorded when any part of the burning edge reaches the 38 cm gage mark. The time will also be noted if the flame is extinguished. If the flame is extinguished, the apparent cause of such extinguishment is noted and the length in centimeters to the point of extinguishment is noted. The test must be repeated as described in the standard.

The data from the test are then used to calculate burning rate and average time of burning. Since the results cannot be used to compare different compositions of different materials at different thicknesses, the standard suggests that useful information may be obtained by plotting burning rate versus thickness for any one material of identical composition. It is not implied that the results of this test can be correlated with the burning characteristics of the tested materials in actual use conditions.

If excessive shrinkage of the sample occurs, the test may be invalidated. Other tests may be more useful under these circumstances, including ASTM Test Method D 635, which tests the burning characteristics of self-supporting plastics; ASTM Specification D 4549, for polystyrene molding and extrusion materials; or ASTM Test Method D2863, which measures the minimum oxygen concentration required to support candlelike combustion of plastics.

D 635 (ASTM)
Standard Test Method for Rate of Burning and/or Extent of Time of Burning of Self-Supporting Plastics in a Horizontal Position

This test is under the jurisdiction of ASTM Committee D-20 on Plastics and is the direct responsibility of Subcommittee D 20.30 on Thermal Properties. The stated purpose of this test method is to develop a procedure "for comparing the relative rate of burning and/or extent and time of burning of self-supporting plastics in the form of bars, molded or cut from sheets, plates or panels, and tested in the horizontal position." Self-supporting plastics are defined by this standard as "those plastics, which, when mounted with the clamped end of the specimen 10 mm above the horizontal screen, do not sag initially so that the free end of the specimen touches the screen."

Inside a test chamber, which may be a totally enclosed laboratory hood or a chamber within such hood, samples will be burned with the flame from a natural gas–supplied (or any other laboratory gas) burner, with no draft across the specimens during the burning process. The samples, 125 mm (± 5 mm) long by 12.5 mm (± 0.2 mm) wide and supplied in the thickness normally found in actual use (or, alternatively, 3 to 12 mm thick), are marked at a distance of 25 mm and 100 mm from one end. Each sample is then held by a clamp on the end nearer the 100 mm mark and held with its longitudinal axis horizontal and its transverse axis at 45° to the horizontal.

When the test is run under the prescribed conditions, the time will be recorded as the flame reaches the marks (or when combustion ceases). If the specimen has not burned, or if the flame not reached the 100 mm mark, measurements are made of the unburned area and recorded. Repeat tests are required, and average time of burning (ATB) and average extent of burning (AEB) are calculated according to formulas provided. Specimens that so warp and/or do not burn because of dripping, flowing, or falling burning particles must be reported, as well as the reignition of the sample by fallen material burning on the wire gauze under the clamped specimen. Once the specimen has ignited, the flame from the burner must be withdrawn a prescribed distance (450 mm) and the hood closed.

For materials that do not burn to the gage mark, ASTM Test Method D 757 may produce additional information. When thin sheets or films of flexible plastics are to be tested, reference should be made to ASTM Test Method D 568. For sheet and plate materials used for electrical insulation, tests are covered in ASTM Test Method D 229.

D 757 (ASTM)
Standard Test Method for Incandescence Resistance of Rigid Plastics in a Horizontal Position (Discontinued)

This test provided for laboratory comparisons of the resistance of rigid plastics to and incandescent surface at 950°C (± 10°C). It was under the jurisdiction of ASTM Committee D-20 on plastics but was discontinued in 1987.

D 876 (ASTM)
Standard Methods of Testing Nonrigid Vinyl Chloride Polymer Tubing Used for Electrical Insulation

This test is under the jurisdiction of Committee D-9 on Electrical Insulating Materials and is the direct responsibility of Subcommittee D 09.07 on Flexible and Rigid Insulating Materials. The stated purpose of this test method is to cover "the testing of general-purpose (Grade A), low-temperature (Grade B), and high-temperature (Grade C) nonrigid vinyl chloride polymer tubing, or its copolymers with other materials, for use as electrical insulation."

The D 876 standard methods cover a dimensional test, a flammability test, a tension test, the effects of elevated temperatures, an oil resistance test, a brittleness temperature test, a penetration test, a volume resistivity test, a dielectric breakdown voltage test, a strain relief test, and corrosion tests. They include, for each test, a statement of significance and use, a description of apparatus to be used, the preparation of test specimens, procedures to be followed, the form of the report, and a statement of precision and bias.

D 1000 (ASTM)
Standard Test Methods for Pressure-Sensitive Adhesive-Coated Tapes Used for Electrical and Electronic Applications

This test is under the jurisdiction of Committee D-9 on Electrical Insulating Materials and is the direct responsibility of Subcommittee D 09.07 on Flexible and Rigid Insulating Materials. The stated purpose of this test method is to cover "procedures for testing pressure-sensitive adhesive-coated tapes to be used as electrical insulation" it includes both elastic and nonelastic backings.

Among the tests required for these tapes is a flammability test. The apparatus and materials needed to run the test are enumerated, in addition to specimen preparation, procedures to follow, contents of the report, and a statement of significance and bias.

The test method references ASTM Standards D 876, which standard includes its own flammability test.

D 1230 (ASTM)
Flammability of Apparel Textiles

This standard is under the jurisdiction of Committee D-13 on Textiles and is the direct responsibility of Subcommittee D 13.52 on Flammability. The stated purpose of this test method is to cover "the evaluation of the flammability of textile fabrics as they reach the consumer for or from apparel other than children's sleepware or protective clothing."

This standard has not established correlation of test results with actual performance, and therefore it cannot be recommended for acceptance testing of commercial shipments of apparel fabrics.

The standard contains a summary of the flammability test method, a statement of use and significance, a description of the apparatus and materials used, the method of sampling, a method of preparation of samples, the procedure to be used, the method of calculation of values, the interpretation of results obtained, the content of the report, and a statement of precision and bias.

This standard contains a caution in the Scope section, stating that this test method is not identical to 16 CFR (Code of Federal Regulations) 1610, Flammability of Clothing Textiles, and that any fabrics that are introduced into commerce must meet the CFR requirements.

D 1433 (ASTM)
Standard Test Method for Rate of Burning and/or Extent and Time of Burning of Flexible Thin Plastic Sheeting Supported on a 45° Incline (Discontinued)

This test provided for the determination of the relative rate of burning and/or extent and time of burning of flexible thin sheeting. It was under the jurisdiction of Committee D-20 on Plastics, but was discontinued in 1988 and replaced by Specification D 4549, for Polystyrene Molding and Extrusion Materials (PS).

D 1929 (ASTM)
Standard Test Method for Ignition Properties of Plastics

This standard is under the jurisdiction of Committee D-20 on Plastics and is the direct responsibility of Subcommittee D 20.30 on Thermal Properties (Section D 20.30.03). The stated purpose of this test method is to establish "a laboratory determination of the self-ignition and flash-ignition of plastics using a hot-air ignition furnace."

The standard describes a hot-air ignition furnace to be used in the test method. Thermoplastic materials may be tested in the form usually supplied by the manufacturer for injection molding, which is as pellets. If sheet or film is the only form in which the material is available, 20 mm by 20 mm squares may be bound together by fine wire to reach the required weight of 3 g (± 0.5 g). The samples must be conditioned in accordance with Procedure A of ASTM 618.

The test devises procedures and an apparatus to determine "the lowest initial temperature of air passing around the specimen at which a sufficient amount of combustible gas is evolved to be ignited by a small pilot flame." This temperature is designated the minimum flash-ignition temperature. The test also determines "the lowest initial temperature of air passing around the specimen at which, in the absence of an ignition source, the self-heating properties of the specimen lead to ignition or ignition occurs of itself, as indicated by an explosion, flame, or sustained glow." This temperature is designated the minimum self-ignition temperature.

In some cases, a self-ignition by temporary glow occurs. This is described by the test method as "the slow decomposition and carbonization of the plastic [which] results only in a glow of short duration at various points in the specimen without general ignition actually taking place."

There are two procedures to test method D 1929, with Procedure B being designated the short method.

The test method suggests that a reference, "A Method and Apparatus for Determining the Ignition Characteristics of Plastics" by N. P. Steckin, appearing in *Journal of Research,* Volume 43, Number 6, December 1949, p. 591, may be of interest in connection with D 1929.

D 2633 (ASTM)
Standard Methods of Testing Thermoplastic Insulations and Jackets for Wire and Cable

These standards are under the jurisdiction of ASTM Committee D-9 on Electrical Insulating Materials and are the responsibility of Subcommittee D 09.18 on Elastomeric Insulation on Cables. The stated purpose of this standard is to cover "procedures for the testing of thermoplastic insulations and jackets on insulated wire and cable. The insulation and water-absorbtion tests do not apply to the class of products having a separator between the conductor and the insulation."

This standard contains a section (38) describing a vertical flame test. It claims that when "properly interpreted [the flame test] may be used to measure and describe the properties of products, materials, or systems in response to heat and flame under controlled laboratory conditions." The test is applicable to wire sizes less than 0.25 inches (6.4 mm) in outside diameter. A test chamber is described, plus a means for holding the sample taut, a burner fueled by natural gas with a prescribed flame size, an adjustable steel jig to locate the burner properly, a timing device, flame indicators, and dry surgical grade cotton.

The object of the test is to determine whether or not the specimens convey flame. The flame is applied to a specimen in the prescribed manner for 15 seconds and removed for 15 seconds. This sequence is repeated four times. Observations are then made concerning whether or not more than 25 percent of the extended portion of the flame indicator has burned after the five applications of flame or whether drops that fell during the test or within 30 seconds of the final flame application ignited the cotton. If either of these two occurrences are present, the insulation is considered to have conveyed flame.

D 2671 (ASTM)
Standard Test Methods for Heat-Shrinkable Tubing for Electrical Use

This standard is under the jurisdiction of Committee D-9 on Electrical Insulating Materials and is the direct responsibility of Subcommittee D 09.07 on Flexible and Rigid Insulating Materials. The stated purpose of this test method is to cover "the testing of heat-shrinkable tubing for electrical insulation. Materials used include polyvinyl chloride, polyolefins, fluorocarbon polymers, silicone rubber, and other plastic or elastomeric compounds."

This standard includes several sections covering a flammability test. It includes a statement of scope (to measure the resistance of the tubing to ignition and propagation of flame after ignition under prescribed conditions of test), a statement of significance and use, and a description of three different procedures (A, B, and C) to be carried out. Each procedure carries its own instructions, with Procedure A being set out in ASTM Test Methods D 876.

D 2843 (ASTM)
Standard Test Method for Density of Smoke from the Burning or Decomposition of Plastics

This standard is under the jurisdiction of ASTM Committee D-20 on Plastics, and is the direct responsibility of Subcommittee D 20.30 on Thermal Properties (Section D 20.30.03). The stated purpose of the test is to cover "a laboratory procedure for measuring and observing the relative amounts of smoke produced by the burning or decomposition." The standard states further that it "is to be used for measuring the smoke producing characteristics of plastics under controlled conditions of combustion or decomposition. Correlation with other fire conditions is not necessarily implied."

The test prescribes the exposure to flame of specimens 25.4 mm (± 0.3 mm) long by 25.4 mm (± 0.3 mm) wide by 6.2 mm (± 0.3 mm) thick in a specifically constructed smoke chamber. If a thickness of greater than 6.2 mm is used, this must be reported. Thinner samples may be used in their normal use thickness, or they may be stacked to reach the 6.2 mm thickness. They then must be conditioned for a specified time at a specified humidity in accordance with Procedure A of Methods D 618.

The specimen is placed on a stainless steel screen and burned using a propane burner at a specified gas pressure. The chamber is closed during the test and the smoke produced by the burning or decomposing plastic is captured in the chamber. A light source, a photoelectric cell and a meter to measure light absorption are used in the tests. Percentage of light absorbed is read every 15 seconds for 4 minutes. The 15-second interval readings for the three samples in each group are averaged, and this average light absorption is plotted against time. The maximum smoke density is read as the highest point on the curve, and the total smoke produced is determined by measuring the area under the curve. The rating for smoke density represents the total amount of smoke present in the chamber for the 4-minute interval.

Other observations, such as time to ignition, flame extinguishment, specimen consumption, melting, dripping, foaming, and charring are also recorded. The test is conducted with the exhaust fan in the smoke chamber or hood turned off. The test must be run in triplicate.

D 2859 (ASTM)
Standard Test Method for Flammability of Finished Textile Floor Covering Materials

This standard is under the jurisdiction of ASTM Committee D-13 on Textiles and is the direct responsibility of Subcommittee D 13.21 on Pile Floor Coverings. Its stated purpose is to cover the "determination of the flammability of finished textile floor covering materials when exposed to an ignition source under controlled laboratory conditions. It is applicable to all types of textile floor coverings regardless of the method of fabrication or whether they are made from natural or man-made fibers." This test method provides a procedure that will identify finished textile floor coverings that can be rated as flame resistant under specific controlled laboratory conditions. It does not specify the use of an underlayment material.

The standard describes a test chamber, a frame, a dessicating cabinet, a circulating air oven, a laboratory fume hood, a vacuum cleaner, and the methenamine reagent tablet to be used. A sample large enough to be cut into eight test specimens is taken according to applicable material specifications, or a sample that is representative of the lot to be tested, conditioned, and cleaned. A methanamine tablet is placed in the center of the hole in the steel plate that is placed on the specimen, and carefully ignited with a match. The test is terminated if the ignition flame or any propagated flame burns out. Otherwise the material is allowed to burn until flame or glowing combustion reaches any point along the edge of the hole in the steel frame. A specimen passes the test if the charred portion of the test specimen does not extend to within 1 inch of the edge of the hole in the steel frame at any point.

D 2863 (ASTM)
Standard Test Method for Measuring the Minimum Oxygen Concentration to Support Candle-like Combustion of Plastics (Oxygen Index).

This standard has been called the "Limiting Oxygen Index" test. It is under the jurisdiction of Committee D-20 on Plastics and is the direct responsibility of Subcommittee D 20.30 on Thermal Properties (Section D 20.30.03). Its stated purpose is to describe "a procedure for measuring the minimum concentration of oxygen in a flowing mixture of oxygen and nitrogen that will just support flaming combustion."

The oxygen index, the desired outcome of the test method determined for a particular plastic, is defined as "the minimum percent of oxygen, expressed as volume percent, in a mixture of oxygen and nitrogen that will just support flaming combustion of a material initially at room temperature under the conditions of this test method."

The test method describes the apparatus to be used and specifies sample specimen dimensions for physically self-supporting plastics (rigid), and alternates for self-supporting flexible plastics, cellular (foamed) plastics, and film or thin sheet (non-self supporting) plastics. If sample sizes other than specified in the test method are used, a difference in the Oxygen Index may result. Additionally, in the case of cellular materials, comparisons of Oxygen Index between identical plastics can be made only if the densities of the materials are also identical.

After the sample specimens are conditioned (according to Procedure A of D 618) and the flow-measuring system is calibrated according to instructions in the test method, the specimen is clamped in the apparatus, the gas mixture flow is started, and the test sample is ignited. The concentration of oxygen in the gas mixture is determined to be too high if measurements of the burning of a particular type sample determines that it burns in accordance with specified criteria. The volume percentage of oxygen must be adjusted downward until the lowest concentration of oxygen in the gas mixture at which the criterion is met is found. At the next reduction of 0.2% of oxygen in the mixture, the criterion in the test should not be met, and the last previous concentration is used to calculate the Oxygen Index.

Warnings are given in the test method concerning cellular specimens, soot-generating specimens, and film or thin sheet specimens that shrink excessively. Special procedures may have to be incorporated to test these materials successfully.

D 3014 (ASTM)
Standard Test Method for Flame Height, Time of Burning, and Loss of Weight of Rigid Thermoset Cellular Plastics in a Vertical Position

This standard is under the jurisdiction of Committee D-20 on Plastics and is the direct responsibility of Subcommittee D 20.30 on Thermal Properties (Section D 20.30.03). Its stated purpose is to describe a method that "covers a small-scale laboratory screening procedure for comparing relative extent and time of burning and loss of weight of rigid thermoset cellular plastics." The method has also been used to test flexible cellular (foamed) plastics and other materials, but there has been no attempt to determine the applicability of this test method to those materials.

The density of the material to be tested is to be determined in accordance with Test Method D 1632. Then, six samples, measuring 254 mm by 19 mm by 19 mm, are to be conditioned as described in the test method, weighed, placed on a specimen support (which must also be weighed), and placed in a test chimney of specified dimensions. The burner may be either a propane burner using propane gas or a natural gas burner using natural gas. After the chimney has been lined with aluminum foil, the specimen is ignited. During the test, the maximum flame height is measured and the time from ignition to extinguishment is recorded. After the specimen has cooled, the specimen support is removed and is weighed without removing the specimen. The test is to be repeated for all six specimens.

The weight percentage of the specimen remaining after burning is calculated, and the average of the six specimens is reported, in addition to average flame height and average time to extinguishment. Also reported is a description of the material, average density of the specimens, air temperature and relative humidity during storage and flame testing, and the number of specimens that produce flaming drips.

D 3675 (ASTM)
Standard Test Method for Surface Flammability of Flexible Cellular Materials Using a Radiant Heat Energy Source

This standard is under the jurisdiction of ASTM Committee D-11 on Rubber and is the direct responsibility of Subcommittee D 11.17 on Flammability. The stated purpose of this standard is to develop a flame-spread index to provide "measurements on materials whose surfaces may be exposed to fire and is intended for research and development purposes only. It provides a laboratory test procedure for measuring and comparing the surface flammability of materials when exposed to a prescribed level of radiant heat energy."

The apparatus needed and described by the standard includes a radiant heat panel with air and gas (acetylene) supply, a specimen holder and framework support, a pilot burner, stack, thermocouples, automatic potentiometer recorder, hood, radiation pyrometer, portable potentiometer, and timer. Preparation and conditioning of the test samples, the procedure for running the test, calculations, and contents of the report are described.

The orientation of the sample is designed so that ignition of the sample will be forced near its upper edge and flame spread is downward. The apparatus is heated for 30 minutes. The sample is then placed within and the pilot flame is ignited. Measurements are taken of the time for flame spread to reach 75-mm marks on the specimen, the stack temperature, and any other observations of the behavior of the specimen during the test. The test is run for 15 minutes but is ended if the flame front has progressed over the entire length of the sample before that.

The prescribed calculations determine the flame-spread index of a material as a product of its flame-spread factors and heat evolution factor. Methods are described to include a "flash potential" in cases where the specimen produces a flash of flames across part or all of the sample, whether or not that flash produces a continuous flaming condition. Specimens that run and/or drip may also be accounted for.

D 3713 (ASTM)
Standard Method for Measuring Response of Plastics to Ignition by a Small Flame

This standard is under the responsibility of Committee D-20 on Plastics and is the direct responsibility of Subcommittee D 20.30 on Thermal Properties (Section D 20.30.03). The stated purpose of the standard is to describe a method that "covers a small scale procedure for characterizing the response of a plastic to an ignition source consisting of a small flame of controlled intensity applied to the base of a standard sample being held in a vertical position."

A flame test chamber is described in the standard, including a bunsen or Tirrill type burner using natural gas or technical grade methane, a timer, a thermocouple, untreated absorbent medical cotton, and a flow meter. The test specimen size is described, with a warning not to use samples thicker than 12.7 mm. (ASTM Method D 3801, unless otherwise specified, uses the same specimens for measuring the comparative extinguishing characteristics of solid plastics when tested in a vertical position.) The samples and the cotton must be conditioned using Procedure A of D 618. The flame height must be preset and the flame temperature measured and recorded. The profile of the flame must meet specifications in the standard. Positioning of the burner in the event that dripping of the sample occurs is also described.

The time is recorded from the time the burner is placed under the specimen until the instant burning or glowing combustion, or both, ceases. A set of the specimens will be exposed to the flame in uniformly increasing 5 second increments to a maximum of 60 seconds, using a new specimen for each sequence of increased increments. The mode of response is noted, with the endpoint of the test being "the point at which the burning time of the sample exceeds 30 seconds after the burner is removed; the material drips, with or without ignition of cotton, either during the flame application or within 30 seconds after the flame is removed; or the combustible material in the specimen is totally consumed with 30 seconds or less after the flame is removed." Other modes of response during the test for the sample listed in the standard include "consumed," "dripping ignition," "dripping, no ignition," and "burning." The test must be performed on ten specimens, tested consecutively.

The end result of this standard is the determination of the ignition response index (IRI), which measures "the response of a sample of specified thickness or shape to the thermal energy produced by a small flame. It consists of the maximum flame inpingement time withstood by the sample without being totally consumed (except for the material in the clamp), or burning or glowing, or both, more than 30 seconds after removal of the ignition source, or producing droplets that ignite cotton. Thus the IRI is composed of time in seconds or letter(s) denoting mode of response, or both, and the sample thickness."

This standard has been proposed as a useful method to detect material substitution, material deterioration during use, or the control of manufacturing processes. The theory is that once an IRI is assigned to a particualr material, a proper sampling plan can be used to select samples for testing leading to acceptance or rejection of that material. Materials that drip cannot be further evaluated to their final ignition characteristics using this standard, and test data may be used to compare results only if such results were obtained using samples of equal dimensions and under equivalent conditions.

D 3801 (ASTM)
Standard Test Method for Comparative Extinguishing Characteristics of Solid Plastics in a Vertical Position

This standard is under the jurisdiction of Committee D-20 on Plastics and is the direct responsibility of Subcommittee D 20.30 on Thermal Properties (Section D 20.30.03). The stated purpose of the standard is to describe a method that "covers a small-scale laboratory procedure for determining comparative extinguishing characteristics of solid plastic material, using a small flame of controlled size and intensity applied to the base of specimens held in a vertical position."

The test chamber may be a laboratory hood or other enclosure with no drafts during the test. A bunsen or Tirrill burner using technical grade methane gas is specified (natural gas with a specified energy density may be used). The test specimens must be of a specified size (ASTM Test Method 3801, unless otherwise specified, uses the same specimens for determining the comparative resistance of solid plastics to ignition by a small flame), and must be conditioned as described in this standard. The burner must be positioned as described in this test method in the event that the test specimens drip during the test.

The test consists of the application of a specified, measured flame to each set of five test samples in two 10-second periods. Flaming time before extinguishment is recorded after the first flame application, and the times of flaming and glowing extinguishment are recorded after the removal of the second flame application. Whether or not the materials drip flaming drops (onto specified dry absorbent surgical cotton) during the test is also recorded.

The results of the test are to be reported as the arithmetical mean of the burning times (flaming, and flaming plus glowing) for each specimen set and their standard deviations, and the duration of the flaming (and flaming plus glowing) times after the first and second flame impingements. Additional information to be reported includes whether the sample was totally consumed (except for material held in the clamp) and whether or not any of the specimens dripped flaming particles that ignited the cotton.

D 3874 (ASTM)
Standard Test Method for Ignition of Materials by Hot Wire Sources

This standard is under the jurisdiction of Committee D-9 on Electrical Insulating Materials and is the direct responsibility of Subcommittee D 09.12 on Electrical Tests. The stated purpose of this method is to "differentiate, in a preliminary fashion, among materials with respect to their resistance to ignition because of their proximity to electrically heated wires and other heat sources."

The test applies to materials in molded and sheet form that are rigid at room temperature. It is intended to test how materials behave when electrical equipment, wires, other conductors, resistors, and other parts of electrical apparatus become abnormally hot and affect those materials near the source of heat.

The standard specifies a No. 24 AWG, Nichrome V wire manufactured by a specific source, a supply circuit, a test fixture, a test chamber, and a specimen-winding fixture. The test specimens are prepared from samples as set out in D 3636 (Method for Sampling and Judging Quality of Electrical Insulating Material), and then they are conditioned. The test is conducted after the wire is wrapped around the test specimen in the prescribed manner. The test circuit is energized by a specific power density, and heating continues until the specimen ignites. Once ignition occurs, the time to ignite is recorded and the power discontinued. The test is to be discontinued if ignition does not occur in 2 minutes. It is also discontinued if the specimen melts through the wire and is no longer in intimate contact with all five turns of the wire.

The contents of the report are enumerated, and a statement of precision and bias concludes the standard. An annex to the standard describes the apparatus, equipment, and procedures for calibrating the wire.

D 3894 (ASTM)
Standard Method for Evaluation of Fire Response of Rigid Cellular Plastics Using a Small Corner Configuration

This standard is under the jurisdiction of Committee D-20 on Plastics and is the direct responsibility of Subcommittee D 20.30 on Thermal Properties (Section D 20.30.03). The stated purpose of this method is to describe "a small-scale laboratory procedure used to observe and measure the response of rigid cellular plastics, under specified exposed or protected conditions, to a standardized heat and flame exposure. Measurements of flame travel, maximum temperature, and material damage are made and reported. Mechanical failures from cracking, spalling, loss of adhesion, and other fire response factors useful in screening for protective purposes may so be noted."

This method uses a scaled-down version of a large-scale corner test, ostensibly for the sake of economy. ASTM claims that this method shows a minimum 75 percent correlation with the Factory Mutual Corner Wall Test.

The standard describes construction of the test fixture, as well as the burner, fuel, and air supplies, gas flow metering, temperature measurement, ignition, temperature versus time control, and smoke measurements.

The results of the test to be reported include visual observation of flame travel distance from the test corner outward on the horizontal plane at the start of the test and every 15 seconds during the test; visual observation of smoke, sample fractures, ignition, and any other significant event; and smolding, afterglow, and other observances as the sample cools, such as damage to the surface and underlying foam. The test must be terminated at 20 minutes, or sooner if the flame becomes "uncontrollable." Anytime the test compartment has become engulfed in flames, no useful experimental data have been found.

Results are based on maximum flame travel and the time, in seconds, at which the maximum flame travel occurred; the maximum temperature and the time to reach maximum temperature; and a chronological description of visual events occurring during the test run in conjunction with a photographic record of the maximum physical damage done to the walls or to the walls and ceiling, including the square area burned.

D 4100 (ASTM)
Standard Test Method for Gravimetric Determination of Smoke Particulates from Combustion of Plastic Materials

This standard is under the jurisdiction of ASTM Committee D-20 on Plastics and is the direct responsibility of Subcommittee D 20.30 on Thermal Products. This is not a fire test but a method for determining the amount of smoke generated by burning plastic materials under specified conditions. Its stated purpose is to "be used to measure and describe the properties of materials, products or assemblies in response to heat and flame under controlled laboratory conditions" and it should not be used to describe or appraise the fire hazard or fire risk of materials, products or assemblies under actual fire conditions."

D 4205 (ASTM)
Standard Guide for Flammability and Combustibility of Rubber and Rubber-Like Materials

This standard is under the jurisdiction of ASTM Committee D-11 on Rubber and is the direct responsibility of Subcommittee D 11.17 on Flammability. Its stated purpose is to cover the "present state-of-the-art test methods for determining the flammability and combustion properties of rubber and rubber-like materials." It was developed to gather together all standard test methods for rubber and other rubber-like materials developed by different sources into one standard. It includes standard test methods developed and promulgated by ASTM, NFPA, ANSI, UL, trade associations, and governmental agencies.

The standard includes a compilation of 54 test methods and standards developed by the foregoing groups for the automotive industry, aviation, belting, electrical and power cables, floor covering, hoses, mattresses, and upholstered furniture; smoke tests; a toxicity test; and miscellaneous additional tests.

D 4549 (ASTM)
Standard Specification for Polystyrene Molding and Extrusion Materials (PS).

This standard is under the jurisdiction of ASTM Committee D-20 on Plastics and is the direct responsibility of Subcommittee D20.15 on Dynamic Mechanical Properties. Its stated purpose is to "be a means of calling out plastic materials used in the fabrication of end items or parts. Material selection should be made by those having expertise in the plastics field after careful consideration of the design and the performance required of the part, the environment to which it will be exposed, the fabrication process to be employed, the inherent properties of the material other than those covered by this specification, and the economics."

The specification covers both crystal and rubber-modified polystyrene intended to be injection molded or extruded, and the basic properties of these materials are included in a table within the standard. A system for setting up the specifications of a polystyrene material is given, incorporating values provided in tables in the standard. Material compositions must be uniform throughout the samples, and such samples must conform to the requirements as they are enumerated in the tables in the standard. Sampling, specimen preparation, conditioning, and the test methods to be used are all designated ASTM standards, guides, and practices and are prescribed in this standard. Quality Assurance Provisions for government/military procurement are also listed, with reference to MIL-STD-105.

D 4723 (ASTM)
Standard Index of and Descriptions of Textile Heat and Flammability Test Methods and Performance Specifications

This standard is under the jurisdiction of ASTM Committee D-13 on Textiles and is the direct responsibility of Subcommittee D 13.52 on Flammability. Its stated purpose is to provide "reference tables of test methods and performance specifications used in the United States of America and Canada for measuring and describing the properties of textiles and textile products or assemblies in response to heat and flame under controlled laboratory conditions."

D 4804 (ASTM)
Standard Test Methods for Determining the Flammability Characteristics of Nonrigid Solid Plastics

This standard is under the jurisdiction of ASTM Committee D-20 on Plastics and is the direct responsibility of Subcommittee D 20.30 on Thermal Properties. The stated purpose of the standard is to cover "small-scale laboratory procedures for determining the extinguishing characteristics (Test Method A) and the relative rate of burning (Test Method B) of solid plastic materials that, due to specimen thinness and nonrigidity, may distort or sag when using ASTM Test Method D 635."

Test Method A uses a flame applied to a specimen held in the vertical position, whereas Test Method B uses a larger flame applied to a specimen in the horizontal position. ASTM Test Methods D 635 and D 3801 are referenced in this standard.

The standard defines certain terms, gives a summary of the test methods and a statement of significance and use, and describes, for both Test Method A and Test Method B, apparatus, sampling procedures, preparation of test specimens, conditioning of the specimens, procedures used to run the tests, calculations used in arriving at the results, the content of the report, and a statement of precision and bias. The size of the test specimens, the size of the applied flame, and the angle at which the flame is applied differ between the two methods. A warning is given to consider the different result that will be attained with different thicknesses of specimens.

Test Method A will produce results that represent the flaming and glowing times, in seconds, for a material tested according to procedures. Test Method B will produce results that represent the rate of burning of the specimen, in centimeters per minute.

E 84 (ASTM)
Standard Test Method for Surface Burning Characteristics of Building Materials

This standard is under the jurisdiction of ASTM Committee E-5 on Fire Standards and is the direct responsibility of Subcommittee E 05.22 on Surface Burning. Its stated purpose is "to determine the relative burning behavior of the material by observing the flame spread along the specimen."

The test apparatus is a tunnel 25 feet long, about 18 inches wide, and about 12 inches deep measured from the bottom of the tunnel to the ledge of the inner walls on which the specimen is supported. The test specimen is 24 feet long by 20 inches wide (or at least 2 inches wider than the width of the tunnel below the ledge) and sits on a ledge about 7.5 inches above the gas ports. A piece of 16-gage uncoated sheet steel 14 inches long is placed on the ledge under and to the front of the test specimen at the upstream end of the tunnel. The inside of the test apparatus must be lined with insulating masonry material, and observation windows of specified composition, size, and placement must be built into one side of the tunnel. The top of the apparatus must be a removable, insulated, noncombustible structure and must completely cover the test chamber and samples. Specifications for the apparatus are available in the standard as published by ASTM.

One end of the tunnel contains natural gas (or methane) ports of specific dimensions, and the other end is vented. Air turbulence is created by positioning six fire bricks of a specified make at predetermined places along the tunnel, with air being moved by an induced draft system. The system is equipped with a photometer, a draft regulating device, thermocouples, and piping and valving for the natural gas or methane used. Methods are given in the standard for calibration of the device with red oak and reinforced cement board.

When the test is run, observations are made concerning the distance and time of maximum flame front travel. Methods for interpreting the results are given in the standard. Methods for supporting specimens that are not self-supporting are also listed.

Quoting ASTM, "This test method . . . is applicable to exposed surfaces, such as ceilings or walls, provided that the material or assembly of materials, by its own structural quality or the manner in which it is tested and intended for use, is capable of supporting itself in position or being supported during the test period. These tests are conducted with the material in the ceiling position."

The test procedure states that if the material is tested with supporting materials on the underside of the test specimen, the flame spread index may be lower than if the material were tested without such support. Under these conditions, the flame spread index obtained in this manner might not relate to indices obtained by testing the material alone.

The test procedure might not produce "valid" results if the test specimen melts, drips, or delaminates during the test, which would probably produce lower flame spreads. Also, the test procedure states that the test "may not be appropriate for obtaining comparative surface burning behavior of some cellular plastic materials."

ASTM E 84 is used primarily to measure surface flame spread, although it also measures smoke generation by relating light absorption by the smoke to values produced by red oak, as well as total heat release (not rate of heat release). Values for the flame spread (and a subsequent Flame Spread Index) of materials will be between 0, assigned to noncombustibles such as cement board and 100, which has been assigned to red oak. It is possible to get higher ratings for some other combustible materials.

A "Guide to Mounting Methods" has been added to ASTM E 84 as Appendix X1.

E 108 (ASTM)
Standard Methods of Fire Tests of Roof Coverings

This standard is under the jurisdiction of ASTM Committee E-5 on Fire Standards and is the direct responsibility of Subcommittee E 05.14 on Roofing. The stated purpose of this standard is to "cover the measurement of the relative fire characteristics of roof coverings under simulated fire originating outside the building. [It] is applicable to roof coverings intended for installation on either combustible or noncombustible decks, when applied as intended for use."

The tests measure "the surface spread of flame and the ability of the roof covering material or system to resist fire penetration from the exterior to the underside of a roof deck under the conditions of the exposures." They also are intended to determine "if the roof covering material will develop flying burning material, identified as flying brands, when subjected to a 12 mph wind during the simulated fire exposure tests." Also included is a test to determine whether the roofing material will be adversely affected by prolonged exposure to rain.

This standard includes four fire tests and a rain test. The fire tests are the (1) an intermittent flame exposure test, (2) a spread of flame test, (3) a burning brand test, and (4) a flying brand test. Each of these tests will be performed on test decks under three classes of fire test exposure, which are designated Class A, Class B, and Class C. These tests are applicable to roof coverings that are effective against severe, moderate, and light test exposure respectively, and they afford high, moderate, and light degrees of fire protection, respectively, to the roof decks. All three tests specify that the tested roof coverings should not slip from position and should not represent flying brand hazards.

The construction, storing, and conditioning of the test decks are specified for each type of test and each class of exposure. The application of the flame, the position of the test deck, the velocity of air moving over the test decks, and the required temperature are prescribed for each of the tests. Specific tests are also described for roof coverings over certain types of supports. Conditions of classification state how the test roof covering material shall behave during and after the test, and what the condition of the test deck shall be after each class of test—A, B, or C. Materials are observed and their behavior in relation to the different classification tests is recorded.

Because no full-scale fire tests for certain house siding material exist, this test has been modified to include the testing of such siding material. In the so-called Modified E 108 Test, a test deck is constructed, and siding is applied over the deck. This construction is then placed in a vertical position and a flame application similar to the regular E 108 test is applied for a specified time. Observations are made and reported relevant to the burning characteristics and flame spread of the siding material under test.

E 119 (ASTM)
Standard Methods of Fire Tests of Building Construction and Materials

This standard is more simply known as the Standard Fire Tests. It is under the jurisdiction of ASTM Committee E-5 on Fire Standards and under the direct responsibility of Subcommittee E 05.11 on Building Construction. These methods were prepared by Sectional Committee A2 on Fire Tests of Materials, under the joint sponsorship of NBS, the ANSI Fire Protection Group, and ASTM, functioning under the procedure of ANSI.

It has been recognized that "the fire resistive properties of materials and assembles [must] be measured and specified according to a common standard expressed in terms that are applicable alike to a wide variety of materials, situations, and conditions of exposure." E 119 tries to describe performance in terms of "the period of resistance to standard exposure elapsing before the first critical point in behavior is observed."

E 119 is really a collection of methods used to test different parts of a building. The standard is written to measure—in walls, partitions, or floor and roof assemblies—the transmission of heat, the transmission of hot gases through assembly sufficient to ignite cotton waste, and the load-carrying ability of the specimen being tested. For individual load-bearing assemblies it measures load-carrying ability under the test exposure for the end support conditions. The methods are also applicable to "other assemblies and structural units that constitute permanent integral parts of a finished building."

E 136 (ASTM)
Standard Method for Behavior of Materials in a Vertical Tube Furnace at 750°C.

This standard is under the jurisdiction of ASTM Committee E-5 on Fire Standards and is the direct responsibility of Subcommittee E 05.23 on Combustion. Its stated purpose is to cover "the determination under specified laboratory conditions of combustion characteristics of building materials. It is not intended to apply to laminated or coated materials." This method does not duplicate actual building fire conditions but is intended to "assist in indicating those materials which do not act to aid combustion or add appreciable heat to an ambient fire. Materials passing the test are permitted limited flaming and other indications of combustion."

The standard describes the vertical tube furnace, heated by electrical heating coils and having a transparent cover, thermocouples or other temperature measuring devices, the time intervals at which temperatures are read, and controlled air flow. The number and size of the test specimens are specified, and the test procedure is given. A material passes the test if at least three of the four test specimens meet specified conditions: (1) The temperature of both the surface and interior as recorded by the thermocouples must not rise more than 30°C (54°F) above the furnace temperature at the beginning of the test; (2) there is no flaming combustion after the first 30 seconds; and (3) when the weight loss of the specimen during the test is greater than 50 percent, the recorded temperature does not rise above the furnace air temperature at the beginning of the test and there is no flaming combustion of the specimen.

E 152 (ASTM)
Standard Methods of Fire Tests of Door Assemblies

This standard is under the jurisdiction of ASTM Committee E-5 on Fire Standards. Subcommittee E 05.12 has been established to investigate the precision and bias of these tests. The stated purpose of this standard is "that tests made in conformity with these test methods will develop data to enable regulatory bodies to determine the suitability of door assemblies for use in locations where fire resistance of a specified duration is required. These methods of fire test are applicable to door assemblies of various materials and types of construction, for use in wall openings to retard the passage of fire. These methods are intended to evaluate the ability of a door assembly to remain in an opening during a predetermined test exposure."

The standard specifies furnace temperatures, unexposed surface temperatures, construction and size of the test door assembly, the mounting procedure, the time of testing, the fire endurance test, and the hose stream test. The report must contain the materials and construction of the door and frame, the performance under the desired exposure period, the temperature measurements of the furnace and the unexposed side, any flaming on the unexposed surface of the door leaf during the first 20 minutes of the fire test, the amount of movement of any portion of the edges of the door, pressure measurements in the furnace, and all observations having a bearing on the performance of the test assembly.

The door will be considered acceptable if it remains in the opening under specified conditions listed in the standard.

E 162 (ASTM)
Standard Test Method for Surface Flammability of Materials Using a Radiant Energy Heat Source

This standard is under the jurisdiction of ASTM Committee E-5 on Fire Standards and is the responsibility of Subcommittee E 05.22 on Surface Burning. The stated purpose of this standard is to cover "the measurement of surface flammability of materials." It provides "a laboratory test procedure for measuring and comparing the surface flammability of materials when exposed to a prescribed level of radiant heat energy."

It is intended to be used for research and development purposes, as opposed to being used "as a basis of ratings for building code purposes." Its intended use is in establishing a method of measurement on materials whose surfaces may be exposed to fire.

The apparatus consists of a radiant panel made of a porous refractory material with an exposed surface of 12 by 18 inches (300 by 460 mm) heated by burning acetylene gas in an appropriate air mixture. It includes a blower, aspirator, filter, and pressure regulator; control valves; a specimen holder on a prescribed framework; a pilot burner; a stack, hood and timer; thermocouples; and a radiant pyrometer, a potentiometer, and a recorder.

The test specimen shall be 6 by 18 inches (150 by 460 mm) and a maximum of

1 inch (25 mm) thick. Thicker specimens may be tested if the holder is modified. Provisions are made to prepare finish material for which the intended application has not been specified. Finish materials may include sheet material opaque to infrared radiation, paints, other liquid films, fabrics, cellular elastomers and plastics, aluminum, foil–backed materials, tiles, and others.

At least four specimens of each sample type must be tested, after conditioning as prescribed in the standard.

The fuel–air mixture is started, ignited, and adjusted, and the recording potentiometer is activated. The pilot flame is ignited and moved into position. The timer is started when the specimen holder is placed into the framework, and the tester records the time of the arrival of flame at each of the 3-inch (76 mm) marks on the specimen holder or on the specimen itself. Simultaneously, observations must be recorded concerning the "flash" (if it occurs) and any dripping or other behavior of the specimen during the test. The test is over when the flame reaches the 15-inch (380 mm) mark or 15 minutes has elapsed, whichever occurs first, provided the maximum temperature of the stack thermocouple has been reached.

The flame index is then calculated in the manner prescribed in the standard. Records must include a complete description of the sample; the type of test specimens used (cores, molded parts, skin, or slabs) and whether or not the specimen was backed with aluminum foil or otherwise laminated; the number of specimens tested; the exposure time; whether the specimen was completely destroyed in, or exposed for, the full 15 minutes; the average flame spread index and range for each set of specimens; and any other characteristics of the specimen.

A calibration procedure for the apparatus is given in Annex A1 of the standard.

E 163 (ASTM)
Standard Methods of Fire Tests of Window Assemblies

This standard is under the jurisdiction of ASTM Committee E-5 on Fire Standards and is the responsibility of Subcommittee E 05.12 on Protection of Openings. The stated purpose of this standard is to "to evaluate the ability of a window or other light transmitting assembly to remain in an opening during a predetermined test exposure of 45 minute duration." The test involves exposing a window assembly to a controlled standard fire to and then applying a specified standard fire hose stream to the fire. The tests are designed to provide measurements of fire performance of those assemblies against which all tested assemblies may be compared.

The objective of the test is to expose window assemblies to a standard time–temperature curve produced by a specified furnace and read by thermocouples placed as prescribed. The test windows are to be representative of the design for which approval is sought, and the square area of test assembly and mounting instructions are given in the standard. Immediately following the fire endurance test, the exposed side of the assembly is exposed to a defined stream from a standard hose applied in the prescribed manner.

The final report must include at least: a description of the test assembly and the wall in which it is mounted; the temperature measurements of the furnace; all observations of behavior during the fire test and the application of the hose stream; a description of the test assembly after the test; any loosening or movement of any part of the test assembly; the condition of the hardware attached to the test assembly; and the ability to open the test assembly after the test. The requirements for passing the test are listed in the standard. They concern themselves with the ability of the test assembly to remain in the opening during the test, with minimal movement, separation, and glass breakage.

E 286 (ASTM)
Standard Test Method for Surface Flammability of Building Materials Using an 8 Foot (2.44 m) Tunnel Furnace

This standard is under the jurisdiction of ASTM Committeee E-5 on Fire Standards and is the responsibility of Subcommittee 05.22 on Surface Burning. Its stated purpose is to "provide comparative measurement of surface flame spread of building materials when exposed to thermal radiation and natural draft conditions." The test method "also includes a photoelectric measurement of the light attenuation produced by the smoke in the vertical furnace stack during the burning of the test specimen."

The end result of this test is the production of a flame spread index. This flame spread index will be different than the one produced by ASTM E 84 (see above). E 84 is used to determine the relative burning behavior of material, and uses natural gas (or methane) burners in a 25-foot tunnel, as compared with this test, which uses radiated heat from a heated metal plate and a small ignition natural gas flame at the lower end of the specimen and the 8-foot tunnel.

The apparatus consists of a combustion chamber 13.75 inches (349 mm) wide by 8 feet (2.44 m) long inside a gas-heated furnace about 10.5 feet long, a gas supply line, a clamp to hold the specimen, a main burner, a hood to collect combustion gases, a stack, and a photoelectric cell to measure smoke density. The construction materials for the test apparatus are listed in the standard.

The gas is adjusted to produce a specific flow relating to the 19 minutes (± 1 minute) required for the flames to travel the length of the standard red oak specimen. With the furnace preheated to a prescribed temperature range, the test specimen is laid on the angle iron frame, and the specimen cover is closed and clamped. The tester watches progress of the flames through the observation holes in the apparatus and notes the time as the flame passes each observation hole. Temperatures and smoke densities are also noted during this period.

The flame spread is expressed as in index-relative to the rate on red oak with an index of 100 and on asbestos board with an index of 0, with the calculation formula presented in the standard. The final report should include a precise description of the specimen, including composition (including any chemicals added in any way to the specimen), moisture content, and origin; how the test specimen was mounted; indices of flame spread, fuel contributed, and smoke (formulas also provided); and observations concerning the appearance of the flame, the char pattern, intumescence, or disintegration of the specimen. If any deviation from the standard method occurred, this must also be reported.

E 603 (ASTM)
Standard Guide for Room Fire Experiments

This standard is under the jurisdiction of ASTM Committee E-5 on Fire Standards. Its stated purpose is to cover "full-scale compartment fire experiments that are designed to evaluate the fire characteristics of materials, products, or systems under actual fire conditions. It is intended to serve as a guide for the design of the experiment and for the interpretation of its results." It attempts to establish "a basis for conducting full-scale experiments for the study of preflashover aspects of compartment fires."

Unlike many of the other standards discussed, this guide does attempt to correlate the results of standard test methods currently being used to evaluate the fire performance of materials with the actual behavior of the materials in a full-scale room fire. It is obvious that materials will behave differently in a full-scale room fire scenario than in a small-scale laboratory fire test. The behavior of a material in a room fire experiment (or in an actual, unexpected, accidental fire within a room) depends on many things, not the least of which are the total fire load in the room, the shape of the materials that make up that fire load, and their arrangement within the room.

This guide does not purport to define a standard room test. Instead, it suggests that specifications be set for most of the variables that exist in a room fire. That is, it suggests specifications for room size, room shape, ventilation, a description of the specimens to be placed within the room, the source of ignition, the instruments required to make the necessary measurements, and safety precautions to be followed when running the actual test.

The guide suggests three possible criteria that can be useful in evaluating the fire performance of materials in the room experiment: the time to flashover of the room, the ignition source required to produce this flashover, and the optical density and carbon monoxide level of the smoke produced in the experiment. The guide also sets out the experimenter's choices in designing the experiment. These are:

1. The size and shape of the test room (suggesting a "standard room")
2. The properties of "compartment linings" that will influence heat buildup
3. The effects of ventilation
4. Recommendations for (and definitions of) the specimens to be included in the test
5. The location of the specimens in the room
6. Type, size, and location of various ignition sources (including comments about each type)
7. The instrumentation to be used for measuring:
    a. The rate of heat release
    b. Heat flux
    c. Temperature
    d. Air velocity
    e. Fire propagation
    f. Smoke
    g. Gas concentration
    h. The potential toxicity of combustion products
8. Selection of instruments and their location during the test
9. Safety precautions prior to, during, and after the experiment
10. Observations and data gathering
11. Analysis and reporting of results

The guide cautions that the experimenter is usually interested in designing the most stringent test possible, and in cases where the most stringent test possible has not been selected, the experimenter should explain why.

E 648 (ASTM)
Standard Test Method for Critical Radiant Flux of Floor-Covering Systems Using a Radiant Heat Energy Source

This standard is under the jurisdiction of ASTM Committee E-5 on Fire Standards and is the responsibility of Subcommittee E 05.22 on Surface Burning. The stated purpose of this standard is to describe "a procedure for measuring the critical radiant flux of horizontally mounted floor-covering systems exposed to a flaming ignition source in a graded radiant heat energy environment, in a test chamber . . . The test method was developed to simulate an important fire exposure component of fires that may develop in corridors or exitways of buildings and is not intended for routine use in estimating flame spread behavior of floor covering in building areas other than corridors or exitways." The test has been devised to simulate the fire behavior of floor coverings observed in full-scale corridor experiments.

The standard describes the construction of a flooring radiant panel tester apparatus, and lists safety precautions to be followed in conducting the test (including the warning of the possibility of a gas–air explosion in the test chamber). It also covers selection of test specimens, standardization of the radiant heat flux profile, conditioning of the test specimens, the test procedure itself, calculation of results, and the final report. The report should include a description of the specimen flooring system, procedure used in assembling the system, the number of specimens tested, the average critical radiant flux (including the standard deviation and coefficient of variation), and observations of the burning characteristics of the specimens during the test exposure.

This standard makes the basic assumption that the critical radiant flux is an important measurement of the sensitivity to flame spread of floor-covering systems in a building corridor. It is intended to provide a method of rank ordering floor-covering systems in their fire spread behavior.

An annex to the 1988 edition of the standard includes instructions for calibrating the radiation instruments used in the test procedure, and an appendix includes a commentary on critical radiant flux and a guide to methods for mounting specimens.

E 662 (ASTM)
Standard Test Method for Specific Optical Density of Smoke Generated by Solid Materials

This standard is under the jurisdiction of ASTM Committee E-5 on Fire Standards and is the responsibility of Subcommittee E 05.21 on Smoke and Combustion Products. Its stated purpose of this standard is to cover "determination of the specific optical density of smoke generated by solid materials and assemblies mounted in the vertical position in thicknesses up to and including 1 inch (25 mm)."

The test is complex and so sensitive to very small variations that great care must be taken to follow the procedure and sample sizes exactly. The test method is sensitive to any unusual fire behavior of the specimen, as outlined in the standard. It is also very sensitive to the geometry of the sample; to its thickness, which is specified; and to the surface orientation, weight, and composition of the specimens. It is not possible to determine mathematically the specific optical density of a specimen of one thickness, when the specific optical density is known of another specimen of exactly the same material, but of a different thickness.

All specimens tested in one test must be the same as far as the above conditions are concerned. It must be placed in the specimen holder the same way, with the same amount of surface exposed.

The test specimens are exposed to an electrically heated radiant energy source that is mounted within an insulated ceramic tube and positioned to produce a specific irradiance level. They are exposed to both flaming and nonflaming conditions within a closed chamber. Varying light transmissions are measured as the smoke builds up, and these measurements are used to calculate specific optical density of the smoke generated. Maximum rate of smoke accumulation, time to a fixed optical density level, and a smoke obscuration index are described in the appendix to the standard.

The standard describes the rather complex apparatus, including the test chamber, the radiant heat furnace, the photometric system, the internal framework for holding the specimen, the burner, the light source, and the instrumentation used for measurements. The standard also describes the size, orientation, and conditioning of the specimens, the specimen assembly and the mounting procedure, and, of course, the procedure for running the tests.

Each report must completely describe the manufacturer, composition, conditioning, shape, dimensions and dimensions of specimen; any special preparation, mounting, and specimen orientation; and number of specimens used, in addition to describing the test itself. All observations concerning behavior of the specimen during the test, and of smoke generation, including color and nature of any particulates, must be reported. A tabulation of curve of time versus either percent transmission or specific optical density must be shown, and test results may be rounded to two significant figures. Maximum specific optical density and maximum specific optical density (corrected) must both be reported.

In addition to appendixes at the end of the standard, annexes covering calibration of test equipment and construction details are provided.

E 800 (ASTM)
Standard Guide for Measurement of Gases Present or Generated During Fires

This standard is under the jurisdiction of ASTM Committee E-5 on Fire Standards and is the responsibility of Subcomittee E 05.21 on Smoke and Combustion Products. Its stated purpose is to describe techniques to determine the concentration of a specific gas in the total sample taken in the test. The guide describes "analytical methods for the measurement of carbon monoxide, carbon dioxide, oxygen, nitrogen oxides, sulfur oxides, carbonyl sulfide, hydrogen halides, hydrogen cyanide, aldehydes, and hydrocarbons." It does not contain enough information for an experimenter to use a particular procedure but provides enough references so that researchers may select the alternative procedure with which they feel most comfortable.

The guide spends considerable time and space on sampling because "more errors in analysis result from poor and incorrect sampling than from any other part of the measurement process." It covers such individual sampling topics as making sure a representative sample is truly representative, collection and transport of the sample, sampling frequency, sampling sites, sampling probes, sampling volume and rate, sample pretreatment, sample maintenance, the test systems themselves, and the reactivity of the fire gases themselves.

The guide suggests different analytical procedures for different gases. Not all of the methods listed may work for every gas, but the chemist or other researcher skilled in analytical techniques will know which tests are the best for a particular gas, based on the chemistry of that gas. For example, carbon monoxide, carbon dioxide, oxygen, and nitrogen may be analyzed by gas chromatography, infrared spectrophotometry, oxidation in an electrolytic cell, galvanic cells, polarographic analyzers, or paramagnetic analyzers. (*Note:* Not all of these tests may be used to analyze all the gases in this group.)

For the quantitative detection of the halogen halides (hydrogen fluoride, hydrogen chloride, hydrogen bromide, and hydrogen iodide), a difficult group of gases to analyze because of their reactivity, the guide recommends proton detection devices, anion detection devices, and hydrogen halide detection devices (such as gas filter-correlation analysis).

For the detection of hydrogen cyanide the report suggests electrochemical methods, infrared and colorimetric methods, and gas chromatographic methods and lists several of each type.

The nitrogen oxides, sulfur oxides, and carbonyl sulfide are treated as a single group for analysis. For the detection of the nitrogen oxides (nitric oxide and nitrogen dioxide) the guide suggests chemiluminesence and ion chromatography. For the sulfur oxides (sulfur dioxide and sulfur trioxide), nondispersive infrared (NDIR) spectrometry is suggested, and for carbonyl sulfide, a gas chromatograph mass spectrometer is recommended.

It is suggested that the tester measure "total hydrocarbons" instead of trying to identify every organic compound evolved during pyrolysis and combustion. For this analysis, a flame ionization detector used alone or in conjunction with a catalytic converter or gas chromatograph is recommended. An additional method suggested is based on nondispersive infrared techniques. For analysis of aldehydes, gas chromatographic and colorimetric approaches are suggested.

The presence of water vapor may be analyzed by infrared techniques or other techniques that depend on the physical properties of water.

Three annexes at the end of the guide cover calibration, gas detector tubes, and a summary table of the test methods. One hundred references are also listed to help researchers select tests.

E 814 (ASTM)
Standard Test Method for Fire Tests of Through-Penetration Fire Stops

This standard is under the jurisdiction of ASTM Committee E-5 on Fire Standards and is the responsibility of Subcommittee E 05.11 on Building Construction. Its stated purpose is to "develop data to assist others in determining the suitability of the fire stops for use where fire resistance is required." Fire stops are materials used to fill openings in walls through which penetrating items such as cables, conduits, ductwork, pipes, and trays pass. The walls must be evaluated as fire resistive as described in Test Methods E 119.

The standard describes the method to determine the performance of fire stops in response to a standard time–temperature fire test and a hose stream test. The hose stream test may be carried out on a fire stop identical to the one used in the fire test, or, at the wish of the sponsor and the concurrence of the testing body, the hose stream test may be carried out on the specimen subjected to the fire test, immediately following such fire test.

The standard sets out the methods to be used in the tests and the rating criteria for pass/fail. Both an "F" rating and a "T" rating are given, after both the fire test and the hose stream test. The ratings are based on the ability of the fire stop to resist the passage of flame and its ability to keep the increase in temperature of any thermocouple on the unexposed side at less than 325°F above its initial temperature. Also, during the hose stream test, the fire stop shall not develop any opening through which water from the stream might pass. The standard also sets out the pertinent items that must be contained in any report of the test.

The standard also contains annexes containing a standard temperature–time curve for control of fire tests, the requirements for thermocouples, and the differential pressure selection. An appendix contains an extended discussion of several items in the test method.

E 906 (ASTM)
Standard Test Method for Heat and Visible Smoke Release for Materials and Products

This standard is under the jurisdiction of ASTM Committee E-5 on Fire Standards and is the responsibility of Subcommittee E 05.21 on Smoke and Combustion. Its stated purpose is to "determine the release rates of heat and visible smoke from materials and products when exposed to different levels of radiant heat using the test apparatus, specimen configurations, and procedures described in this test method."

The test method provides for exposure of a specimen to radiant thermal energy both with and without a pilot flame. Smoke and heat release are measured continuously from the time the specimen is put into the controlled exposure chamber, through ignition, progressive flame involvement to the end of the test. The external heat flux may be varied from 0 to 100 kilowatts per square meter (kW/m^2). The specimen may be tested either vertically or horizontally, and nonpiloted, piloted, or point (impingement) ignition may be used. Heat and visible smoke release rates are calculated by monitoring the changes in temperature and optical density of the gas exiting the chamber. The test allows materials and their behavior to be examined under different heat fluxes.

Several cautions are given. Heat release values depend on the mode of ignition. Release rates depend on several factors, some of which may not be controlled. Heat release values obtained by this test are for the specimens tested, and there is no straight-line relationship between specimens of the same composition but different sizes. The method is limited to the specimen sizes given in the standard. No general relationships can be drawn between horizontal and vertical values.

The standard describes the apparatus to be used, calibration of the equipment, sample preparation, the procedure for running the test, and calculation of the heat release rate (RHR) and the smoke release rate (SRR). The report shall include a description of the specimen, including its orientation and mounting; the radiant heat flux in kW/m^2; the type of ignition; all data giving the release rates as a function of time; and a description of the behavior of the specimen during the test.

Appendices to E 906 include information concerning release rate tests, release rate calibration, flux distribution, a radiation reflector for horizontal specimens, and a commentary.

E 970 (ASTM)
Standard Test Method for Critical Radiant Flux of Exposed Attic Floor Insulation Using a Radiant Heat Energy

This standard is under the jurisdiction of ASTM Committee E-5 on Fire Standards and is the responsibility of Subcommittee E 05.22 on Surface Burning. The stated purpose of this standard is to describe "a procedure for measuring the critical radiant flux [at the point at which the flame advances the farthest] of exposed attic floor insulation subjected to a flaming ignition source in a graded radiant heat energy environment in a test chamber." The heat energy simulates the amount of heat received by the insulation by radiation from a roof or other upper surface that is heated by the sun or by flames of a fire in the attic.

The critical radiant flux is defined as the distance to the farthest measured advance of flaming and converted to kilowatts per square meter (kW/m^2) from a previously prepared radiant flux profile graph. It is assumed that the critical radiant flux is a valid measure of the surface burning characteristics of exposed attic floor insulation. The test is designed to simulate behavior observed in attic experiments.

The standard describes the apparatus to be used in the test, the hazards faced in running the test, sampling, test specimens, radiant heat energy flux profile standardization, conditioning of the specimens, the procedures used in running the test, and the calculation of results. The report must contain a description of the specimens and the procedures used to prepare them, the density and critical radiant flux of each specimen tested, and the average critical radiant flux with its statistics. The standard contains an annex on the procedure for calibration of the radiation instrumentation, and an appendix with a commentary on critical radiant flux test for exposed attic floor insulation.

E 1321 (ASTM)
Standard Test Method for Determining Material Ignition and Flame Spread Properties

This standard is under the jurisdiction of ASTM Committee E-5 on Fire Standards and is the responsibility of Subcommittee E 05.22 on Surface Burning. The stated purpose of this fire test response standard is to determine the "material properties related to piloted ignition of a vertically oriented sample under a constant and uniform heat flux and to lateral flame spread on a vertical surface due to an externally applied radiant-heat flux."

The test is designed to provide data that can be used to predict the time to ignition and the velocity of lateral flame spread on a vertical surface of tested materials under a specified external heat flux.

The standard references certain ASTM documents, defines terminology used, presents a summary of the test method, gives a statement of significance and use, describes the apparatus used, and describes hazards of the test, the size and number of test specimens, the calibration of the apparatus, conditioning of the specimens, the procedures to be used in running the test, calculation of results, content of the report, and a statement of precision and bias.

An annex to the standard provides instruction on the assembly of the test apparatus, and the appendices provide information on ignition theory and flame spread theory. A commentary gives historical aspects and the scope of the test method.

F 777 (ASTM)
Standard Method for Resistance of Electrical Wire Insulation Materials to Flame at 60°

This standard is under the jurisdiction of ASTM Committee F-7 on Aerospace and Aircraft and is the direct responsibility of Subcommittee F 07.06 on Flammability. Its stated purpose is to determine "the resistance of insulated wire and cable to sustaining and propagating flame."

Described in the standard are the draft-free enclosure in which the specimen will be burned, a Tirrill or bunsen burner fueled with 99 percent pure methane gas, a weight to be attached to the specimen to keep it taut, a holder with clamp to hold the specimen in the proper position, a scale, one or more timers, and a conditioning chamber, if necessary, to condition the samples.

Specimens 760 mm long are positioned in the holder, and a flame of specified parameter is brought into contact with it in a prescribed manner at the specified point on the sample. The timer is started when the flame contacts the specimen and is kept in contact for a defined time. The flame is removed after the defined time and the timer is allowed to run until the flame of the sample ceases. If any flaming, dripping particles continue to flame after striking the floor of the enclosure, the flame duration time must be recorded. When the flame has ceased and the timer stopped, the time (in seconds to the nearest 0.2 seconds), less the defined flame application time, is recorded as the time of afterburn.

When the sample has cooled, the burn length (in inches) must be recorded. Burn length is defined as "the greatest distance from the original location of the test mark to the farthest point showing evidence of damage due to that area's combustion including areas of partial combustion, charring or embrittlement but not including areas sooted, stained, warped, or discolored nor areas where material has shrunk or melted away from the heat." Also, the burn time of flaming, dripping particles, if any; the defined or specified flame application time (in seconds); the number of specimens tested; and the time of afterburn (in seconds) must all be recorded and reported.

FTMS 191A, Method 5903.1 (Federal Test Method Standard)

This test is for the flame resistance of cloth in a vertical position. It was developed by the U.S. Army Natick Research and Development Center to measure the char length, afterflame time, and afterglow time of sample specimens. It is used for all fabrics but particularly for flame resistant fabrics.

FTMS 406, Method 2021 (Federal Test Method Standard)

This method has been replaced by ASTM D 635, Standard Test Method for Rate of Burning and/or Extent and Time of Burning of Self-Supporting Plastics in a Horizontal Position. ASTM D 635 is approved for use by agencies of the Department of Defense and is approved for listing in the Department of Defense Index of Specifications and Standards.

IEC 707 (International Electrotechnical Commission)

Methods of testing for the determination of the flammability of solid electrical insulating materials when exposed to an igniting source.

IEC 829 (International Electrotechnical Commission)

Methods of testing for the determination of the flammability of solid electrical insulating materials when exposed to electrically heated wire sources.

ISO 871 (International Standards Organization)

Plastics—ignition temperature of evolution of flammable gases from a small sample of pulverized material.

ISO 1182 (International Standards Organization)

Noncombustibility test for building materials.

ISO 1326 (International Standards Organization)

This test is the same as ASTM D 635, Standard Test Method for Rate of Burning and/or Extent and Time of Burning of Self-Supporting Plastics in a Horizontal Position.

ISO 5657 (International Standards Organization) Fire Tests, Reaction to Fire, Ignitability of Building Products

Quantitative information from this test includes ignition temperatures for a variety of radiation exposure levels. Extrapolation may provide values for the minimum level of impressed flux to cause ignition.

Motor Vehicle Safety Standard 302 (U.S. Department of Transportation)

This standard prescribes the tests to be run to determine the flammability of materials used in the interior of passenger cars, multipurpose passenger vehicles, and trucks and buses. It measures the horizontal burn rate of sample specimens 4 inches by 14 inches, using a 1.5-inch natural gas flame. The burn rate must be not more than 4.0 inches per minute or the material should self-extinguish before burning 2.0 inches past the start of the timing zone.

NFL Full-Scale Test (National Fire Laboratory of Canada) National Fire Laboratory Full-Scale Test

This test is under the jurisdiction of the National Research Council of Canada (NRCC). Its purpose is to burn a full-size sided wall to "develop meaningful data to respond to perceived concerns of code writers, building officials, and others, and to justify needed changes in the building codes."

A wall assembly is described, including the placement of thermocouples. It is mounted on the front wall of the burn building. Propane burners are used to simulate a fire exposure equivalent to 1485 pounds of 2-inch by 4-inch wood cribs within the rooms, and the fire is allowed to propagate out through a window of specified size and placement. The burners are ignited, and the total time of the fire includes "about 5 minutes of development, 15 minutes of high fire exposure, and 5 minutes decay." Temperature and radiant heat flux measurements are taken in the room of fire origin, at the window, and at several heights on the test wall. Observations are to include the progress of the fire, the highest extent of flaming on the wall, the appearance of the studs and gypsum sheathing behind the siding being tested, and any existing burn patterns.

The test can be used to demonstrate that rigid PVC siding will burn as long as there is a supporting flame, but that it will not propagate the fire horizontally.

NFPA 701 (National Fire Protection Association) Standard Methods of Fire Tests for Flame-Resistant Textiles and Films

This test covers specifications for curtains and drapes, films, other textiles, tents, tarpaulins, and other protective coverings. It includes small- and large-scale tests and specifies the maximum burning time. Observations must be made concerning afterflame time, char length, and flaming melt drip.

NFPA 702 (National Fire Protection Association)
Standard for Flammability of Wearing Apparel

This test is similar to 16 CFR 1610. It covers clothing except for hats, gloves, footwear, and interlinings. The test sets up criteria that include the minimum time to ignition for different classes of materials. It measures flame spread and ease of ignition.

NFPA 1971 (National Fire Protection Association)
Standard for Protective Clothing for
Structural Firefighting

This standard covers all fabrics used in the manufacture of protective clothing for firefighters. It measures char length and afterflame time, with a maximum of 4 inches in char length and 2 seconds in afterflame time. It uses FTMS 191 Method 5903.

UL 94 (Underwriters Laboratories, Inc.)
Standard for Tests for Flammability of Plastic
Materials for Parts in Devices and Appliances

This standard is under the jurisdiction of Underwriters Laboratories, Inc. Its stated purpose is to cover "tests for the flammability of plastic materials used for parts in devices and appliances. They are intended to serve as a preliminary indication of [the materials'] acceptability with respect to flammability for a particular application."

The apparatus described includes a test chamber or laboratory hood free of drafts and a bunsen or Tirrill burner fueled by regulated technical grade methane gas, natural gas of a specific energy density, propane, butane, or acetylene. Although a choice may be made among fuels, the methane is preferred and shall be used to settle a dispute. Also described is a wire gauze, a ring stand with clamps, a stopwatch or other timer, a metal support fixture, and a conditioning chamber or room. The sample specimens are described as to size, markings, and position in the test chamber. The composition of the samples in regard to color, melt flow and reinforcement is also considered, with samples of the same plastic but differing in these characteristics, also being tested.

Horizontal Burning Test

94HB Classification. The flame of the burner is adjusted to specifications and applied to the sample in a prescribed manner. Instructions are given in the event the specimen burns to a specific mark before the flame application has reached 30 seconds. Instructions are also given in the event the specimen continues to burn after the removal of the flame, with the time to burn from one mark to the next determined and the rate of burning calculated.

A material shall be classified 94HB if (A) it does "not have a burning rate exceeding 1.5 inches (38.1 mm) per minute over a 3.0 inch (76.2 mm) span for specimens having a thickness of 0.120–0.500 inches (3.05–12.7 mm); *or* (B) if it does not "have a burning rate exceeding 3.0 inches (76.2 mm) per minute over a 3.0 inch span for specimens having a thickness less than 0.120 inches (3.05 mm)"; *or* (C) if it ceases to burn before the 4.0 inch (102 mm) reference mark."

Vertical Burning Tests

Apparatus and specimen composition, size, thickness and position in the test are described, including the adjustment and multiple applications of the test flame.

94V-0 Classification. Under this test, a material shall be classified 94V-0 if it does (A) "not have any specimens that burn with flaming combustion for more than 10 seconds after either application of the test flame"; *and* (B) does "not have a total flaming combustion time exceeding 50 seconds for the 10 flame applications for each set of 5 specimens"; *and* (C) does "not have any specimens that burn with a flaming or glowing combustion up to the holding clamp"; *and* (D) does "not have any specimens that drip flaming particles that ignite dry surgical cotton located 12 inches (305 mm) below the test specimen; *and* (E) does "not have any specimens with glowing combustion that persists for more than 30 seconds after the second removal of the test flame."

94V-1 Classification. A material shall be classified 94V-1 if it (A) does "not have any specimens that burn with flaming combustion for more than 30 seconds after either application of the test flame;" *and* (B) does "not have a total flaming combustion time of 250 seconds for the 10 flame applications for each set of 5 specimens;" *and* (C) does "not have any specimens that burn with flaming or glowing combustion up to the holding clamp"; *and* (D) does not have any specimens that drip flaming particles that ignite the dry surgical cotton located 12 inches (305 mm) below the test specimen"; and (E) does "not have any specimens with glowing combustion that persists for more than 60 seconds after the second removal of the test flame."

94V-2 Classification. For a material to be classified 94V-2, it must meet all the specifications of a material classified as 94V-1 with the exception that section D of the specifications says that the material should "be permitted to have specimens that drip flaming particles that ignite the dry absorbent surgical cotton located 12 inches (305 mm) below the test specimen."

A retest of specimens is described in the standard in the event that one (and only one) of the specimens fails the required test. Where the test requires a minimum total of flaming time, a retest is allowed if the total number of seconds of flaming time is no more than 5 seconds more than prescribed.

Vertical Burning Tests for Thin Materials

These tests are designed to be performed on samples that "[owing] to their thinness, distort, shrink, or are consumed up to the holding clamp when tested using the tests for 94V-0, 94V-1, and 94V-2." The apparatus, specimen size and flexibility, thickness, and position in the test are described, including adjustment of the flame and multiple applications of the test flame. The method of behavior during the test is discussed for materials to be classified 94VTM-0, 94VTM-1, and 94VTM-2, with the differences being similar to the vertical burning tests for materials that do not shrink or distort.

Flame Spread Index

This standard calls for the flame spread index of a material to be determined in accordance with ASTM E 162, Standard Test Method for Surface Flammability of Materials Using a Radiant Heat Source.

UL 723 (Underwriters Laboratories, Inc.)

Underwriter's Laboratory Standard UL 723 is essentially the same test as ASTM E 84, Test Method for Surface Burning Characteristics of Building Materials. It has been designated as the Steiner Tunnel Test in honor of Albert J. Steiner, who spent a great deal of time developing this and other fire tests.

9

Toxicity Testing

Testing for the toxicity of combustion or decomposition products of plastics is an issue that is very highly charged emotionally, even more so than the fire testing controversy. Several different researchers have developed different tests and test methods, with each researcher claiming his or her results to be the most accurate. Researchers have each selected their own test parameters, equipment, and procedures and have interpreted their own results.

Some tests use animals moving unrestrained within the test chamber, claiming that this freedom of movement is more natural, since in the "wild" they would be moving freely, perhaps even frantically during exposure to a fire. Other tests have the animals restrained, exposing only the head, while some expose the entire body in the restrained mode. The rationale here is that if the animals were unrestrained, they would huddle together in fear and perhaps breathe air filtered through fur, thus removing the toxicant or smoke particles. This, it is reasoned, gives the test animal an extra chance of survival.

Some tests use no animals at all. Researchers have developed a method using a fractional effective dose (FED) whereby calculations are made determining the amount of toxic combustion product that would have killed the animals had they been present. When more than one toxic combustion product is present, fractional doses are added together to determine whether or not the sum total will cause lethality. This is the N-gas model, and when the fractional effective dose sums up to 1.0, the animal is presumed to have died from the lethal effects.

Some tests use moving air at a constant temperature, passing it over the test sample and then into the animal chamber. Other tests use air whose temperature is constantly increased, and some tests, while controlling the temperature of the air, dilute it by 50 percent with "outside" air. Still other tests expose the sample specimens to direct flame whereas others use radiant heat or some other heat source and then move the combustion or decomposition products into the animal chamber.

Time of exposure to the products of combustion or decomposition varies from one test to another, as does the post-test observation period, if the test includes one.

Reporting and interpretation of results varies widely among tests. Criteria range from the concentration of gases required to bring about incapacitation (measured in different manners in different tests) of half the test animals, to the concentration of gases required to kill half the test animals. Some researchers are interested in determining disorientation rather than total incapacitation. Disorientation may be recorded as time to intoxication as evidenced by staggering or spinning around, or in another test, the determination may be based on whether the animals can complete certain tasks. Some tests do not measure the amount of gases required to bring about an undesirable circumstance but instead measure how much of a particular material must burn or decompose to bring about the desired measurable endpoint.

With such a confusing array of endpoints and the subsequent measurement of those endpoints, is it any wonder that there is little understanding or agreement about the exact toxic combustion or decomposition products produced in a fire? It is even more difficult to determine what the results of these tests mean to humans.

It is relatively easy to determine the final combustion products of a particular polymer, once its molecular structure and composition are known. That is, it is relatively easy using the same test, the same samples (in the same configuration and weight), and the same apparatus under the same test conditions—or at least it should be. It is true that all organic materials, not just plastics, produce different combustion and decomposition products under different conditions. But the only conditions for testing that will make *any* sense are those conditions in which the plastic (or any other organic material) finds itself in a real-world, unplanned, unexpected, and uncontrolled fire. In other words, the researchers working on identifying combustion and decomposition products of plastics have the same problems as researchers who are working on studying on how plastics burn.

Once the test sample has been determined to be standard (according to established scientific methods), and the apparatus, the procedures, and the conditions have also been standardized, one can burn the test specimen according to those set procedures and determine the combustion products of the sample. The really big question must then be posed: "What does all this testing have to do with what really happens in a real, uncontrolled, accidental, and unexpected fire in the real world?" Certainly, the test animals will have been killed by the combustion or decomposition products of the specimen example, whether the sample was plastic or not (so long as it is organic in composition). *It just doesn't matter of what material the sample is made,* because there *will* be a toxic combustion product that *will* kill the animals.

The researchers counter this statement by saying that the combustion or decomposition products of one material are more toxic than those of another material, and that by using their tests they can rank the toxicity of those combustion and degradation products. They say they can do that by measuring how fast the animals are killed or otherwise affected by determining how much of a particular material is required in the test to produce deadly results.

Again the question must be asked: "How are these tests relevant to what really happens in a real uncontrolled and unexpected fire in the real world?" The researchers again answer that by saying that by controlling the materials used in the construction of occupancies (by sharing their data on which materials are "deadly" with architects and engineers) they can control the number of fire deaths in those occupancies. The assumption would have to be made that they can also control the exact dimensions of each room and the furnishings and personal belongings present in each room. They also assume that they can control method of ignition, the length of time the fire smolders, the assurance that no one will discover the fire and no warning will be given, the ventilation system, the methods of escape, and indeed, whether anyone is present when the fire begins. This is a tremendous amount of assuming.

The fallacy is, of course, that there has never been, nor will there ever be, *in the real world,* an unexpected, uncontrolled, accidental fire of exactly the same nature and composition as produced by a researcher in a laboratory, or in any other test fire, small, medium, or large in scale.

There may seem to be one set of circumstances—fuel, air flow, conditions, and situations—that researchers *can* duplicate, because fires that occur in this situation usually happen in a compartment with a contained fire load that is more or less the same most of the time (even in this case, it is not *always* the same). The apparent exception is fires that occur in airplanes, particularly commercial airliners. Each type of commercial aircraft is very similar to another aircraft of the same model. The passenger compartment is nearly always the same (it may vary from one airline's favorite configuration to that of another airline). The furnishings are nearly always the same, given some airline-to-airline variation. The load of passengers and their belongings is the major variable, and this changes with every flight. So, even when one would expect all the parameters to be the same, there are many differences. Consider the many variations on a fire scenario in an airplane (including midair collisions and crashes), and even the exception seems to follow the rule that no two fires are alike.

In more precise language, until researchers can duplicate *exactly* all the fuel (including size, shape, configuration, and so on) and *all* conditions (humidity, atmospheric pressure, temperature, wind movement, smoke and heat transport, and so on) of a real-life fire, they will *not* be able to assess differences in the toxicity of combustion or decomposition products—*except* for the one toxic combustion product that is predictably present whenever organic materials burn. This one constantly present, deadly combustion product, which is present in lethal quantities in every uncontrolled organic material fire in the real world, is carbon monoxide. Over and over again, carbon monoxide is the killer when fire victims die from breathing combustion gases. Yet even though carbon monoxide is always present in lethal quantities in every uncontrolled (and in most controlled) fire in the real world, not everyone who is exposed to it is killed or harmed by it.

As reiterated elsewhere in this book, I recognize that researchers cannot duplicate real-world, uncontrolled fires. I also recognize that since researchers cannot hope to duplicate those fires, they have to make assumptions, assign constant values to unknowns, and include in (or exclude from) their calculations occurrences over which they have no control. After all these adjustments, the actual experiment necessarily becomes so contrived that it has no relationship to what actually happens. But it is on the basis of these contrived experiments that researchers want their findings to be accepted!

Certain facts remain constant. Those plastics with the same elements in their molecules that are present in the cellulose that makes up the majority of the composition of wood produce the same final combustion products as wood—namely, carbon, carbon monoxide, carbon dioxide, and water. It is true that some of the intermediate combustion products of plastics may be somewhat different than those of wood, but they are either mostly consumed in the fire or are carried away by the thermal column.

Those plastics that contain only carbon and hydrogen in their molecules evolve the same final combustion products as wood, as do other hydrocarbon solids, liquids, and gases. Again, the intermediate combustion products may be very slightly different from those of wood.

Those plastics containing a halogen in their molecule have the added final combustion product hydrogen chloride or hydrogen fluoride, depending on which halogen (chlorine of fluorine) is part of the molecule. These gases are upper respiratory irritants, not mysterious "supertoxicants." If a human being (or some other animal) breathes enough hydrogen chloride or hydrogen fluoride, he or she will die. However, the victims of toxic combustion products in fires where hydrogen chloride or hydrogen fluoride has been produced were found to have died of carbon monoxide poisoning. A correlation between the levels of carboxyhemoglobin in the blood and hydrogen chloride or hydrogen fluoride in the atmosphere would suggest that the presence of these irritant gases causes a person to die of lower levels of carboxyhemoglobin. This substance, you will recall, is the compound in the red blood cell caused by the combination of carbon monoxide with hemoglobin. It is this combination that renders the red blood cell incapable of "picking up" oxygen in the lungs for distribution to the rest of the body. No such correlation has been shown. There *have* been studies, however, that show how a fire victim's age, physical condition, and use of alcohol or other drugs can influence what level of carboxyhemoglobin causes death.

And then there are the plastics that contain nitrogen in their molecules. Some of these, when burning, liberate hydrogen cyanide. There is no argument about the toxicity of hydrogen cyanide. However, this gas, like carbon monoxide, is flammable, and the vast majority of it is consumed in the fire itself. Victims of fires in which both "natural" and synthetic materials burned and have evolved hydrogen cyanide were found to have died from carbon monoxide poisoning. Some of the fire victims did have hydrogen cyanide in their lungs, but they died from elevated carboxyhemoglobin levels in their blood. Researchers are currently trying to establish a link between hydrogen cyanide and elevated carboxyhemoglobin levels of fire victims.

Regarding decomposition products that evolve from plastics at temperatures below their ignition temperatures, not all plastics behave in the same manner. Some plastics, such as polyurethane and polymethyl methacrylate, depolymerize at high temperatures, but below the ignition temperature of the plastic. Depolymerization is the reverse of polymerization, and the reaction yields the monomers used to make the polymer. In the case of depolymerization, the decomposition products are known, and they are different from the combustion products of those polymers.

Other polymers, like polyvinyl chloride, liberate decomposition products similar to their combustion products, assuming oxygen is present. For this total decomposition to occur, temperatures must be reached that would make human life untenable. Therefore, anyone in an environment hot enough to decompose these plastics to the point that large amounts of the decomposition products are present will not be able to survive the elevated temperature, the decomposition products of the polymer notwithstanding. If the thermal degradation is happening behind a wall or in an adjoining room (or even one far removed), and the products of decomposition are entering the room through openings or a ventilation system, the irritating properties of hydrogen chloride will cause a person (even one asleep) to hurriedly leave for fresh air. Any other decomposition products will resemble the decomposition products of wood, paper, and related products.

It is true that some plastics, particularly polyvinyl chloride (PVC), will degrade over a long periods of exposure to sunlight or elevated temperatures (over 100°F). However, this particular scenario presents no harm to humans, since the degradation is so slow (probably taking months or years to happen) that the degradation products are not present in quantities sufficient to be picked up by delicate air-testing equipment. The argument that polyvinyl chloride decomposes over time in normal living conditions in a manner that will harm humans is just not true.

It is true that some liquid thermosetting systems can be compounded incorrectly (probably deliberately to save money) and then are improperly "cured" or crosslinked. In this situation, the improper crosslinking of some urea-formaldehyde foams could cause a release of formaldehyde gas. The odor of formaldehyde is so strong that it can be detected by humans in very low concentrations, and steps can then be taken to correct the problem. But here one should not blame the plastic but the person who had economic reasons for the incorrect formulation.

In summary, my major argument against current tests of combustion or decomposition product toxicity is the isolation and distance from reality in which these tests are carried out. The combustion and decomposition products produced in tests come from isolated materials or assemblies and ignore the other materials present at the scene of a fire that would also be producing harmful products of combustion and decomposition. Testing just does not duplicate conditions that exist in the everyday real world.

The solution to the problem of fire safety is not a simple one. It is simplistic to say "If we didn't have fires, we wouldn't have to worry about how safe or how dangerous a material is when it is exposed to fire or excessive heat." Even though I personally believe that all fire are preventable, it is not reasonable to assume that none will occur.

Even so, the most important aspect of fire safey *is* fire prevention. It *is* possible to educate people about means of preventing accidental fires. Education has already helped reduce the chances of accidental fires occurring, and education has also helped reduce the number of fire-related deaths. Work must continue in both these areas until accidental fires are reduced to the absolute minimum possible, and deaths from fires are eliminated.

Second, *all* occupancies, including residences, must be equipped with efficient detection and suppression systems so that *every* fire is detected early, suppressed quickly, and at worst, gives all occupants plenty of time to escape.

Third, all existing building and fire codes must be strictly enforced, and codes must be enacted where they do not currently exist. The practice of granting "grandfather" exemptions to buildings in existence when the codes or other regulations were enacted must stop. All existing structures must be brought up to current code requirements and then improved again as codes are tightened in the future. The same should be true for products and raw materials. All raw materials should have to face the same specifications for use. Only when existing buildings are covered by modern model building and fire codes will we quit reading about multiple deaths that occur in buildings not equipped with the latest safety systems. Only then will we no longer hear the criminal excuse that the building wasn't properly protected "because it wasn't required by the code." That is, the building was already in existence when the current code was adopted, and it didn't have to be as safe as new buildings. It happens over and over, and it has to stop.

Fourth, *all* products, materials, articles, and assemblies that burn should be made as fire-safe as possible, including "natural" products and materials. This can be (and is being) done responsibly by the manufacturers. What is absolutely *not* needed is the emotion-charged half-truths that are uttered by individuals or groups with hidden agendas. Often this includes an economic interest to the complainant in one manner or another. These false charges get consumers and governmental agencies all worked up, often for no good reason. The money spent by companies defending themselves from these charges could better be spent on product improvements, or even returned to the consumer in the form of lower prices.

Finally, a spirit of cooperation is sorely needed. This spirit should bring together manufacturers of products made from *all* materials, members of the Fire Service, everyone interested in fire safety, and the consuming public. Let the free enterprise system work again and let competing products fight it out over consumer benefits, and not half-truths.

INDIVIDUAL TESTS

Following are descriptions of individual small-scale tests that have been recognized at one time or another as tests that can be duplicated by a competent laboratory. Theoretically, this competent laboratory can then duplicate the results of previous tests. They are presented without comment. Many additional tests are in existence, and many new tests are in varying degrees of development.

Some of the tests not discussed are medium-scale or large-scale tests. Some small-scale tests have been done and results correlated with large-scale tests. Two studies have been carried out in conjunction with fire departments responding to real-world fires; these have been done in Boston and San Antonio.

Other researchers are working on computer models that will predict the growth and spread of fires in certain materials or room configurations, and the transport of heat and smoke from the room where a fire originates to other parts of the building. Others are trying to predict, without ever having a fire, the effect of smoke and other products of combustion and decomposition of that fire on humans without ever exposing anyone, including testers, to the actual danger.

DIN 53 436

This test uses a quartz tube 1,300 mm in length with a movable electric oven that surrounds the tube. The sample is placed within the tube and the oven moves along the tube at a rate of 10 mm per minute. Air is drawn through the tube at 100 liters per hour. The samples are prepared so that they have an equal weight per unit length, and the furnace heats the tube at constant temperatures, selected between 392°F and 1112°F. Rats are exposed to the air from the furnace after it has been cooled and diluted. Usually 20, but as few as 5 rats may be used in the test, usually held in the head-only configuration. Death of the animals is the end point.

The test is used to determine the relative toxicity of combustion products of the materials at different applied temperatures. At the end of the test, blood carboxyhemoglobin (COHb) of the animals is measured. Additional measurements gained by the test include the number of dead animals after the 30-minute exposure, the ratio of dead exposed rats to live exposed rats after the test, or the amount of material necessary to kill half the animals during the test period.

This test will provide a measurement of the **critical temperature**, T_c°C, defined as the pyrolysis temperature that just fails to produce a level of lethal concentration of decomposition products. **Pyrolysis temperature**, $T(LC_{50})$°C, is defined as the temperature at which half the animals die during exposure. **Critical concentration**, C_c, is defined as the concentration of decomposition gases that just fails to produce a lethal concentration. It also yields three additional measurements.

Dome Chamber Toxicity Test

This test is a modification of the University of San Francisco method. It uses a horizontal tube furnace in which a 1-gram sample is placed, and temperatures are maintained between 200°C (392°F) and 800°C (1472°F) at fixed 100°C (180°F) intervals. The test is run for 30 minutes (or less if all the mice die before the 30 minutes is up) on four mice. The mice, assuming any survive, are then tested and observed for 14 days.

The mice are not restrained in the exposure chamber. Levels of oxygen and other selected gases (such as hydrogen cyanide, hydrogen chloride, sulfur dioxide, or carbonyl fluoride) in the exposure chamber are recorded during the test, and carbon monoxide is measured at the end of the test.

This test measures time to staggering and time to death (t_d). It allows calculation of a concentration-time product for lethality, or death-product (DP).

Dow Test Method

The Dow test uses the same furnace as the National Bureau of Standards test (see below). Samples are tested in both the combustion and pyrolysis modes, using 2 grams of material in the pyrolysis mode and 3 grams in the combustion mode. Each test is run twice. The combustion products are chemically analyzed for acrolein, carbon monoxide, carbon dioxide, formaldehyde, hydrogen cyanide, oxygen, nitrogen, the nitrogen oxides, and other gases. Seven rats are used, and the test includes a 14-day postexposure observation period.

Measurements taken include lethality, signs of toxic effects, and body weights. An autopsy is also conducted.

Federal Aviation Administration Method

This test uses a tube furnace in which a 0.75-gram sample is placed, and the sample is heated to 1112°F. A closed recirculating system is used to move air at a rate of 4 liters per minute through the combustion chamber and into the exposure chamber, where three rats are exposed in a rotating cage. The air is continually analyzed to measure carbon monoxide, hydrogen cyanide, and oxygen content during the exposure.

The heating of the sample continues for 10 minutes, after which the furnace is shut off, and the recirculating system disconnected. The animals are exposed in the chamber for 30 minutes, and any survivors are observed for 14 days. Survivors then undergo pathological examination to determine effects of the test.

The test measures time to incapacitation (t_i) and time to death (t_d). It is usually run three times.

Harvard Medical School Test Method

The Harvard test uses an oven as the heating chamber for specimens. The animals may run on exercise wheels during their exposure to products of combustion. The purpose of this construction is to measure a Time of Useful Function (TUF). Measurements are made of time to when the animals quit running and also time-to-death (T_d). Combustion products are continually analyzed for carbon monoxide, carbon dioxide, and oxygen content of the air in the exposure chamber.

McDonnell-Douglas Test Methods

These are two separate tests that have been integrated into a single process using animals exposed to the combustion products of test specimens. One test connects an external electrode belt around the body of the rat to measure electrocardiogram (EKG) impulses and respiration rates of the test animals. This method measures the beginning of cardiac arrhythmia (C_a).

The second test allows an animal to run free on a rotating rod until it could no longer walk or run, which is then recorded as time to incapacitation (T_i).

National Bureau of Standards (now the National Institute for Standards and Technology) Test

This test uses a stainless steel crucible-type furnace in which a one liter beaker is placed that will hold the sample specimen. Tests are conducted in the flaming and nonflaming modes, with temperatures adjusted to be 45°F above and below the autoignition temperature of the specimen. For the flaming mode, a piloted ignition source is used to ignite products of decomposition.

The test is run for 30 minutes (with a 14-day postexposure observation period on any surviving animals) on six rats in the animal exposure chamber. Samples of the air are withdrawn and analyzed for carbon monoxide, carbon dioxide, and oxygen content of the circulated air. The primary end point measurement of the test is the lethal concentration that kills half the animals, or LC_{50}.

National Institute of Building Sciences Test Protocol

This test has been submitted to ASTM Committee E-5 in the hope that it will be adopted as a standard for determining the hazard index of building products.

This test uses a combustion cell connected to an exposure chamber, but usually no animals are used. Animals were used to develop the test, but a computer model now determines the toxic hazard index of a product under a specific set of conditions. Animals may be used in a supplementary test, or where a regulation calls for it. In the case of the use of live animals, six rats are restrained in the heads-only exposure mode. The exposure is for 30 minutes, less if all the animals die. Survivors are observed for 14 days. Radiant heat is used as the source of energy, and an exposure of 50 kW/m^2 will be produced.

Continuous gas analyzers monitor for carbon monoxide, hydrogen chloride, hydrogen cyanide, and oxygen. Carboxyhemoglobin (COHb) levels will be determined for many animals that die. Test specimens used must be representative of the end use of the material being tested. The test specimen may be a homogeneous sample or it may be a complete assembly, or something in between, but a maximum size is specified.

This test seeks to establish a method of measuring toxic potency and to establish a Toxic Hazard Index (IT_{50}). It will measure irradiation time (IT) and a concentration-time product (C_t) to lead to its other calculations. It seeks to determine ease of ignition expressed in time, the rate of smoke generation as measured by mass loss, and toxic potency (the quality of the gases produced).

The Toxic Hazard Index will be calculated by combining the results of testing for ignition properties, gas generation, and toxic potency of the material tested.

National Institute for Standards and Technology Test

This organization, formerly the National Bureau of Standards, is developing a test essentially similar to the National Institute of Building Sciences Test Protocol (see above). The combustion chamber and exposure chamber are the same, as are the size of the samples and the time and level of exposure. The major difference between this test and the National Institute of Building Sciences Test Protocol is that ignition properties, gas generation, and toxic potency are calculated separately and individually, and no Toxic Hazard Index (IT_{50}) will be determined. Instead, an LC_{50} is determined.

Another difference between this test and the National Institute of Building Sciences Test Protocol is that in determining the LC_{50}, the test sample surface is blackened with a thin layer of carbon black to give samples a uniform surface.

PITT Test (University of Pittsburgh)

This test has been submitted to ASTM Committee E-5 in the hope that it will be adoped as a standard for determining the hazard index of building products.

This test uses a furnace in which the temperature can be increased in a constant manner, usually 36°F per minute. The sample is placed in the furnace, and as it decomposes the products are diluted by air and led to the animal exposure chamber where four rats are held in the heads-only position. The exposure begins when the sample loses 0.2 percent of its weight and continues for 30 minutes.

This test measures the concentration of smoke that produces a 50% decrease in respiratory rate (RD_{50}) of the animals, a concentration–response relationship for sensory irritation, a stress index (SI), asphyxiant effects on the animals, the sample weight that produces death in 50% of the animals (LC_{50}), and an acute lethal hazard (ALH). During the test, the circulating air is analyzed on a continuing basis for carbon monoxide, carbon dioxide, oxygen, hydrogen chloride, hydrogen cyanide, and formaldehyde content. Information from this test is used to calculate the relative toxicity on a time–response basis (LT_{50}) and on a concentration–time basis (LCT_{50}).

Radiant Heat Test Method

This test uses a radiant heat source in its combustion furnace capable of delivering a heat flux of 5 watts per square centimeter (5 W/cm^2) to the sample, which is suspended in the combustion zone. Tests are run for 30 minutes on six rats, and air is analyzed on a continuing basis for carbon monoxide, carbon dioxide, and oxygen content of the circulated air.

The test measures time to incapacitation (t_i). Since the materials used in this test are exposed to heat in the manner in which they would be exposed in a fire, the test can express toxic potency on the basis of surface area of the material exposed rather than weight.

Southwest Research Institute Primate Studies

These are tests sponsored by the Federal Aviation Administration and the U.S. Department of Transportation to determine the possible impairment of escape activities of passengers from airplane crashes when a fire occurs. The tests try to determine how combustion gases will impair or prevent escape from a burning plane, and they attempt to validate results from behavioral tasks performed by rodents in laboratory combustion toxicity tests.

The tests are carried out with young baboons to determine the level of concentration of irritant gases that will produce escape impairment. Results are compared with tests of equal complexity administered to rodents in laboratory smoke tests.

Stanford Research Institute International Test Method

This is a test designed to generate and deliver products of combustion of test specimens to rats in a special behavioral response cage. After exposure to products of combustion, an electrical shock is provided to the cage bottom, and to escape the shock, the rat must climb a pole. Measurements include the content of carbon monoxide, carbon dioxide, and oxygen content of the air in the exposure cages, and the time until the rats can no longer climb the pole to escape the shock.

University of San Francisco Method

See Dome Chamber Toxicity Test.

University of Tennessee Test Method

In this test, four rats are exposed to the combustion products of the sample specimens at temperatures 90°F above the sample's degradation temperature.

Surviving animals are observed for 14 days, whereas blood removed from the dead rats is measured for carboxyhemoglobin content. Varying weights of the sample material are used to determine the initial weight of the material that kills half the test animals (the LC_{50}).

University of Utah Test Method

This test uses a laboratory-size conductive heat furnace and tests both the flaming and nonflaming modes. The test lasts 30 minutes, and the air is monitored for carbon monoxide, carbon dioxide, hydrogen cyanide, and oxygen. Results are reported in terms of incapacitation (IC_{50}) and lethality (LC_{50}). Other determinations of the toxicity can be made from examination of the surviving animals and pathological examination of those that die.

U.S. Testing Company Test Method

This test uses a radiant heater to decompose sample specimens, and the products of decomposition are delivered to an exposure chamber where six rats have access to exercise wheels. The test is for 20 minutes, with a 14-day postexposure observation period for animals that survive. Test equipment monitors the amounts of carbon monoxide, carbon dioxide, hydrogen chloride, hydrogen cyanide, hydrogen sulfide, and oxygen in the circulated air. Measurements include time to incapacity and time to death. Any additional toxic effects that occur during the test are also recorded.

THE LATEST TECHNOLOGY

Additional tests are in different stages of being developed and "perfected." Some tests measure toxic potency, some measure potential toxic hazard, whereas others address risk assessment. Toxic potency tests try to determine only the quantity of the tested material that must burn or decompose to create enough toxic products to kill half the animals in the test ("quality of the smoke"). A test of potential toxic hazard takes into account the ease of ignition (or time to ignition), the rate of burning, the quality of the smoke, and the quantity of the smoke. A risk assessment involves the ease of ignition, the rate of burning (or rate of mass loss), flame spread, fuel load, the toxic potency of the combustion or decomposition products, the rate of heat release, and other factors relating to the combustion characteristics of the sample specimen. A risk assessment also takes into the consideration the probability that the fire *will* occur. Any of the tests may or may not use a finished article or assembly.

The latest combustion toxicity tests include "HAZARD I," a computer model that is "a set of procedures combining expert judgment and calculations to estimate the consequences of a fire resulting from changes in such aspects as: materials fire properties, fire mitigation strategies, types of people present, and elements of building design." HAZARD I was developed by the National Institute for Standards and Technology (NIST). The Consumer Products Safety Commission (CPSC) then asked NIST for a simplified, shortened version of HAZARD I, which was dubbed "QUICK HAZARD I." Some believe that NIST does not put a lot of stock in Quick Hazard I and is in fact developing upgraded versions of the regular Hazard I program.

As this is being written, the National Fire Protection Research Foundation is in the process of submitting to ASTM a proposed fire hazard assessment standard, the purpose of which is to provide a standardized procedure for compiling information that will accurately describe properties of a product, article, or assembly relevant to the fire hazard of that product. This fire hazard assessment is the most complete to date, encompassing ease of ignition, rate of flame spread, rate of heat release, total heat released, fire growth rate, rate of smoke generation, opacity and corrosivity of the smoke, the total irritant and toxic products produced, the rate at which they are produced, the toxic potency of those products, the thermal decomposition rate of the specimen sample, a full profile of its behavior under fire conditions, including structural integrity, thermal conductivity and mechanical response to the fire, its ease of extinguishment, and finally, the quantity of product in use as compared with the size and type of occupancy.

The International Electrotechnical Commission (IEC) has proposed a second edition of its fire hazard assessment, Standard 695-1-1 (1982). It provides guidance for the preparation of specifications and test requirements for assessing fire hazards of electrotechnical products. It concerns itself primarily with minimizing the risk of fires (and the propagation thereof) caused by electrically induced ignition within electrotechnical products. The standard concerns itself primarily with combustibility factors but gives consideration to heat release and the opacity, toxicity, and corrosivity of smoke generated. The tests are run on products.

Other tests not described above include the following:

Center de Recherches Toxicologiques Studies
GE/Southern Research Institute Lab and Full Scale Comparisons
Huntington Research Centre Primate Studies
ISO/TR 6543—The Development of Tests for Measuring Toxic Hazards in Fires
ISO/DTR 9122—Toxicity of Fire Effluents
Japanese Combustion Toxicity Tests
Olin Corporation Studies

University of Michigan Test Methods
University of Pittsburgh/PPG Industries Studies
Vesicol/U.S. Testing Study

SUMMARY

- Universities, standards-setting organizations, and governmental agencies are trying to develop a comprehensive toxicity test that will once and for all predict what toxic products are formed when any particular material burns, where these toxic combustion or decomposition products are transported, who will be killed by them, and how long will it take.
- This task is even more unlikely to be accomplished than the development of a comprehensive predicting fire test, since there are even more variables in the toxicity tests than there are in the fire tests.
- Because of the greater number of variables, the researchers must make more and more assumptions in their tests, thus removing the tests farther and farther from reality.
- The toxicity problem should not be addressed by determining where the killer gas is originating: Carbon monoxide is being liberated by every burning substance that contains carbon.
- Toxicity must be addressed by preventing fires, developing quick acting-detection and suppression systems, developing and enforcing safety codes uniformly, making all articles that enter the stream of commerce as fire safe as possible, and promoting cooperation among industry, government, and consumers to work out a system whereby everyone contributes to the solution of the fire problem.

Appendix A

Abbreviations for Plastics

The following abbreviations are used for polymers, copolymers, and plastics. Some are formally recognized and have been adopted by standards-setting organizations, and some have been adopted by other references. Alternative abbreviations exist for specific compounds and copolymers, but it is not possible to capture all the variations. There may even be some arguments where a particular abbreviation stands for another material not covered by this list. Some abbreviations may stand for more than one polymer, copolymer, or plastic compound. Also, one of these materials may have more than one abbreviation. This list is provided for the benefit of readers who may come across such an abbreviation and want to know what polymer is represented by it.

ABA	Acrylonitrile–butadiene–acrylate
ABFA	1,1-Azobisformamide
ABS	Acrylonitrile–butadiene–styrene
ACS	Acrylonitrile–chlorinated polyethylene–styrene
ACM	Polyacrylate
ADC	Allyl diglycol carbonate
AES	Acrylonitrile–ethylene–styrene
AMMA	Acrylonitrile methyl/methacrylate
AMS	Alpha methyl styrene
AN	Acrylonitrile
ANM	Acrylate–acrylonitrile copolymer
APAO	Amorphous polyalphaolefin
APET	Amorphous polyethylene terephthalate
APO	Amorphous polyolefin
ARP	Aromatic polyester, for example poly(arylterephthalate)

ASA Acrylate–styrene–acrylonitrile
AU Polyester–urethene

BCB Benzocyclobutene
BMC Bulk molding compound
BMI Bismaleimide
BOPP Biaxially oriented polypropylene
BR Polybutadiene

CA Cellulose acetate
CAB Cellulose acetate butyrate
CAP Cellulose acetate propionate
C/C Carbon/carbon composite
CE Cellulosic plastics
CF Cresol–formaldehyde
CFC Chlorofluorocarbon
CM Chlorinated polyethylene
CMC Carbon matrix composite
CMC Carboxy methyl cellulose
CN Cellulose nitrate
CO Polychloromethyloxirane (epichorohydrin polymer)
CP Cellulose propionate
CPE Chlorinated polyethylene
CPET Crystallized polyethylene terephthalate
CPVC or PVCC Chlorinated polyvinyl chloride
CR Polychloroprene
CS Casein
CSM Chlorosulfonated polyethylene
CTA Cellulose triacetate
CTFE Chlorotrifluoroethylene

DAC Diallyl chlorendate
DAIP Diallyl isophthalate
DAM Diallyl maleate
DAP Diallyl phthalate
DGEBA Diglycidol ether of bisphenol A (epoxy)
DVA Dynamically vulcanized alloy

EA Ethylene acid copolymer
EAA Ethylene acrylic acid
EBA Ethylene butyl acrylate

EC	Ethyl cellulose
ECO	Epichlorohydrin copolymer
ECTFE	Ethylene–chlorotrifluoroethylene
EEA	Ethylene–ethyl acrylate
EMA or EMMA	Ethylene–methyl acrylate copolymer or ethylene methacrylic acid
EMAA	Ethylene–methacrylic acid
EMAC	Ethylene–methyl acrylate copolymer
EMI	Electromagnetic interference
EP	Epoxy
EPD	Ethylene/propylene/diene
EPDM	Ethylene–propylene diene modified rubber
EPM	Ethylene–propylene copolymer
EPR	Ethylene–propylene rubber
EPS	Expanded polystyrene
ESD	Electrostatic dissipation
ETE	Engineering thermoplastic elastomer
ETFE	Ethylene–tetrafluoroethylene
ETP	Elastomeric thermoplastic
ETP	Engineering thermoplastic
EU	Polyether urethane
EVA	Ethylene vinyl acetate
EVAL	Ethylene–vinyl alcohol
EVOH	Ethylene vinyl alcohol
FEP	Fluorinated ethylene propylene
FF	Furan–formaldehyde
FKM	Carbon chain fluoropolymer
FP	Fluoropolymers
FRP	Fiberglass reinforced plastic
FTIR	Fourier transform infrared
FVMQ	Fluorosilicone rubber
GMT	Glass mat thermoplastic
GP	General purpose
GPO	Propylene oxide copolymer
GRP	Glass reinforced plastic
HCPP	High crystallinity polypropylene
HDPE	High density polyethylene
HIPS	High impact polystyrene
HMF	Heavy metal free
HMV	High melt viscosity

HMWHDPE High molecular weight high density polyethylene
HTD Heat distortion temperature

IPS Impact-modified polystyrene
IIR Butyl rubber
IR Synthetic polyisoprene

LDPE Low density polyethylene
LLDPE Linear low density polyethylene

MBS Methacrylate butadiene styrene
MDPE Medium density polyethylene
MF Melamine–formaldehyde
MIPS Medium impact polystyrene
MMC Metal matrix composite
MQ Silicone elastomer
MPR Melt processible rubber

NBR Nitrile rubber or acrylonitrile rubber
NR Natural rubber

OSA (also AES) Olefin-modified styrene–acrylonitrile
OPP Oriented polypropylene
OPS Oriented polystyrene
OPVC Oriented polyvinyl chloride

PA Polyamide
PAA Poly(acrylic) acid
PAE Polyarylether
PAEK Polyaryletherketone
PAI Polyamide–imide
PAK Polyester alkyd
PAN Polyacrylonitrile
PAni Polyaniline
PARA Polyarylamid (polyaramide)
PASU Polyarylsulfone
PAT Polyarylate [poly(arylterephthalate)]
PB Polybutylene (polybutene)
PBAN Polybutadiene acrylonitrile
PBI Polybenzimidazole
PBS Polybutadiene–styrene

PBT or PBTP	Polybutylene terephthalate
PC	Polycarbonate
PCDP	Polydicyclopentadiene
PCTFE	Polychlorotrifluoroethylene
PDAP	Poly(diallyl phthalate)
PDCPD	Polydicyclopentadiene
PE	Polyethylene
PEBA	Polyether block amide
PEEK	Polyetherether ketone
PEI	Polyetherimide
PEK	Polyetherketone
PEN	Polyethylene naphthalate
PEO or PEOX	Polyethylene oxide
PES	Polyether sulfone
PESV	Polyether sulfone
PET	Polyethylene terephthalate
PETE	Polyethylene terephthalate
PETG	Polyethylene terephthalate glycol
PETP	Polyethylene terephthalate
PF	Phenol–formaldehyde
PFA	Perfluoroalkoxy or perfluoroalkoxy alkane
PFB	Perfluoroalkoxy resin
PFF	Phenol–furfural
PI	Polyimide
PIB	Polyisobutylene
PIR	Polyisocyanurate
PISU	Polyimidesulfone
PMC	Polymer matrix composite
PMCA	Poly (methyl–alpha–chloroacrylate)
PMMA	Polymethyl methacrylate
PMP	Polymethylpentene
PMQ	Silicone elastomer
PMS	Paramethyl styrene
PMVQ	Silicone elastomer
PO	Polyolefin
POB	Poly(p-oxybenzoate)
POM	Polyoxymethylene, polyformaldehyde, or acetal
POP	Polyphenylate oxide
PP	Polypropylene
PPE	Polyphenylene ether

PPO	Polyphenylene oxide
PPOX	Polypropylene oxide
PPS	Polyphenylene sulfide
PPSS	Polyphenylene sulfide sulfone
PPSU	Polyphylene sulfone
PS	Polystyrene
PSO	Polysulfone
PSU	Polysulfone
PTFE	Polytetrafluoroethylene
PUR	Polyurethane
PVA	Polyvinyl alcohol
PVAC	Polyvinyl acetate
PVAL	Polyvinyl alcohol
PVB	Polyvinyl butyral
PVC	Polyvinyl chloride
PVCA	Polyvinyl chloride–acetate
PVCC or CPVC	Chlorinated polyvinyl chloride
PVDC	Polyvinylidene chloride
PVDF	Polyvinylidene difluoride
PVF	Polyvinyl fluoride
PVFM	Polyvinyl formal
PVK	Polyvinyl carbizole
PVP	Polyvinyl pyrrolidone
RMC	Resin molding compound
RP	Reinforced plastic
RPBT	Reinforced polybutylene terephthalate
RPET	Reinforced polyethylene terephthalate
RTM	Resin transfer molding
RTPU	Rigid thermoplastic polyurethane
RTPUR	Rigid thermoplastic polyurethane
SAN	Styrene–acrylonitrile
SB or S/B	Styrene–butadiene
SBR	Styrene–butadiene rubber
SEBS	Styrene/ethylene–butylene/styrene
SF	Structural foam
SI	Silicone (all polymers)
SLS	Selective laser sintering
SMA	Styrene-maleic anhydride
SMC	Sheet molding compound

SMS	Styrene–alpha–methylstyrene
SP	Straight polyimide
SRIM	Structural reaction injection molding
SRP	Styrene–rubber plastics
STM	Structural thermoplastic composite
TEEE	Thermoplastic elastomer, ether–ester
TEO	Thermoplastic elastomer, olefinic
TES	Thermoplastic elastomer, styrenic
TPE	Thermoplastic elastomer
TPEL	Thermoplastic elastomer
TPES	Thermoplastic elastomers
TPO	Thermoplastic olefins
TP	Thermoplastic
TPU	Thermoplastic polyurethane
TPUR	Thermoplastic polyurethane
TPV	Thermoplastic vulcanizate
TS	Thermoset
TSBR	Thermoplastic styrene-butadiene copolymer
UF	Ureaformaldehyde
UHMWPE	Ultra high molecular weight polyethylene
ULDPE	Ultra low density polyethylene
UP	Unsaturated polyesters
VCE	Vinyl chloride–ethylene
VCEMA	Vinyl chloride–ethylene–methyl acrylate
VCM	Vinyl chloride monomer
VCMA	Vinyl chloride–methyl acrylate
VCVAC	Vinyl chloride–vinyl acetate
VCVDC	Vinyl chloride–vinylidene chloride
VDC	Vinylidene chloride
VMQ	Silicone elastomer

Appendix B

Plastics Trade Names

Selecting trade names that will mean something to a person seeking information about the type of plastic represented by a particular brand name is arbitrary at best and misleading at worst. Information of this kind is included in this book to enable an interested persons to identify the type of plastic present once they have discovered a brand name. There is no guarantee that the information given here is completely accurate in terms of the type of resin, alloy, or blend. Possible reasons for inaccuracy are explained below. If life or property is at stake, or if other serious consequences might follow an error in identification, you will have to consult a number of sources to make sure the information is correct.

The tracking of trade names used for any product, whether a plastic resin, compound, alloy, or other specialty use or any other article of commerce, is a very difficult task. The specific trade name for a specific product may be too new for inclusion as this book goes to press, or it may be as old as the plastics industry itself and included here. The trade name may cover one particular product, or it may encompass an entire line of resins, compounds, blends, or alloys. It may be limited to a particular family of chemically similar plastics, or it may cover all plastics with a similar additive, such as carbon fibers. It may not be a name, as an ordinary person might recognize a name, but a series of numbers or letters, or both. It may not include the manufacturer's name as part of the tradename—and then again, it might. Trade names of this kind are not listed here, because the opportunity to build in confusion is too great. Some trade names for products (as opposed to names for plastics) are included if the polymer from which that product is made can be identified.

Moreover, what is true today about the type of plastic represented by a trade name may not be true tomorrow, since the manufacturer may want to add other materials or change the name altogether. The information provided below is as accurate as the author can make it, but there are many chances for incorrect information to creep in. And just because only one type of plastic is given after a particular trade name, do not infer that another type of material may not also be represented by that name. Please keep these cautions in mind when you use the following information.

Abasafil	Glass-filled acrylonitrile–butadiene–styrene
Ablebond	Epoxy and polyimide compounds
Ablefilm	Epoxy and polyimide compounds
Abletherm	Epoxy and polyimide compounds
Ablex	Polyolefin blends
Absinol	Glass-filled acrylonitrile–butadiene–styrene
Abson	Acrylonitrile–butadiene–styrene
ACCO	High crystallinity polypropylene
Acetron	Filled and reinforced acetal
Aclar	Polytrifluorochloroethylene
Aclon	Polytrifluorochloroethylene
Aclyn	Ionomer
Acpol	Unsaturated polyester
Acridur	Acrylic sheet
Acriglas	Cast acrylic sheet
Acrilan	Polyacrylonitrile
Acrivue	Acrylic
Acrycal	Acrylic
Acrylafil	Styrene–acrylonitrile copolymer
Acrylite	Acrylic
Acrylivin	Polymethyl methacrylate/polyvinyl chloride
Acrysol	Acrylic
Acrysteel	Impact resistant acrylic sheet
Adiprene	Polyurethane
Adpro	Polypropylene; thermoplastic olefin
A-Fax	Atactic polypropylene
Airex	Polyvinyl chloride foam
Akulon	Nylon
Alathon	Polyethylene
Aldyl	Polyethylene pipe
Alcryn	Thermoplastic elastomer
Alflow	Glass-filled polypropylene
Alftalat	Alkyd resins
Algoflon	Polytetrafluoroethylene
Allaco	Epoxy compounds
Alloprene	Chlorinated rubber

Almatex	Acrylic
Alpha	Polyvinyl chloride and compounds
Alstamp	Glass-filled polypropylene
Altek	Polyester
Alton	Polytetrafluoroethylene/polyphenylene sulfone alloy
Altulite	Acrylic
Amilan	Nylon
Amodel	Polyphthalamide
Andrez	Polystyrene–butadiene resins
Antron	Nylon
Apec	Polyester carbonate; polycarbonate
Apical	Polyimide film
Arakote	Polyester
Araldite	Epoxy
Ardel	Polyarylate
Arimax	Acrylamate
Aristech GPA	General purpose acrylic sheet
Arloy	Styrene maleic anhydride; copolymer/polycarbonate alloy
Armorthane	Polyurethane
Arnite	Polybutylene terephthalate, polyethylene terephthalate
Arnitel	Thermoplastic elastomer
Arofene	Phenolic
Aropol	Unsaturated polyester
Aroset	Acrylic
Arpak	Expanded polyethylene beads
Arpro	Expanded polypropylene beads
Arylon	Polyarylate
Ashlene	Nylon
Aspect	Polyethylene terephthalate; thermoplastic polyester compounds
Astyrn	Polypropylene
Atryl	Sheet molding compound
Attane	Ultra low density polyethylene
Aukuthane	Polyurethane
Autofroth	Polyurethane
Avdel	Polyphenylene sulfide–based line of prepregs
Azamide	Polyamide
Azdel	Polypropylene, filled
Azloy	Polycarbonate
Azmet	Polybutylene terephthalate/thermoplastic composite
Azrez	Unsaturated polyester

Bapolan	Styrene, acrylonitrile–butadiene–styrene, and styrene–acrylonitrile
Bapolene	Polyethylene and polypropylene
Barex	Polyacrylonitrile
Bayblends	Polycarbonate/acrylonitrile–butadiene–styrene blends
Baydur	Polyurethane
Bayester	Polyester
Bayfide	Polyphenylene sulfide
Bayfit	Polyurethane
Bayflex	Polyurethane
Bayflon	Fluoroplastics
Baylon	Nylon
Baytal	Acetal
Baytec	Polyurethane
Baytherm	Polyurethane
Beckosol	Alkyd resins
Beetle	Nylon; urea; ureaformaldehyde
Beta	Nylon; polypropylene
Betathane	Polyurethane
Bexloy	Thermoplastic copolyester elastomer; polybutylene terephthalate
Blendex	Acrylonitrile–butadiene–styrene modifying resins
Boronol	Polyolefins with boron
Bruschcal	Polyester
Budd	Urea molding powder
Budene	Polybutadiene
Bur-A-Loy	Nitrile–polyvinyl chloride compounds
Butacite	Polyvinyl butyral sheeting
Butvar	Polyvinyl butyral
Cadon	Engineering thermoplastic: styrene maleic anhydride terpolymer; styrene maleic anhydride/acrylonitrile–butadiene–styrene alloy
Calibre	Polycarbonate
Capran	Nylon film
Capron	Nylon
Captima	Nylon
Captiva	Nylon
Castethane	Urethane elastomers
Castomer	Thermoset urethane elastomers
Celanex	Thermoplastic polyester: polyethylene terephthalate/polybutylene terephthalate alloy

Celazole	Polyetherketone
Celcon	Acetal
Cellbond	Polyester core fiber
Cellidor	Cellulose acetate butyrate
Cellobond	Phenolics
Cellulite	Polystyrene foam
Cellulex	Polystyrene
Celstran	Reinforced acrylonitrile–butadiene–styrene; polyphenylene sulfide; filled polypropylene; thermoplastic elastomer
Celtec	Foamed polyvinyl chloride
Centrex	Acrylic–styrene–acrylonitrile
Ceramathane	Ceramic-polyurethane composite
C-Flex	Thermoplastic elastomer
Chemigum	Thermoplastic elastomers; nitrile rubber powder
Chem-O-Sol	Polyvinyl chloride plastisol
Chem-O-Thane	Urethane elastomers or foams
Chromspun	Acetate
Clarene	Ethylene vinyl alcohol copolymer
Cleartuf	Polyethylene terephthalate
CoExcel	Polyvinyl chloride pipe
Colo-Fast	Polyurethane
Compodic	Nylon; ethylene vinyl acetate copolymer
Comtuf	Nylon; polybutylene terephthalate
Conacure	Polyurethane polyol
Conapoxy	Epoxy compounds
Conathane	Polyurethane resins and elastomers
Coro-foam	Polyurethane foam
Corrolite	Polyester
Corton	Polypropylene, filled
Corvic	Polyvinyl chloride
Cosmic	Diallyl phthalate; diallyl isophthalate
Crystalor	Polymethylpentene
Cyanacryl	Acrylic elastomer
Cyanaprene	Polyurethane
Cycolac	Acrylonitrile–butadiene–styrene
Cycolin	Acrylonitrile–butadiene–syrene/polybutylene terephthalate alloy
Cycovin	Acrylonitrile–butadiene–styrene/polyvinyl chloride alloy
Cycoloy	Acrylonitrile–butadiene–styrene/polycarbonate alloy; acrylonitrile–butadiene–styrene/polyvinyl chloride alloy
Cyglas	Unsaturated polyester
Cymel	Melamine

Cyracure	Epoxy resin
Cyro	Acrylic sheet
Cyroflex	Polycarbonate sheet
Cyrolite	Acrylic copolymer
Cyrolon	Polycarbonate sheet
Dacron	Polyester
Dalvor	Polyvinylidene fluoride
Dapex	Diallyl phthalate
Daran	Polyvinylidene chloride emulsions
Daratak	Polyvinyl alcohol adhesive emulsion
Darex	Styrene–butadiene emulsion
Delrin	Acetal; acetal/elastomer alloy
D.E.N.	Epoxy novolac resin
Denflex	Polyvinyl chloride plastisols; polyurethanes; epoxies
D.E.R.	Epoxy resins
Derakane	Vinyl ester resins
Desmocoll	Polyurethane resins
Desmodur	Aliphatic polyisocyanate resins
Desmopan	Thermoplastic polyurethane
Desmophan	Saturated polyester resin
Devcon	Epoxy
Dexcarb	Polycarbonate; polycarbonate alloys
Dexflex	Thermoplastic olefin; thermoplastic elastomer
Dexlon	Nylon
Dexpro	Polypropylene and polypropylene alloys
Diaron	Melamine liquid
Dielectrite	Unsaturated polyester
Diene	Polybutadiene
Dimension	Nylon/polyphenylene ether alloy
Driscopipe	Polyethylene pipe
Dolphon	Epoxy and polyester resins
Dominus	Polyvinyl chloride; polyurethane elastomer
Dowlex	Linear low density polyethylene
Duracap	Polyvinyl chloride compound
Duracryn	Thermoplastic elastomer
Duradene	Styrene–butadiene copolymers
Duralex	Polybutylene
Dural	Polyvinyl chloride; polyvinyl chloride alloy
Duraflex	Polyvinyl chloride alloy; thermoplastic elastomer
Duramid	Thermoplastic polimide
Durane	Polyurethane

Duraplex	Alkyd
Duraloy	Acetal/elastomer alloy
Durayl	Acrylic
Durel	Polyarylate
Durethan	Nylon
Durez	Alkyd; diallyl phthalate; phenolic or unsaturated polyester
Durocryl	Acrylic emulsion
Duroset	Ethylene/vinyl acetate emulsions
Durvil	Flame resistant rayon
Dyalon	Polyurethane elastomer
Dylark	Styrene maleic anhydride copolymer
Dylene	Polystyrene
Dylite	Polystyrene
Dyphene	Phenolic resins
Dytherm	Expandable styrene maleic anhydride
Dytron	Thermoplastic elastomer
Ecdel	Thermoplastic elastomer
Eco-Foam	Cornstarch with 5% synthetic additive
Ektar	Polyethylene terephthalate glycol, polycarbonate/polyester alloy; polypropylene; poly 1-4-cyclohexylenedimethylene terephthalate
Elastalloy	Thermoplastic elastomer
Elastocell	Thermoplastic polyurethane
Elastoflex TF	Polyurethane foam sheet
Elastolit	Polyurea/amides
Elastollan	Thermoplastic polyurethane
Elastoloc	Thermoplastic elastomer
Elemid	Polyester alloy
Elexar	Thermoplastic elastomer
Elite	Acrylonitrile–butdiene–styrene
El Rexene	Polyethylene, polypropylene
Elvace	Vinyl acetate ethylene emulsions
Elvamide	Nylon
Elvanol	Polyvinyl alcohol
Elvax	Ethylene vinyl acetate copolymer
Empee	Polyethylene; polystyrene
Enathene	Ethylene–normal butyl acrylate copolymers
Endure-C	Polyurethane
Enpnite	Nylon; polypropylene
Envex	Polyimide
Envirez	Fire retardant polyester

Epacron	Epoxy
Epalex	Polypropylene
Epocast	Epoxy or polyimide
Epolene	Polyolefin wax
Epon	Epoxy
Eponex	Epoxy
Eponol	Epoxy
Epotuf	Epoxy
Epoxyglas	Epoxy, glass-filled
Epsyn	Ethylene propylene elastomer
Eraelrene	Polyethylene
Escalloy	Polybutylene terephthalate
Escorene	Polyethylene; polypropylene; ethylene vinyl acetate
Escoform	Polypropylene
Estamid	Thermoplastic elastomer
Estane	Thermoplastic polyurethane
Esterlene	Woven nylon coated with polyethylene
Estralon	Thermoplastic polyester
Estron	Acetate
Ethafoam	Polyethylene foam
Ethocel	Ethyl cellulose
Ethofil	Polyethylene, filled
Ethomid	Polyethoxylated amides
Euatane	Ethylene vinyl acetate
Eurecryl	Cyanoacrylates
Eurelon	Polyamides
Euremelt	Polyamides
Eurepox	Epoxy
EVAL	Ethylene vinyl alcohol
Exothane	Polyurethane
Extron	Polypropylene, filled
Eymyd	Polyimide
Eypel	Polyphosphazene elastomer
FaRez	Polyurethane polyol
Fastcast	Polyurethane
Ferrene	Polyethylene, filled
Ferrex	Polypropylene, filled
Ferropak	Polypropylene/high density polyethylene alloy
Ferroflo	Polystyrene
Ferroflex	Thermoplastic elastomer
Fiberite	Epoxy, melamine; phenolic; vinyl ester; polyimide or silicone

Fiberloc	Polyvinyl chloride, filled
Fibresinol	Phenol/glass fiber molding compound
Fina	Polypropylene, polystyrene
Finaprene	SBS elastomer
Flexclad	Thermoplastic copolyester resin
Flexel	Low smoke, low flame polyvinyl choride compound
Flexobond	Polyurethane
Flexomer	Polyolefin copolymer resins
Fluon	Polytetrafluoroethylene
Fluorel	Fluoroelastomer
Florocomp	Fluoropolymer
Foamol	Line of polyester polyols
Foraflon	Polyvinylidene fluoride
Formaldafil	Glass-filled acetal
Formica	Melamine-formaldehyde
Formion	Ionomer
Formolon	Polyvinyl chloride
Formvar	Polyvinyl formal
Fortiflex	Polyethylene
Fortilene	Polypropylene
Fortron	Polyphenylene sulfide
Froth-Pak	Polyurethane

Gardglas	Acrylic glazing sheets
Gecet	Polyphenylene oxide/polystyrene foam beads
Gedex	Polystyrene
Gedexel	Polystyrene, expandable
Gelon	Nylon
Geloy	Acrylate–styrene–acrylonitrite and alloys
Gelva	Polyvinyl acetate
Gemax	Polyphenylene oxide/polybutylene terephthalate alloy
Gen-Flo	Styrene butadiene latex copolymer
Geolast	Thermoplastic elastomer
Geolite	Polyurethane
Geon	Polyvinyl chloride
Glaskyd	Alkyd
Gore-Tex	Expanded polytetrafluoroethylene products
Granlar	Aromatic or liquid thermoplastic polyester
G-Resin	Polyethylene
Grilamid	Nylon; thermoplastic elastomer
Grilesta	Polyester powder coatings

Grilon	Nylon
GTX	Polyphenylene oxide/nylon blend
Halar	Ethylene–chlorotrifluoroethylene compound
Haloflex	Polyvinyl chloride, polyvinyl chloride, acrylic terpolymer emulsions
Halon	Polytetrafluoroethylene
Halon-ET	Ethylene–tetrafluoroethylene copolymer
HDR	Biaxially oriented polypropylene film
Hercuprene	Thermoplastic elastomer
Hetron	Unsaturated polyester; vinyl ester
Hicond	Filled polyethylene
HiFax	Thermoplastic elastomer
HiGlass	Glass-reinforced polyethylene
Hilene	Polyethylene; polypropylene
Hiloy	Nylon; polybutylene terephthalate
Hi Styraclear	High-impact polystyrene
Hostaflex	Polyvinyl chloride copolymer
Hostaflon	Polytetrafluoroethylene
Hostaform	Acetal
Hostalen	Polyethylene
Hostaphan	Polyester film
Hostaprime	Styrene maleic anhydride
Hostatec	Polyether ketone
HTX	High temperature polyvinyl chloride compound
Hycomp	Polyimide
Hydlar	Nylon; thermoplastic elastomer
Hydrex	Polyester
Hyflon	Perfluoroalkoxy
Hylozene	Blended polyethylene film
Hypalon	Synthetic rubber
Hypol	Polyurethane isocyanate
Hytrel	Engineering thermoplastic (polyester) elastomer
Hyzod	Polycarbonate sheet
Impet	Reinforced polyethylene terephthalate
Incoblend	Conductive polymer blend
Incopol	Polyaniline
Insulcast	Epoxies; silicones
Iotek	Ionomers
Isonate	Polyurethane isocyanate
Isonol	Polyurethane polyol

Isoplast	Thermoplastic polyurethane, nonelastomeric
Isothane	Flexible polyurethane foams
Iupilon	Polycarbonate
Iupital	Acetal
Jenite	Polyvinyl chloride shapes
J-Plast	Thermoplastic elastomer
Kadel	Polyetherketone
Kalar	Butyl rubber
Kalene	Elastomeric compounds
Kalex	Polyurethane elastomers
Kalrez	Tetrafluoroethylene/perfluoromethylvinylether
Kamax	Acrylic/imide copolymer
Kaneace	Methacrylate–butadiene–styrene copolymer
Kanevinyl	Chlorinated polyvinyl chloride
Kapton	Polyimide film
Karlex	Glass-filled polycarbonate
Kel-F	Polychlorotrifluoroethylene
Keltan	Ethylene–propylene diene modified
Kemid	Polyetherimide film
Kevlar	Aromatic polyamide (aramid)
K F Polymer	Polyvinylidene difluoride
Kinel	Polymide
Klemite	Melamine molding compound
Kodacel	Cellulose acetate
Kodabond	Copolyester
Kodapak	Polyethylene terephthalate
Kodar	Polyethylene terephthalate glycol
Kodel	Polyester
Kohinor	Polyvinyl chloride
Korad	Acrylic film
Korton	Fluoropolymer films
Kraton	Thermoplastic elastomer (styrenic)
K-Resin	Styrene–butadiene
Kryptane	Polyurethane elastomers
K-Tac	Polypropylene
Krystallite	Polyvinyl chloride shrink film
Kydex	Acrylic/polyvinyl chloride sheet
Kynar	Polyvinylidene fluoride
Ladene	Polystyrene; melamine; polyvinyl chloride; polyethylene
Lamgard	Polyester

Leguval	Unsaturated polyester
Lekutherm	Epoxy
Lennite	Crosslinked polyethylene
Lexan	Polycarbonate
Lexorez	Line of polyester polyols
Lin-foam	Expandable polystyrene
Locod	Ethylene vinyl acetate copolymer
Lomond	Engineering thermoplastic elastomer
Lotrene	Low density polyethylene
Lotrex	Low density polyethylene
LP	Liquid polysulfide polymer
LR	Thermoplastic elastomer
Lubmer	Polyethylene
Lubrilon	Nylon
Lucite	Acrylic
Lupranate	Polyurethane isocyanate
Luran	Styrenic copolymer; acrylate–styrene–acrylonitrile
Lustran	Acrylonitrile–butadiene–styrene/styrene–acrylonitrile; acrylonitrile–butadiene–styrene; styrene–acrylonitrile copolymer
Lustrofilm	Extruded polyethylene terephthalate
Lutanol	Polyvinyl ether
Luxis	Nylon
Lytex	Epoxy
Magnacomp	Nylon, polypropylene
Magnex	Acrylonitrile–butadiene–styrene; Polyphenylene oxide/ polyphenylene ether (PPO/PPE) alloy; conductive polyvinyl chloride compound
Magnum	Acrylonitrile–butadiene–styrene
Makroblend	Polycarbonate blends
Makrofol	Polycarbonate film
Makrolon	Polycarbonate
Malon	Polyethylene terephthalate
Maloy	Thermoplastic polyester injection molding alloys
Mamax	Thermoplastic polyester
Maranyl	Nylon
Marblette	Phenolic resin, cast
Mark-A-Cite	Polystyrene shapes
Marlex	Polyethylene; polypropylene
Marvaloy	Acrylic modified styrene alloy
Matrimid 5218	Polyimide

Maxcel	Polyurethane
Meldin	Polyimide and reinforced polyimide
Melinar	Polyethylene terephthalate
Melinex	Polyethylene terephthalate
Melthene	Ethylene vinyl acetate copolymer
Merlon	Polycarbonate
Metacast	Epoxy
Methafil	Polymethylpentene
Methocel	Cellulose ethers
Metre Foam	Polyurethane
Metton	Polydicyclopentadiene
Microsol	Polyvinyl chloride plastisol
Microthene	Polyethylene; polyolefin powders; ethylene vinyl acetate
Migralube	Tetrafluoroethylene and silicone lubricated thermoplastics
Milastomer	Thermoplastic elastomer
Mindel	Polysulfone; polysulfone/acrylonitrile–butadiene–styrene alloy
Minlon	Mineral-filled nylon
Mirrex	Rigid polyvinyl chloride film
Mistafoam	Polyurethane
Mistapox	Epoxy
Mondur	Polyurethane isocyanate
Monocast	Nylon
Monothane	Castable polyurethane
Moplen	Polypropylene
Morthane	Thermoplastic polyurethane
Movilith	Ethylene-vinyl acetate; polyvinyl acetate
Moviol	Polyvinyl alcohol
Movital	Polyvinyl butyral
Mulrath	Polyurethane isocyanate
Multranl	Polyurethane polyol
Mylar	Polyester film
Mylex	Polyester film
Mytex	Engineered polypropylene
Naftex	Linear low density polyethylene film
NC Proofboard	Polyurethane foam
NeoCryl	Acrylic emulsion
Neo Pac	Polyurethane–acrylic copolymer emulsions
Neopolen	Polyethylene foam
NeoRez	Polyurethane
Neo Tac	Polyurethane and acrylic adhesive polymers

Neulon H	Sheet molding compound
Nevchem	Alkylated petroleum resins
Nevex	Modified hydrocarbon resins
Nevillac	Hydroxy modified resins
Nevtac	Aliphatic hydrocarbon resins
New-TPI	Polyimide
Nivionplast	Nylon
Norcast	Epoxy casting compound
Norex	Polyvinyl chloride foam
Norchem	Polyethylene; polypropylene
Norguard	Polyethylene shrink covers
Norlene	Polyethylene film
Norsoflex	Very low density polyethylene
Norsorex	Thermoplastic rubber polynorborene
Nortuff	Polyethylene; polypropylene
Noryl	Modified polyphenylene oxide; polyphenylene ether
Noryl GTX	Engineering thermoplastic
Nourcryl	Acrylic
Novablend	Polyvinyl chloride
Novaloy	Acrylonitrile–butadiene–styrene/polyvinyl chloride alloy
Novapol	Polyethylene
Novatemp	Acrylonitrile–butadiene–styrene/polyvinyl chloride alloy
Novatron	Polyurethane
Novon	Starch-based polymer
Novophalt	Asphalt containing recycled polyolefin
Nucrel	Ethylene methyl acrylate
Nupol	Thermosetting acrylic
Nybex	Nylon-based compound
Nycast	Nylon
Nycoa	Nylon
Nydur	Nylon
Ny-kon	Nylon
Nylafil	Glass-filled nylon
Nylamid	Nylon
Nylatron	Nylon
Nylomet	Metallized nylon
Nynel	Ethylene vinyl acetate copolymer
Nypel	Nylon
Nyrim	Nylon
Nysyn	Nitrile butadiene elastomer

Olehard	Polypropylene
Oletron	Polypropylene
Omnifilm	Polyvinyl chloride film
Ontex	Thermoplastic olefin alloy
Opalesque	Nylon tricot
Oppanol	Polyisobutylene
Optema	Ethylene methyl acrylate
Optene	Family of barrier polymers containing vinyl alcohol groups
Opticite	Polystyrene film
Optix	Acrylic
Orevac	Polyolifin binder resins
Orthane	Polyurethane
Oxyblend	Polyvinyl chloride
Oxycast	Epoxy casting compounds
Oxyclear	Polyvinyl chloride
Panalane	Hydrogenated polybutene
Panlite	Polycarbonate
Pantalast	Polyvinyl chloride/ethylene vinyl acetate copolymer
Parachlor	Chlorinated polyethylene elastomer
Paracril	Butadiene–styrene elastomer
Paraglas	Acrylic panels
Paraloid HT-510	Imide copolymer
Parshield	Silicone
Paxon	Polyethylene
PBI	Polybenzimidazole
Pebax	Polyether block amide; thermoplastic elastomer
Pelaspan-Pac	Expandable polystyrene
Pellathane	Thermoplastic polyurethane
Pennlon	Molecularly oriented polyolefin
Pentathane	Cast polyurethane
Permapol	Polyether and polythiol ether polymers
Petlon	Polyethylene terephthalate
Petra	Reinforced polyethylene terephthalate
Petrothene	Polyethylene, polypropylene
Petsar	Polyethylene terephthalate; polycarbonate/polyester alloy
Philjo	Polyolefin films
Piccodiene	Polydicyclopentadiene
Piccolastic	Polystyrene
Piccopale	Aliphatic-type hydrocarbon resins

Piccotex	Polyvinyl toluene copolymer
Piccovar	Alkyl–aromatic resins
Pioloform	Polyvinyl butyrate
Plastbau	Expandable polystyrene boards
Plaskon	Epoxy
Plastalloy	Polysulfone
Platilon	Polyurethane film
Platamid	Nylon copolymers; hot melt adhesives
Plenco	Melamine/phenolic; phenolic or unsaturated polyester
Plenex	Polyvinyl chloride
Plex	Acrylic sheets
Plexar	Polyethylene; ethylene vinyl acetate copolymer
Plexiglas	Line of acrylic resins
Pliolite	Styrene–butadiene
Pliothene	Polyethylene–rubber blends
Pliovic	Polyvinyl chloride
Pluracol	Polyurethane polyol
Pluronic	Polyethers
Plyophen	Phenolics
Pocan	Polybutylene terephthalate; polyester/elastomer alloy
Polane	Polyurethane coatings
Polybac	Epoxy; silicones
Polyblend	Polyphenolene sulfide alloy
Polycast	Polyester-based casting compounds
Polycin	Polyurethane polyol
Polycomp	Polyphenylene sulfide
Poly/Con	Epoxy
Polycryl	Acrylic
Polycure	Polyethylene
Poly-Dap	Diallyl phthalate
Polydene	Polyvinyl chloride alloys
Poly-Eth	Polyethylene
Poly-FAB	Fluorinated polyacrylate
Polyfine	Modified polyolefin
Polyflex	Polyvinyl chloride and polyvinyl chloride alloys; oriented polystyrene
Polyflon	Polytetrafluoroethylene
Polyfort	Polypropylene; polyethylene
Polygard MR	Allyl–polycarbonate sheet
Polylac	Acrylonitrile–butadiene–styrene
Polylite	Polyurethane; polyesters

Polyman	Acrylonitrile–butadiene–styrene/polyvinyl chloride alloy; styrene–acrylonitrile copolymer
Polypur	Thermoplastic polyurethane
Polyrex	Polystyrene
Polyrite	Unsaturated polyester
Polysar	Polystyrene
Polyset	Epoxy
Polytac	Polypropylene
PolyTHF	Polytetrahydrofuran
Polytone	Polystyrene
Polytron	Polyvinyl chloride; unsaturated polyester
Polytrope	Polypropylene, filled; thermoplastic elastomer
Polytuff	Polypropylene
Polyviol	Polyvinyl alcohol
Polywax	Polyethylene
PremiGlas	Unsaturated polyester
Prevail	Thermoplastic polyurethane/ABS blend
Prevex	Modified polyphenylene ether
Primacor	Ethylene acrylic acid copolymer
Prime-Impax	Polystyrene sheet
Prism	Polyurethane
Procond	Polypropylene, filled
Profax	Polypropylene
Profil	Polypropylene
Prolastic	Thermoplastic elastomer
Pro-Pac	Polypropylene strapping
Propak	Polypropylene, filled
Propiofan	Polyvinyl propionate
Propylux	Polypropylene
Prostat	Polypropylene
Pulse	Acrylonitrile–butadiene–styrene/polycarbonate alloy
Pynothane	Polyurethane elastomer
Q-Thane	Polyurethane
Quadrol	Polyurethane polyol
Qualitool	Epoxy
Quatrex	Epoxy
Racoplast	Phenolic molding compound
Radel	Polyarylsulfone
Renflex	Thermoplastic elastomer
Renoprop	Polypropylene

Reny	Nylon
Repete	Recycled polyethylene terephthalate
Resimene	Melamine–formaldehyde/ureaformaldehyde
Resin 18	Poly–alpha–methyl styrene
Resinoid	Phenolic
Resinol	Polyolefins
Resipol	Polyester molding compound
Retpol	Polypropylene, filled
Revinyl	Recycled polyvinyl chloride sheet
Rexene	Polyethylene; polypropylene
Rextac	Poly–alpha–olefin
Rhoplex	Acrylic polymers; vinyl acetate emulsions
Riblene	Polyethylene; ethylene vinyl acetate
Ricon	Vinyl polybutadiene homopolymers and copolymers with styrene
Rilsan	Nylon
RIMline	Polyurethane
Rimpact	Polyurethane elastomer
Rimplast	Nylon; thermoplastic vinyl acetate; silicone
Riteflex	Thermoplastic polyester elastomer
Rolox	Epoxy
Rotoflame	Flame retarded polyethylene
Rotoflex	Polyvinyl chloride powder compound
Rotothene	Polyethylene
Rotothon	Polypropylene
Rovel	Acrylic–styrene–acrylonitrile; olefin modified styrene–acrylonitrile
Royalene	Ethylene propylene elastomer
Royalex	Acrylonitrile–butadiene–styrene foam core sheet
Royalite	Acrylonitrile–butadiene–styrene sheet; polyvinyl chloride sheet; acrylonitrile–butadiene–styrene/polyvinyl chloride sheet; polyvinyl chloride/acrylic sheet
Royaltherm	Silicone modified ethylene propylene elastomer
Royaltuf	Modified ethylene propylene elastomer
Rubinate	Polyurethane isocyanate
Rucoplex	Polyesters
Rucothane	Polyurethane
Rynite	Reinforced polyethylene terephthalate
Ryton	Engineering thermoplastic; polyphenylene sulfide
Sabre	Polycarbonate/polyester alloy
Salox	Filled polytetrafluoroethylene resin

Santoprene	Polypropylene/ethylene propylene diene modified thermoplast rubber
Saran	Polyvinylidene chloride
Sarlink	Thermoplastic elastomer
Schulamid	Nylon alloys and compounds
Sclair	Polyethylene
Sclairlink	Crosslinkable polyethylene
Scotchpar	Polyester film
Selar	Nylon; polyethylene terephthalate, ethylene vinyl alcohol copolymer; nylon/polyethylene alloy
Silastic	Silicone
Silgranit	Acrylic/granite–silicate material
Silite	Silicone rubber
Sinkral	Acrylonitrile–butadiene–styrene
Sinvet	Polycarbonate
Skybond	Polyimide resins
Smoke Guard	Low smoke, low flame polyvinyl chloride compound
Sniamid	Nylon compounds
Soarnol	Ethylene vinyl alcohol copolymer
Solef	Polyvinylidene fluoride
Solidur	Polyethylene
Sonite	Epoxy
Sparkle	Styrene acrylonitrile
Spectra	Special high strength polyethylene fiber
Spectrim	Polyurethane
Spectrum	Polyolefin sheet
Stanuloy	Polyethylene terephthalate
Stanyl	Nylon
Stat-Kon	Anti-static polystyrene
Stat-Rite	Static dissipative resin
StepanFoam	Polyurethane
Stereon	Styrene butadiene copolymers
Stycond	Styrene copolymer
Stypol	Polystyrene
Styrafil	Styrene maleic anhydride copolymer
Styresol	Styrenated resins
Styrex	Polystyrene films
Styrofoam	Polystyrene foam
Styron	Polystyrene
Styropor	Polystyrene expandable beads
Sulfil	Polysulfone
Sullvac	Polyvinyl chloride/acrylonitrile–butadiene–styrene blends

Sumikon	Polyvinyl chloride
Sumilite	Polyvinyl chloride sheet
Sunprene	Thermoplastic elastomers
Supec	Polyphenylene sulfide
Superkleen	Polyvinyl chloride compound
Sur-Flex	Ionomer film
Spectra	Polyethylene
Styrofoam	Polystyrene
Styron	Polystyrene
Styropor	Polystyrene
Sulfil	Polyetherether ketone; polyphenylene sulfide
Super Beckacite	Phenol–formaldehyde
Surlyn	Ionomer
Synergy	Polyphenylene ether/nylon blend
Synthacryl	Acrylic
Taitalac	Acrylonitrile–butadiene–styrene
Taffen	Polypropylene
Tamcin	Polypropylene, filled
Technigram	Polyethylene
Technyl	Nylon
Tecoflex	Thermoplastic polyurethane
Tedlar	Polyvinyl fluoride
Tedur	Polyphenylene sulfide
Teflon	Polytetrafluoroethylene and other fluoropolymers
Telcar	Thermoplastic olefin elastomer
Tefzel	Ethylene–tetrafluoroethylene copolymer
Telene	Polydicyclopentadiene
TempRite	Chlorinated polyvinyl chloride
Tenamatte	Polyvinyl chloride sheet
Tenaplast	Polyvinyl chloride sheet
Tenite	Lines of resins including cellulosics, polyolefins and polyesters
Terblend S	Acrylic–styrene–acrylonitrile/polycarbonate alloy
Tercuran	Acrylonitrile–butadiene–styrene
Terluran	Acrylonitrile–butadiene–styrene styrenic terpolymers
Terlux	Acrylonitrile–butadiene–styrene
Tetrafil	Reinforced polyethylene terephthalate or polybutylene terephthalate
Tetralon	Polytetrafluoroethylene
Texalon	Nylon

Texin	Thermoplastic polyurethane; polycarbonate/thermoplastic polyurethane alloy
Thanol	Polyurethane polyol
Thermid	Thermosetting polyimide
Thermex	Polybutylene terephthalate
Thermopoxy	Epoxy
Thermx	Copolyester
Tivar	Polyethylene
Tone	Polyurethane polyol
Torayfan	Polypropylene
Torelina	Biaxially oriented polyphenylene sulfide film
Torlon	Polyamide–imide
Toyolac	Acrylonitrile–butadiene–styrene
TPX	Polymethylpentene
Traytuf	Polyethylene terephthalate
Treax	Polypropylene
Trefsin	Thermoplastic elastomer
Triax	Engineering thermoplastic and other alloys; acrylonitrile–butadiene–styrene/polycarbonate alloy; acrylonitrile–butadiene–styrene/nylon alloy
Trimflex	Polyethylene foam
Trogamid	Nylon
Tufel	Silicone
Tuffak	Polycarbonate sheet
Turcite	Acetal
Tuflin	Polyethylene
Typlax	Ethylene vinyl acetate copolymer; polypropylene
Tygothane	Polyurethane tubing
Tyril	Styrene–acrylonitrile copolymer
Tyrin	Chlorinated polyethylene
Ube	Nylon
Ucar	Polyethylene; phenolic; phenoxy; vinyl resins
Udel	Polysulfone
Ultem	Polyetherimide
Ultrac	Polyethylene
Ultradur	Polybutylene terephthalate
Ultra-Ethylux	Polyethylene
Ultraform	Acetal
Ultrafil	Thermoplastic polyurethane, filled
Ultralon	Polytetrafluoroethylene

Ultramid	Nylon
Ultranyl	Polyphenylene ether/nylon blend
Ultrapek	Polyaryletherketone
Ultrason E	Polyethersulfone
Ultrason S	Polysulfone
Ultrathene	Ethylene vinyl acetate copolymers
Ultrawear	Polyethylene
Ultros	Cellulose acetate or butyrate
Unichem	Polyvinyl chloride compounds
Unifoam	Polyurethane foam
Unipol	Polypropylene, polyethylene
Uni-Rez	Polyamides
Unival	Polyethylene
Upilex	Polyimide
Uraflex	Polyurethane
Uralac	Polyesters
Uralam	Polyesters
Urotuf	Polyurethane

Valite	Phenolic
Valox	Polybutylene terephthalate; reinforced polyethylene terephthalate; polyethylene terephthalate/polybutylene terephthalate alloy
Valtec	Polypropylene
Vandar	Polybutylene terephthalate
Varcum	Phenolic resins
Vectra	Liquid crystal polymer
Vedoc	Powder coatings
Versicon	Polyaniline
Verton	Nylon; polyphenylene sulfide; polypropylene styrene copolymer
Vespel	Polyimide
Vestamelt	Thermoplastic polyesters
Vestamid	Nylon
Vestanamer	Polyoctenylene
Vestodur	Polybutylene terephthalate
Vestolen	Polyethylene; polypropylene
Vestolit	Polyvinyl chloride
Vestoran	Polyphenylene oxide
Vestosint	Nylon
Vibrathane	Polyurethane

Victrex	Polyethersulfone; polyetherketone; polyetherether ketone; aromatic or liquid thermoplastic polyester; polyetherether ketone/polyethersulfone alloy
Vinakon	Polyvinyl chloride compound
Vinnol	Polyvinyl chloride
Vintec	Polyvinyl chloride sheet
Vinylex	Polyvinyl chloride films
Vi-Seal	Polyvinyl chloride
Vistaflex	Thermoplastic olefin
Vistalon	Ethylene propylene rubber
Vistel	Polyvinyl chloride
Vitafilm	Polyvinyl chloride film
Vitel	Thermoplastic copolyester resins
Viton	Hexafluoropropylene/vinylidene fluoride/tetrafluoroethylene terpolymer
Vituf	Polyester resins
Vivana Lite	Nylon
Voloy	Nylon; polybutylene terephthalate
Voranol	Polyurethane polyol
Vorite	Polyurethane isocyanate
Vulkollan	Polyurethane
Vydyne	Nylon
Vynathene	Vinyl acetate ethylene copolymers
Vynite	Polyvinyl chloride/nitrile rubber alloy; thermoplastic elastomer
Vyram	Thermoplastic elastomer
Vythene	Polyvinyl chloride/thermoplastic elastomer alloy; thermoplastic elastomer
Wallkyd	Alkyd
Wellamid	Nylon
Wellite	Polyester
Wilflex	Vinyl plastisols
Winlon	Polyethylene
Xenoy	Thermoplastic alloy; polycarbonate/polyester alloy
XT-Polymer	Acrylic copolymer
Xtralife	Polyvinyl chloride
XXCEL	Thermoset olyester
Xycon	Polyester-polyurethane hybrid resin
Xydar	Aromatic or liquid thermoplastic polyester; liquid crystal polymer

Zefran	Acrylic
Zelux	Polycarbonate
Zemid	Ionomer or polyethylene
Zetafax	Polyethylene chlorinated copolymer
Z-Thane	Polyurethane
Zurcon	Polytetrafluoroethylene
Zylac	Hydrocarbon polymer
Zylar	Acrylic; styrene/acrylic copolymers
Zytel	Nylon

Appendix C

Glossary

ABS Resins. A terpolymer of acrylonitrile, butadiene, and styrene.

Abrasion Resistance. Resistance to surface wear.

Acetal Resins. Unbranched polyoxymethylene. Also called *polyformaldehyde*.

Acrylate Elastomers. A family of synthetic elastomers based on polyethyl acrylate copolymerized with a monomer that allows crosslinking.

Acrylic Resins. Polymers prepared from monomers such as acrylic acid and some of its derivatives.

Acrylonitrile. A monomer with the structure $CH_2{=}CHCN$.

Additive. Any material mixed with a plastic or resin to modify its properties.

Aging. The effect of exposure of articles to the environment for an extended period. If the environment is an artificial one for test purposes, or if the test is designed to provide exposure to the article faster than would occur in the environment in which it is to perform, the aging is said to be artificial or accelerated.

Alcohol. An organic hydrocarbon derivative with the general formula R—OH, where the R— is a hydrocarbon radical or backbone and the —OH is the hydroxyl radical.

Aldehyde. An organic hydrocarbon derivative with the general formula R—CHO, where the R— is a hydrocarbon radical or backbone and the —CHO is the carbonyl radical (—C=O) attached to a hydrogen atom.

Alicyclics. Organic compounds whose carbon atoms are arranged in a closed ring but do not contain the hexagonal benzene ring and are therefore not aromatic compounds.

Aliphatics. Organic compounds such as hydrocarbons characterized by straight-chain or branched-chain arrangement of the carbon atoms.

There are three groups of aliphatic hydrocarbons: alkanes (paraffins), alkenes (olefins), and alkynes (acetylenes).

Alkanes. An analogous series of saturated hydrocarbons with the general formula C_nH_{2n+2}, where *n* represents the number of carbon atoms in the chain. The first four alkanes are methane (CH_4), ethane (C_2H_6), propane (C_3H_8), and butane (C_4H_{10}).

Alkyd Resins. Polyester resins made with a fatty acid modifier. The most important surface coating resins.

Alkyl. Pertaining to a radical formed by removing one hydrogen atom from an alkane, for example, methane to methyl and ethane to ethyl.

Alkylated. The introduction of an alkyl radical onto an organic molecule.

Alloy. A composite plastic made by physically blending and melting together different polymers (and sometimes copolymers) to achieve certain properties.

Allyl Resins. Plastics made by the polymerization of monomers containing the allyl group ($CH_2{=}CH{-}CH_2{-}$).

Ambient Temperature. The temperature of the medium (air, water, and so on) surrounding the object under observation.

Amino Plastics. Polymers containing the NH— or NH_2— group. Based on reactions of formaldehyde with melamine or urea.

Amorphous. Noncrystalline; having no formal structure.

Antioxidant. A chemical additive designed to minimize the effects of oxygen on the compound or article.

Antistatic Agent. A chemical additive designed to prevent the accumulation of electrostatic charges on the surface of the article. The chemical may be sprayed on the surface, or it may be incorporated within the plastic compound. The antistatic agent may be a mixture, a chemical compound, another polymer, metal flakes or fibers, or any other conductive material.

Aramid. Generic name for a class of highly aromatic polyamide fibers that are flame retarded.

Aromatic. Pertaining to benzene-based organic compounds. Contrast with aliphatic compounds, which have straight or branched chains.

Ash. The powdered mineral (inorganic) residue produced by the complete combustion of a material.

Atom. The smallest particle of an element able to be identified as the element.

Banbury. An intensive mixing device using counterrotating, sigma-shaped mixing blades inside a closed chamber. Used to masticate, flux, and thoroughly mix plastic and rubber compounds.

Barrier Plastics. A term used to describe a plastic compound that, when processed into an article, will not allow the passage of products through itself. The barrier is intend to stop gases, aromas, and flavors from passing in either direction.

Bi-. A prefix used in chemical nomenclature meaning "two." May also appear as "di-".

Binder. The portion of a compound that holds the active ingredients together. Usually the compound before the additives are added.

Blackbody Temperature. The temperature of a perfect radiator. (ASTM E 648)

Bleeding. The undesirable movement of materials from the plastic compound to the surface of the finished article. Also called *migration*.

Blow Molding. A process by which a hot plastic tube or parison is forced to take shape by being forced against the inside of a mold by a blast of air from inside the tube.

Blown Film. Plastic film produced by extruding film, passing it around a mandrel, and expanding it to the desired dimensions by forcing air against it from within the mandrel.

Branched. Molecular structure characterized by the presence of side chains attached to the main chain. Contrast with **linear**.

British Thermal Unit. *See* BTU.

BTU or Btu. British Thermal Unit. The amount of energy required to raise 1 pound of water 1 degree Fahrenheit (1°F).

Burn. To undergo combustion. (ASTM E 176) Also, as used in the plastics processing industry, the pyrolysis or degradation of a "batch" of material in the equipment. It is said that the batch is "burned" or "burned up."

Burning Behavior. The manner in which a material acts when it is subjected to a specific ignition source, as in a **fire test**.

Burning Velocity. Speed of a plane (two-dimensional) flame front, normal to its surface and relative to the unburned, gaseous and fuel, oxidizer mixture. (ASTM E 176)

Butadiene, 1,3-. $CH_2{=}CH{-}CH{=}CH_2$. The monomer of polybutadiene.

Butyl Rubber. A synthetic rubber made by the copolymerization of isobutylene with isoprene.

Calendering. The forming of a plastic into film or sheet by forcing the plastic through a machine called a calender, a series of rolls.

Calorie. The amount of energy required to raise 1 gram of water 1 degree Celsius (1°C). The Calorie referred to in connection with food is the Kilogram calorie (Kc), which is 1,000 times the amount of a calorie (c).

Caprolactam. The monomer of polycaprolactam, or nylon-6.

Casting. Forming of shapes by pouring a liquid plastic into a mold or onto a moving belt or other surface.

Catalyst. An additive that controls the speed of a reaction but is not consumed by the reaction.

Cellophane. Regenerated cellulose.

Cellular Plastic. A plastic that has been processed to contain many empty cells throughout its mass. Also referred to as *foamed plastic*.

Cellulosic Plastics. Thermoplastics made by replacing the hydroxide groups of natural cellulose with acidic groups. Also called cellulosics. Includes cellulose acetate, cellulose acetate propionate, cellulose acetate butyrate, cellulose nitrate, cellulose propionate, and cellulose triacetate.

Cellulose Acetate. A thermoplastic acetic ester of cellulose.

Cellulose Acetate Butyrate. A thermoplastic acetic and butyric ester of cellulose.

Cellulose Acetate Propionate. A thermoplastic acetic and propionic acid ester of cellulose.

Cellulose Nitrate. A thermoplastic nitric acid ester of cellulose.

Cellulose Propionate. A thermoplastic propionic acid ester of cellulose.

Cellulose Triacetate. A thermoplastic acetic triester of cellulose.

Chain Length. The number of repeating monomer units in a polymer molecule. The degree of polymerization.

Char. Carbonaceous material formed by pyrolysis or incomplete combustion. (ASTM E 176)

Chemical Compound. A chemical combination of two or more atoms, either from the same elements or from different elements, that is electrically neutral.

Chemical Resistance. The ability of a material to resist chemical reaction by active chemicals.

Chlorinated Polyether. The polymer of chlorinated oxetane.

Class A Fire. A fire involving Class A materials, which include wood, paper, plastics, and cloth.

Class A Materials. Wood, paper, plastics, and cloth.

Class B Fire. A fire involving Class B materials, such as flammable and combustible liquids.

Class B Materials. Flammable and combustible liquids.

Class C Fire. A fire involving charged electrical equipment and wires.

Class C Materials. Charged electrical equipment and wires.

Class D Fire. A fire involving metals.

Class D Materials. Metals.

Coextrusion. The process by which two or more layers of an extruded product are produced. Two (or more) extruders are used, and the two (or more) extrudates are laid on each other while still hot.

Combustible. Capable of **combustion**.

Combustion. Any chemical process that produces heat and light as glowing or flaming.

Combustion Products. Airborne effluent from a material undergoing combustion. (May also include **pyrolysates**) (ASTM E 800)

Comonomer. A monomer polymerized with another monomer to form a copolymer. The copolymer will have many of the properties of each polymer.

Compression Molding. The process by which a compound or a sheet of thermoplastic is placed in a molding cavity while heat is applied to the mold and pressure is applied to the sheet by part of the mold moving against it.

Concentrate. An additive that is a compound containing many more times the active ingredient required by the final article. This concentrate is then added to a "natural" compound at a predetermined rate to achieve the desired result. The concentrate can contain colorants, anti-oxidants, flame retardants, or any other active ingredient. Also called *masterbatch*.

Conduction. Transfer of heat *through* a medium.

Convection. Transfer of heat *with* a medium.

Conversion. *See* Process.

Copolymer. Formed when two monomers are polymerized together. The copolymer will have some properties of polymers made from each monomer.

Corridor. An enclosed space connecting a room or compartment with an exit. The corridor may include normal extensions, such as lobbies and other enlarged spaces. (ASTM E 648)

Covalent Bond. A type of chemical bonding characterized by the sharing of one or more pairs of electrons between the bonded atoms. The resulting chemical compound is called a covalent compound.

Critical Radiant Flux. The level of incident radiant heat energy on the floor-covering system at the most distant flameout point. It is reported as watts per square centimeter (W/cm^2) or BTUs per square foot (BTU/ft^2). (ASTM E 648)

Crosslinking. A process whereby chemical bonds are set up between polymer chains. The process occurs mostly in thermoset resins or by crossblending fillers and other additives with a thermoplastic resin, which then becomes a thermosetting resin.

Crystallinity. The molecular structure of some polymers. The term denotes uniformity and compactness of the molecular chains, forming a crystal-like structure. Contrast with **amorphous**.

Cure. The process by which the chemical properties of a material are changed through a chemical reaction. Usually refers to the chemical reaction involving a thermosetting resin, where the polymerization is called *curing* or *setting up*.

Cyclic. An organic compound structure characterized by the presence of one or more closed rings. The number of rings present is designed by the prefix mono-, bi-, tri-, or poly-.

Deflagration. The very rapid combustion of a material, usually a low explosive. An explosion accompanied by a pressure wave moving at subsonic speed. Compare with **detonation**.

Degradation. A deleterious change in the chemical structure, physical properties, or appearance of a plastic. (ASTM D883)

Density. Weight per unit volume, usually expressed in grams per cubic centimeter (g/cc or g/cm^3), or pounds per cubic foot (lb/ft^3).

Depolymerization. When a polymer reverts to its monomer, or a lower molecular weight polymer, usually occurring upon the polymer's exposure to high temperature.

Detonation. The extremely rapid combustion of a material such as a high explosive; an explosion accompanied by a shock wave moving at supersonic speed. Compare with **deflagration**.

Di-. A prefix used in chemical nomenclature meaning "two." May also appear as "bi-".

Dimer. A molecule formed from two molecules of a monomer.

Dispersion. Finely divided particles of one material evenly mixed throughout another material.

Dyes. Natural or synthetic colorants that are soluble in most solvents and usually dissolve in the polymer in which they are mixed. Contrast with **pigments**.

Ease of Ignition. The ease with which a material or product can be ignited under specified conditions.

EC_{50}. The effective concentration of gas or smoke that will produce a response in 50 percent of the test animals within a specified time. It is a general term and may designate any measured response of the animal.

Elastomer. A material that in its final shape and at room temperature may be stretched to at least twice its length and will return to its original shape rapidly and with some force.

Electroplating. The electrostatic deposition of metal on plastics, molds, or both.

Element. A pure substance that cannot be broken down into simpler substances by chemical means.

Elongation. The increase in length of a material stressed under a specified tension.

Endothermic. A chemical reaction in which heat is absorbed. Contrast with **exothermic**, which is a chemical reaction in which heat is liberated by the reaction.

Epichlorohydrin Rubber. A polyether elastomer with very good resistance to oil, heat, and ozone. Polymerized from the monomer epichlorohydrin.

EPDM. Ethylene–propylene diene modified elastomer.

Epoxy Resins. Thermosetting polymers based on ethylene oxide and its derivatives.

Ester. The reaction product of an alcohol and an acid. It is an organic hydrocarbon derivative with the general formula R—COO—R', where the R— is a hydrocarbon radical or backbone, and the —R' is either the same hydrocarbon radical, or a different one.

Ether. An organic hydrocarbon derivative with the general formula R—O—R', where the R— is a hydrocarbon radical or backbone, and the —R' is either the same hydrocarbon radical or a different one.

Ethylene Vinyl Acetate. A copolymer of ethylene and vinyl acetate. It has many of the properties of polyethylene but has increased flexibility, elongation, and impact resistance. Also called *EVA*.

Exothermic. A chemical reaction in which heat is liberated. Contrast with **endothermic**, which is a chemical reaction in which heat is absorbed by the reaction.

Explosion. The phenomenon characterized by the instantaneous release of heat and pressure caused by the oxidation of fuels or the decomposition of molecules. Contrast with **deflagration** and **detonation**.

Extender. *See* Filler.

Extrudate. The end product of the extrusion process. May be a sheet, film, rod, or a formed product (called a *profile*).

Extruder. A machine that forms continuous sheet, film, rods or profiles by the action of a screw rotating (or in an older machine, a ram or reciprocating plunger) in a barrel and carrying forward and forcing the fused plastic mass through a die that imparts the shape to the plastic as it cools.

Extrusion. The process whereby a shape is imparted to a plastic material by forcing the molten mass through a die.

Extrusion Blow Molding. Process wherein a parison, or hollow tube, is extruded and placed in a mold, and a blast of hot air inside the parison forces it to assume the shape of the mold.

Filler. A material added to a resin or plastic to alter its properties—usually to reduce its cost. May also be called an *extender*.

Film. Sheet with a maximum thickness of 0.01 inch (10 mils).

Film Blowing. The extrusion of a hollow tube followed by continuous inflation of the tube by internal air pressure (no mold involved).

Film Casting. The pouring of a fluid plastic compound or resin onto an endless carrier, followed by removal of the solidified film from the carrier.

Fire. A rapid chemical reaction involving the combination of a material with oxygen or another oxidizer, usually giving off heat and light. *See also* Combustion.

Fire Behavior. Any physical changes that occur in a material or product when that material or product burns.

Fire Effluent. The total amount of gases, particulates, and aerosols liberated during **combustion** or **pyrolysis**.

Fire Endurance. A measure of the elapsed time during which a material or assemblage continues to exhibit fire resistance. (ASTM E 176)

Fire Exposure. The heat flux of a fire, with or without direct flame impingement, to which a material, product, building element, or assembly is exposed. (ASTM E 176)

Fire Gases. The airborne products emitted by a material undergoing combustion or pyrolysis, which, at the relevant temperature, exist in the gas phase. (ASTM E 176)

Fire Hazard. The potential for harm (loss of life, injury, or damage to property) associated with a particular fire.

Fire Load. The total amount of combustible (or flammable) material present in a precisely defined area where a fire might occur. When considering fire load, one must include not only the total amount of fuel present but its total heat combustion and the rate of heat release of the fuel.

Fire Performance Characteristic. A response of a material, product, or assembly to a prescribed source of heat or flame under controlled fire conditions. Such characteristics include ease of ignition, flame spread, smoke generation, fire endurance, and toxicity of smoke. (ASTM E 176)

Fire Performance Test. A procedure that measures a response of a material, product, or assembly to heat or flame under controlled fire conditions. (ASTM E 176)

Fire Resistance. The property of a material or assemblage to withstand fire or give protection from it. (ASTM E 176)

Fire Retardant Chemical. A chemical that, when added to a combustible material, delays ignition and combustion of the resulting material when exposed to fire. (ASTM E 176)

Fire-Retarded Treatment. The use of a fire-retardant chemical or a fire-resistant coating. (ASTM E 176)

Fire Risk. The probability that a fire will occur, and the potential for harm to life and damage to property resulting from its occurrence. (ASTM E 176) Should include at least three measurements: the potential for harm; the probability that the fire will occur; and the probability of exposure of people, animals, property, systems, and the environment.

Fire Risk Assessment Standard. (Formerly *fire hazard standard*). A standardized method of assessing fire risk of a material, product, or assembly in a specific environment or application. (ASTM E 176)

Fire Stop. A through-penetration fire stop is a specific construction consisting of the materials that fill the openings around penetrating items such as cables, cable trays, conduits, ducts, and pipes and their means of support through the wall or floor opening to prevent the spread of fire. (ASTM E 814)

Fire Test. A procedure, not necessarily a standard test method, in which the response of materials to heat or flame, or both, under controlled conditions is measured or otherwise controlled. (ASTM E 800)

Fire Triangle. A theory of fire that states that if fuel, energy, and an oxidizer are brought together in proper amounts, a fire *will* occur.

Flame. A hot, usually luminous, zone of gas of particulate matter in gaseous suspension, or both, undergoing combustion. (ASTM E 176)

Flame Front. The leading edge of a flame propagating through a gaseous mixture or across the surface of a liquid or a solid. (ASTM E 176)

Flame Resistance. The ability to withstand flame impingement or give protection from it. (ASTM E 176)

Flame Retardance. The ability of a material to suppress, reduce, or delay the propagation of flame through a material or product.

Flame Retardant. A substance that, when added to another substance, material, or product, will suppress, reduce, or delay the propagation of flame through that substance, material or product. A reactive flame retardant is built chemically into the polymer molecule, whereas an additive flame retardant is added to the plastic after polymerization.

Flame Speed. Propagation of a flame front per unit of time through a gaseous fuel and oxidizer mixture relative to a fixed reference point. (ASTM E 176)

Flame Spread. The propagation of the flame away from the source of ignition across the surface of a liquid or solid. (ASTM E 176) The tendence of a material to spread flame as it burns.

Flame Spread Index. A number or classification indicating a comparative measure derived from observations made during the progress of the boundary of a zone of flame under defined test conditions. (ASTM E 176)

Flaming Combustion. *See* Space Burning.

Flammability. A measure of a material's propensity to burn or, conversely, its resistance to ignition.

Flammable. Subject to easy ignition and rapid flaming combustion. (ASTM E 176)

Flammable Range. The concentration of gas or vapor in air between the **lower** and **upper flammable limits**. It is the range of concentrations in air in which ignition of the gas or vapor will occur.

Flashover. The phenomenon that may occur during a fire, whereby the surface of everything in a compartment appears to break into flame at the same time.

Flash Point. The minimum temperature of a liquid at which it produces vapors sufficient to form an ignitable mixture with the air at the surface of the liquid or near the container.

Fluorinated Ethylene Propylene. A thermoplastic copolymer of tetrafluoroethylene and hexafluoropropylene.

Fluorocarbon Polymers. The family of polymers made from monomers whose molecules contain at least carbon, hydrogen, and fluorine.

Flux. To make fluid by melting or fusing. A fluxed material is not in the liquid state but rather in a "doughy" plastic state.

Flux Profile. The curve relating incident radiant heat energy on the specimen plane to distance of the point of initiation of flaming ignition (0 cm). (ASTM E 648)

Free Radical. A molecular fragment possessing at least one unpaired electron. It is very active chemically and must react with another free radical rapidly to form a compound. The high reactivity of free radicals is useful in the polymerization process.

Fuel. Anything that will burn.

Fuel Load. The amount and distribution of combustibles in an occupancy. (ASTM E 931)

Functional Group. *See* Radical.

Fuse. To make fluid by melting or fluxing. A fused material is not in the liquid state but rather is in a "doughy" plastic state.

Glowing Combustion. The direct combination of an oxidizer with a solid fuel on the surface of the fuel. No flame will be present, but heat and light may be emitted. Also called *surface burning*.

Halocarbon Plastics. Plastics made from monomers containing only carbon and a halogen or halogens.

Halogen. One of the elements of Group VII on the periodic table of the elements: fluorine, chlorine, bromine, and iodine. Fluorine and chlorine are the halogens present in some polymer molecules. Astatine is also a halogen but it is so rare it will not be encountered in plastics.

Halogen Acid. Hydrofluoric acid, hydrochloric acid, hydrobromic acid, or hydiodic acid.

Halogen Acid Gas. Hydrogen fluoride, hydrogen chloride, hydrogen bromide, or hydrogen iodide.

Halogenated Hydrocarbon. A hydrocarbon compound that has had one or more hydrogen atoms replaced by a halogen (fluorine, chlorine, bromine, or iodine).

Hardener. A substance that brings about the curing of a plastic *(thermosetting)*.

Heat Flux. A measure of heat impinging onto a surface per unit of time, usually from a radiant heating device.

Heat of Combustion. The total amount of energy (heat) released during the complete combustion of a material.

Heat Sink. Any system or device that will absorb heat from a thermal system.

Heat Stress. The adverse effects in humans or animals exposed to heated atmospheres or to radiant heat in a fire.

Homopolymer. The polymer produced by polymerization of a single **monomer**.

Hydrocarbon. A covalent chemical compound containing only carbon and hydrogen.

Hydrocarbon Derivative. A covalent compound made up of a hydrocarbon "backbone" and a functional group.

Hydrocarbon Plastics. Plastics made from resins or monomers containing only hydrogen and carbon.

Hydrogenated. The addition of hydrogen to an organic molecule.

Hygroscopic. Having the ability to absorb moisture from the air.

IC_{50}. The incapacitating concentration: The concentration of gas or smoke that will produce incapacitation in 50 percent of the test animals within a specified exposure time.

IDLH. The amount of gas, vapors, fumes or dust in air that is *i*mmediately *h*armful to *l*ife and *h*ealth.

Ignitability. The ease with which a substance will ignite.

Ignition. The initiation of **combustion**. (ASTM E 176)

Ignition Temperature. The minimum temperature to which a material must be raised before combustion will begin.

Impact Modifiers. Materials added to brittle polymers or compounds to improve their resistance to breaking under impact.

Impact Resistance. The relative ease with which plastic parts break under high speed stress applications. Also called *impact strength*.

Inhibitors. Materials used to prevent or slow the polymerization reaction (or any chemical reaction).

Initiators. Chemicals used to begin the polymerization process.

Injection Blow Molding. The process by which a parison (hollow part) is formed by **injection molding** or **extrusion** and is then formed by a blast of air forcing the parison into the shape of the mold.

Injection Molding. The process by which plastic parts are formed by forcing fused or molten resin or compound into a mold by use of a ram or reciprocating screw. See also **extrusion** and **injection molding**.

Inorganic Compound. *Inorganic* refers to all those compounds that are not compounds of carbon, with the exception of carbon disulfide, the carbon oxides, and the carbon-containing oxysalts. *Inorganic chemistry* is usually described as the chemistry of minerals and compounds made from them. The vast majority of inorganic compounds will not burn.

Inorganic Pigments. Natural occurring or synthesized colorants that are ionic in nature. They may be oxides, sulfides, or other salts. May also be synthetics. Compare with **dyes**.

Inorganic Polymer. A polymer without carbon in its backbone.

Intumescence. The property of swelling and producing a cellular surface when exposed to high temperatures.

Ionic Bonding. A type of bonding in which the parts of the compound are electrically charged and are called ions. Ionic bonding differs from **covalent bonding** in that the parts of an ionic compound (the ions) are held together by the electrostatic attraction of the opposite charges on the ions.

Ionomer. A thermoplastic composed mainly of polyethylene that contains both covalent and ionic bonds.

Isocyanate. A compound containing the isocyanate (—NCO) radical.

Isomers. Molecules containing the same kind and number of atoms but have a different molecular structure of these atoms, and therefore possessing different chemical and physical properties.

Isotactic. A polymeric molecular structure having a sequence of regularly spaced asymmetric atoms arranged similarly along the polymer chain.

Ketone. An organic hydrocarbon derivative with the general formula R—CO—R', where the R— is a hydrocarbon radical or backbone, the —R' is either the same hydrocarbon radical or a different one, and the —C=O is the carbonyl radical.

Lamination. A bonding together of two or more layers of materials usually either all plastic or plastic and nonplastic materials.
Latex. A water dispersion of a polymeric material.
Lay-up Molding. Process in which fluid resin is applied to a layer of reinforcing material, cured, and then formed with or without the use of pressure.
LC_{50}. Lethal concentration. In terms of fire, the concentration of gas or smoke that will kill 50 percent of the test animals within a specified exposure and postexposure time.
Limiting Oxygen Index. (LOI). The limiting concentration of oxygen in the atmosphere necessary for sustained combustion. A material with an LOI of more than 21 should not burn in air at room temperature.
Linear Polymer. A polymer that has the appearance of a chain, with little or no side branching.
Lithopone. An inorganic white pigment used in plastics made from barium sulfate, zinc oxide, and zinc sulfide.
Loading Level. The amount of additive added to a plastics compound.
Lower Flammable Limit. (LFL) The minimum percentage of gas or vapor in air below which ignition will not occur (the mixture is too lean). Also called the *Lower Explosive Limit* (LEL).
Lubricant. Materials that reduce friction between two surfaces.

Macromolecule. In plastics chemistry, the giant molecule formed by polymerization.
Mass Burning Rate. Mass loss per unit of time by materials burning under specified conditions. (ASTM E 176)
Masterbatch. A plastics compound containing a very high concentration of an active ingredient. Also called **concentrate**.
Melt Index. The measurement of the amount of a material that can be forced through a particular opening (0.0825 inch) by a specific weight (2160 grams) in a specified time (10 minutes) at a specified temperature (190°C or 374°F). (ASTM D 1238)
Mer. The repeating molecular unit in any polymer.

Metalizing. The application of a thin coating of metal to a nonmetallic surface.

Methyl Methacrylate. The monomer of polymethyl methacrylate (PMMA). Its formula is $CH_2{=}CCH_3COOCH_3$.

Migration. *See* Bleeding.

Modacrylic. Any synthetic fiber containing more than 35 percent but less than 85 percent acrylonitrile ($CH_2{=}CHCN$).

Mold. To shape a plastic by confining it in a closed cavity.

Molecular Weight. The total weight of all atoms in a molecule.

Molecule. The smallest particle of a compound that can still be recognized as the compound. Consists of two or more atoms bound together chemically by covalent bonds and is electrically neutral.

Mono-. A prefix used in chemical nomenclature meaning "one."

Monomer. A very small molecule that has the capability of chemically combining with itself to form a giant molecule called a polymer. *Mono* = one; *mer* = part.

Natural. In referring to a plastics compound, a natural compound is one that has no color or other special property additive included. May contain a plasticizer, stabilizer, filler, or other additive.

Natural Polymers. Those polymers produced in nature, as opposed to synthetic polymers. Natural polymers include cellulose (the major ingredient in wood and plants), cotton (almost pure cellulose), wool, silk, leather, and human skin.

Neoprene. Commercial name for polychloroprene.

Nitrile Rubber. Polybutadiene copolymerized with acrylonitrile.

Nomex™. Trade name for an aramid (highly aromatic polyamide) fiber or fabric.

Noncombustible. Having the property of not being capable of combustion under specified test conditions.

Nonflammability. Having the property of not being capable of burning with a flame under specific test conditions.

Nylon. The generic name for polyamides.

Occupancy. The purpose for which a building or portion therefore is used or intended to be used or, an occupied building. (ASTM E 931)

Olefin Plastics. The polyolefins (polyetheylene, polypropylene, polybutylene, polyisobutylene, and polymethylpentene).

Opacity of Smoke. The ratio of incident luminous flux (I) to transmitted luminous flux (T) through smoke under specified test conditions. (ISO/IEC Guide 52:1989)

Optical Density of Smoke. *D* is a measure of the attenuation of a light beam passing through smoke, expressed as the common logarithm of the ratio of the incident flux, I_o, to the transmitted flux, I. (D = $\log_{10}$ (I_0/I). (ASTM E 176) Also the common logarithm of the **opacity of smoke**.

Organic Compound. Organic pertains to chemical compounds that were once part of living things, but now may be manufactured. These are generally the hydrocarbon compounds and their derivatives.

Organic Peroxides. A group of highly dangerous hazardous materials. Used as initiators for thermoplastics and curing agents for thermosets, they are highly reactive oxidizing agents that burn, and they can start their own decomposition process when contaminated, heated, or shocked.

Organic Pigments. Colorants that are covalently bonded, not including dyes. Organic pigments are generally not soluble in the plastic in which they are dispersed. Contrast with **dyes**.

Oxidation. Originally, oxidation was defined as the combination of any substance with oxygen. Today, the technical definition has been broadened to include any chemical reaction in which electrons have been transferred.

Oxygen Index. *See* Limiting Oxygen Index.

Parison. In blow molding, it is the hollow tube that is extruded or injection molded, placed inside a mold, and is forced to assume the shape of the mold by an internal blast of air.

Peroxide. An organic hydrocarbon derivative with the general formula R—OO—R', where the R is a hydrocarbon radical or backbone, the R' is either the same hydrocarbon radical, or a different one, and the —O—O— is the peroxide radical. An inorganic peroxide is an ionic compound in which hydrogen or metals are attached to the peroxide ion ($—O—O—^{-2}$).

Phenolic Resin. A thermosetting polymer made from the condensation of phenol and formaldehyde. Also called phenol–formaldehyde resin since it is the most important resin in the family.

Phenoxy Resin. A thermoplastic polymer based on the reaction between bisphenol-A and epichlorohydrin.

Pigment. A material that imparts color to a substrate. Usually an inorganic or organic compound not soluble in the medium in which it is dispersed. Pigments that are not compounds include carbon black and finely divided metals. Contrast with **dye**.

Piloted Ignition. Initiation of combustion as a result of contact of a material or its vapors with an external high energy source such as a flame, spark, electrical arc, or glowing wire. (ASTM E 176)

Plastic. A material that contains as an essential ingredient one or more an organic polymeric substances of large molecular weight, is solid in its finished state, and, at some stage in its manufacture or processing into finished articles, can be shaped by flow. (ASTM D 883)

Plasticizer. A material added to a resin or polymer to increase its softness and flexibility.

Plastics Compound. A mixture of resin and the necessary additives to give the resin the required properties of the finished part.

Plastisol. A suspension of polyvinyl chloride in a liquid plasticizer. The resulting compound is a liquid.

Poly-. A prefix meaning many. A polymer means many (poly) parts (mer).

Polyacetylene. The polymer of acetylene that has high intrinsic electrical conductivity.

Polyacrylamide. A water-soluble polymer used as a thickening agent.

Polyacrylates. A family of thermoplastics made from acrylic monomers. Also called *acrylics*.

Polyallomers. A family of crystalline polymers made from two or more olefin monomers.

Polyamide. A family of polymers in which amide groups link the structural units. Also called **nylon**.

Polyarylates. An aromatic engineering thermoplastic.

Polybenzimidazole. A polymer with no melting point. It begins to char at 1200°F. It can be spun into fibers

Polybutadiene. An **elastomer** made from the polymerization of butadiene.

Polybutylene. (PB) A polyolefin thermoplastic made by the polymerization of butylene (butene). Also called *polybutene*.

Polybutylene Terephthalate. (PBT) A thermoplastic polyester.

Polycarbonate. (PC) A thermoplastic made from the reaction of bisphenol-A with phosgene. (PC)

Polychloroprene. A synthetic rubber also known as **neoprene**.

Polyesters. Polymers formed by the reaction between dibasic acids and dihydroxy alcohols.

Polyethers. A large family of polymers containing the ether group (—C—O—C—).

Polyethylene. (PE) A thermoplastic polyolefin made by the polymerization of ethylene (ethene). Also called *polyethene*. The largest volume plastic in the world.

Polyethylene Terephthalate. (PET) A thermoplastic polyester made from the polymerization of ethylene glycol and terephthalic acid.

Polyformaldehyde. *See* Acetal Resins.

Polyimide. (PI) Thermoplastics made by reacting aromatic dianhydrides with aromatic diamines.

Polyisobutylene. The polymer made from the polymerization of isobutylene.

Polymer. Generally, a "giant" molecule made up of thousands of "tiny" molecules called monomers, which are linked together in a long chain-like structure.

Polymerization. The process by which monomers combine with themselves to form giant molecules called polymers in a controlled atmosphere (inside a large vat called a reactor). When the reaction is in a runaway (uncontrolled) mode, the result is *instant polymerization,* which in large quantities is accompanied by a violent and destructive explosion.

Polymethyl Methacrylate. (PMMA) A thermoplastic acrylic polymer made from the polymerization of methyl methacrylate.

Polyol. An alcohol containing three or more hydroxy groups. A "common" alcohol contains just one hydroxy group and has the general formula R—OH, where the R is a hydrocarbon radical or backbone.

Polyolefin. A polymer made from an alkene (olefin) monomer. The major polyolefins are polyethylene, polypropylene, polybutylene, and polymethylpentene.

Polyoxmethylene. *See* Acetal Resins.

Polyphenylene Oxide. (PPO) An engineering thermoplastic polyether.

Polypropylene. (PP) A polyolefin thermoplastic made by the polymerization of propylene (propene).

Polysilyne. Polymer based on silicon.

Polystyrene. (PS) A hydrocarbon thermoplastic made by the polymerization of styrene.

Polytetrafluoroethylene. (PTFE) A thermoplastic fluorocarbon polymer made by the polymerization of tetrafluoroethylene. Also known as Teflon™.

Polyurethane. (PUR) A family of polymers made by reacting diisocyanate with organic compounds having two free hydrogens.

Polyvinyl Acetal. A family of thermoplastic polymers made by reacting polyvinyl alcohol with an aldehyde.

Polyvinyl Acetate. A thermoplastic polymer produced by polymerizing vinyl acetate. Used widely in adhesives.

Polyvinyl Alcohol. A water-soluble thermoplastic whose monomer is vinyl alcohol.

Polyvinyl Butyral. A member of the polyvinyl acetal family.

Polyvinyl Chloride. (PVC) A halogenated thermoplastic that has more uses than any other plastic. May be rigid, semiflexible, or flexible.

Polyvinyl Fluoride. (PVF) A fluorocarbon thermoplastic made by the polymerization of vinyl fluoride.

Polyvinyl Formal. A member of the polyvinyl acetal family.

Polyvinylidene Chloride. A halogenated thermoplastic whose trademarked commercial name is Saran™.

Premix. A mixture of polyester resin and fillers.

Prepolymer. A polymer of intermediate molecular weight, somewhere between the monomer and final polymer.

Prepreg. A mat of reinforcing fibers that have been impregnated with a thermosetting resin that is partially cured.

Process. The method by which plastics resins, compounds, alloys, and blends are converted from the powdered, liquid, cubed, or pelletized state into a finished or semifinished shape. Processes include extrusion, injection molding, blow molding, calendering, casting, and forming, among others. May be referred to as *conversion* or *conversion process*.

Processing Equipment. Machines and other equipment used to convert plastics resins, compounds, alloys, and blends from the powdered, liquid, cubed, or pelletized state into a finished or semifinished shape. Such equipment includes but is not limited to extruders, injection molders, blow molders, blown film towers, calenders, casting lines, mixers, blenders, driers, metering equipment, mills, intensive mixers, dissolvers, presses, and forming equipment.

Pyrolysis. Irreversible chemical decomposition caused by heat, usually without oxidation. (ASTM E 176) Pyrolysis of polymers can produce shorter-chain (lower molecular weight) polymers or the original monomer.

Pyrophoric. Having the property of rapid or violent combustion produced simply by contact with air.

Pyrolysate. A product of pyrolysis.

Radiation. Pertaining to heat, it is the transfer of heat without use of a medium. Contrast with **convection**.

Radical. A molecular fragment that, when attached to another molecular fragment or a hydrocarbon backbone, will impart specific properties to the new compound. May also be called a *functional group*.

Rate of Heat Release. The amount of heat released over time by a material as it burns. The rate must be high for the material to spread flame to adjoining material.

Rayon. Originally a name for yarns made from regenerated cellulose, but now may include cellulose acetate or cellulose triacetate.

Reactant. Any substance that takes part in a chemical reaction.

Regrind. Rejected or recycled plastic parts and any other parts of the finished molding that are removed from the part (or anything classified as scrap) and ground up into small particles to be reprocessed with virgin (unused) material.

Reinforced Plastic. A plastic resin or compound into which fillers are added to greatly increase the strength or other physical properties of the finished part.

Repeating Unit. The identifiable fragment of the monomer(s) that formed the polymer. When monomers combine with themselves to form a polymer, they line up in clearly distinguishable units that repeat themselves throughout the polymer.

Resin (natural). Solid or semisolid viscous substances that are secretions of certain plants and trees.

Resin (synthetic). A solid, semisolid, or pseudosolid organic material that has an indefinite and often high molecular weight, exhibits a tendency to flow when subjected to stress, usually has a softening or melting range, and usually fractures conchoidally (following an involutely-curved surface). (ASTM D 883)

Retarder. An **inhibitor**.

Rotation Casting. The forming of plastic articles by adding to a rotating mold a fluid plastic material and rotating the mold so the fluid is distributed on the mold walls and heating until the plastic material has hardened.

Rotation Molding. Same as **rotation casting** except that a dry powdered plastic is used.

Rubber. An **elastomer**.

Saturated. Referring to hydrocarbon compounds that contain only single covalent bonds between the carbon atoms.

Self-heating. An **exothermic** reaction within a material resulting in a rise in temperature in the material. (ISO/IEC Guide 52:1989)

Self-ignition. Ignition resulting from **self-heating**. (ISO/IEC Guide 52:1989)

Set. To change from an uncured state to a cured state; **cure**.

Smoke. The airborne solid and liquid particulates and gases evolved when a material undergoes pyrolysis or combustion. (ASTM E 176)

Smoldering. Combustion of a solid without flame, often evidenced by visible smoke. (ASTM E 176)

Solid Casting. The pouring of a liquid resin into a mold and curing the resin to a solid part.

Soot. Carbonaceous material, usually very finely divided particles, produced by the incomplete combustion of carbon-containing materials.

Space Burning. A phenomenon of combustion that occurs when a flame exists. Flames are present when fuel mixes with the air during combustion, releasing heat and light. Also called *flaming combustion.*

Spontaneous Ignition. The process by which oxygen combines slowly with a fuel, usually on its surface, with the slow evolution of heat energy. The heat energy is absorbed by the fuel, raising its temperature slowly, until the fuel's ignition temperature is reached and it breaks into flame.

Spray-up. The application of resin and reinforcement into a mold by use of a spray gun.

Stabilizer. A chemical used in the production of plastics compound to maintain the polymer's original properties. Stabilizers are usually meant to protect against heat, light, and/or oxidation.

STEL. Short Term Exposure Limit. The maximum amount of gas, vapor, dust or fumes to which a person may be exposed for a short time, usually 15 minutes, without being harmed.

Styrene. $CH_2{=}CHC_6H_5$. Technically, the monomer of polystyrene, but polystyrene is often referred to as styrene.

Styrene–acrylonitrile. (SAN) A thermoplastic made from the copolymerization of styrene and acrylonitrile.

Styrene–butadiene. (SB) An **eslastomer** made from the copolymerization of styrene and butadiene.

Surface Burning. *See* Glowing Combustion.

Synergist. A substance that, when added to a compound in conjunction with another substance, causes an enhancement of the properties of the other substance. It is said to cause **synergism**.

Synergism. A phenomenon in which the chemical effect of two materials acting together is greater than the effect that would be produced if each material acted individually and the effects were added together.

Synthesize. To manufacture a molecule duplicating a molecule made in nature.

Synthetic. Manufactured, as opposed to being made in nature.

Synthetic Polymers. Polymers that are manufactured, as opposed to natural polymers. Plastics are synthetic polymers.

Tensile. Pertaining to tension. Capable of being stretched.

Tensile Strength. Resistance of a material to a force tending to tear it by stretching. The maximum stress sustained by a specimen during a tension test. The pulling stress required to break a specimen. Measured in psi.

Test Assembly. The wall or floor into which the test sample(s) is (are) mounted or installed. (ASTM E 800)

Tetrahedron of Fire. A theory of fire stating that four elements are necessary to have a fire: fuel, energy, an oxidizer, and the chain reaction of burning.

Thermoforming. A plastic sheet is formed into a finished part by clamping it onto a frame over a mold, causing it to soften by heating it, and then applying pressure to make it conform to the mold.

Thermoplastics. Resins or compounds that may be formed over and over again upon applying heat and pressure.

Thermosets. Resins or compounds that may be formed and cured once, with any subsequent heat causing the plastic to degrade or "burn."

Transfer Molding. Resins or plastic compounds are placed in a heated vessel above the mold and, once fluxed, can be forced into the mold by the pressure of a ram or plunger. Usually used for **thermosets**.

Trimer. A molecule formed from three molecules of a monomer.

TLV–TWA. Threshold Limit Value–Time Weighted Average. The average (weighted over time) maximum amount of gas, vapor, dust, or fumes to which a person may be exposed for 8 hours a day, 5 days a week, without harm.

Total Flux Meter. Instrument used to measure the level of radiant heat energy incident on the specimen plane at any point. (ASTM E 648)

Toxicity. The harmful effects of a chemical on some biologic mechanism. The ability of a chemical substance to produce injury once it reaches a susceptible site in or near the body.

Toxicology. The study of the harmful effects of chemicals on biologic tissue.

Uncontrolled Polymerization. A rapid chemical reaction in which polymerization occurs without an opportunity for the heat to be dissipated. The result may be an **explosion**.

Unsaturated. Referring to hydrocarbon compounds that contain at least one multiple covalent bond between two carbon atoms.

Upper Flammable Limit. The maximum percentage of gas or vapor in air above which ignition will not occur (the mixture is too rich). Also called the *Upper Explosive Limit* (UEL).

Urea-formaldehyde. A thermosetting polymer made from the reaction of urea and formaldehyde.

UV Stabilizer. An additive that protects materials from the harmful effects of ultraviolet rays by selectively absorbing the rays. The ultraviolet is defined as radiation in the region of the electromagnetic spectrum including wavelengths from 100 to 3,900 angstroms, just below the visible light region. The angstrom is approximately one one-hundred-millionth (10^{-8}) centimeter.

Vacuum Forming. A means of thermoforming in which the pressure applied is that of a vacuum.

Vinyl Resins. Any polymer made from a monomer containing the vinyl group ($CH_2{=}CH{-}$), but usually restricted to polyvinyl acetate, alcohol, butyral, chloride, and formal resins.

Weatherometer. A device that simulates, in one manner or another, conditions to which a finished article would be exposed when used outdoors.

Appendix D

Fire Tests and Miscellaneous Tables

Ignition Temperatures of Various Plastics

Polymer	Ignition Temperature
ABS	780°F to 915°F
acetal (polyoxymethylene)	825°F
acrylics	806°F to 1040°F
cellulose acetate	887°F
cellulose nitrate	286°F
cellulose triacetate	1004°F
ethyl cellulose	565°F
melamine	1153°F to 1193°F
modacrylic	1274°F
nylons	795°F to 990°F
phenolics	1060°F to 1076°F
polyacrylonitrile	896°F
polycarbonate	968°F to 1076°F
polyester (thermoset, glass-filled)	653°F to 752°F
polyethylene	660°F
polyethylene terephthalate	896°F
polyimides	1000°F
polypropylene	806°F to 824°F
polystyrene	910°F to 1000°F
polytetrafluoroethylene	986°F to 1076°F
polyurethanes	780°F
rigid polyvinyl chloride	1035°F
flexible polyvinyl chloride	850°F
polyvinylidene chloride	986°F
rayon	788°F
silicone rubber	860°F to 896°F
styrene–acrylonitrile	690°F
styrene–butadiene rubber	680°F
unsaturated polyesters	900°F to 910°F

Heats of Combustion of Nonplastic Materials

Material	Heat of Combustion, BTU/lb.
acetone	13,225
acetylene	21,450
asphalt	15,800
butane	21,300
coal (average)	13,000
corrugated fiber carton	5,975
cotton	7,150
cottonseed oil	17,100
douglas fir	9,050
ethyl alcohol	12,800
ethyl benzene	22,800
ethyl ether	22,000
fuel oil, No. 1	19,800
fuel oil, No. 6	18,300
hydrogen	60,950
lubricating oil	20,400
methyl ethyl ketone	16,075
newsprint	8,475
octane	22,525
paraffin wax	20,100
propane	21,650
silk	9,800
tallow	17,100
toluene	18,250
wood (average)	8,675
wool	9,800
wrapping paper	7,100

Limiting Oxygen Indices

Polymer	Percentage of Oxygen
ABS	18.0 to 39.0
acetal (polyoxymethylene)	14.7 to 16.0
acrylics	16.6 to 19.4
alkyd	29.0 to 63.4
cellulose	19.0
cellulose acetate	18.0 to 27.0
cellulose acetate butyrate	18.0 to 20.0
cellulose butyrate	18.8 to 19.9
cellulose triacetate	18.0 to 27.0
chlorinated polyethylene	21.1
chlorinated polyvinyl chloride	60.0 to 70.0
epoxy	18.3 to 49.0
melamine	30.0 to 60.0

Limiting Oxygen Indices *(Continued)*

Polymer	Percentage of Oxygen
modacrylic	26.7 to 30.0
natural rubber	17.2
nylons	20.0 to 30.1
phenolics	29.0 to 66.0
polyacrylonitrile	18.2
polyallomer	17.0
polybutadiene, crosslinked	18.3
polycarbonate	22.5 to 44.0
polyetherether ketone	35.0
polyester (thermoset, glass-filled)	22.0 to 46.0
polyethylene	17.4
polyethylene terephthalate	20.0 to 22.7
polyimides	36.5 to 40.6
polyphenylene oxide	15.0 to 34.0
polyphenylene sulfide	43.0 to 52.0
polypropylene	17.0 to 28.0
polystyrene	18.1
polysulfone	30.0 to 51.0
polytetrafluoroethylene	95.0
polyurethanes	17.0 to 21.4
rigid polyvinyl chloride	45.0 to 49.0
flexible polyvinyl chloride	19.0 to 40.0
polyvinyl fluoride	22.6
polyvinylidene chloride	60.0
polyvinylidene fluoride	43.7
rayon	18.9
silicone rubber	25.8 to 41.0
styrene–acrylonitrile	18.0 to 28.0
styrene–butadiene rubber	16.0 to 19.0
unsaturated polyesters	22.0 to 46.0
ureaformaldehyde	30.0

Individual Fire Tests and Standards

Tests and standards are listed here by their best-known name. At time of publication, these names were the accepted designations. In some cases, the designation is not really the final name of the test or standard but may be the name of the committee or working group within an organization charged with developing the test and bringing it to acceptance by the organization. Some of these tests may be obsolete and no longer in use. However, since many of them are still quoted in the literature, they are listed below. For the up-to-date list of tests required by any jurisdiction or standards group, that group or organization should be contacted. Their addresses are listed in Appendix F.

Descriptions of many of the tests and standards listed here have been presented in Chapter 8. Not all the tests and standards listed here are described. Not all of the tests and standards listed are for plastics only. For further information on each of the standards or tests, the sponsoring organization should be contacted.

By no means is this a comprehensive list of all fire tests and standards. This list represents tests and standards in existence in the United States, Canada, and Europe. No attempt has been made to include the tests and standards from the U.S.S.R., Japan, Australia, or other countries or regions that may issue them. ISO (International Organization for Standardization) standards are included because they will probably replace many European, Canadian, and United States standards in the future.

There are many other tests covering many materials other than plastics, some of which are listed here. Some of the tests by one organization may be very close to, or an exact duplicate of, another test. Such similarities are noted. Some of the standards listed do not describe tests but give definitions, suggest vocabulary, or are compilations of other tests and standards.

In the case of ISO (International Organization for Standards), a DP (for Draft Proposal) may appear in the title. At the time of publication, the standard was being considered for adoption.

Standards developed by individual countries in Europe are listed. When the European Economic Community is formed at the end of 1992, ISO standards may prevail—and then again they may not. The countries not scheduled for entry into the European Economic Community may retain their own standards—and then again, they may not.

Standards developed by the Federal Republic of Germany (West Germany) are listed only as Germany. The standards developed by the German Democratic Republic (East Germany) are listed as GDR. At the time of publication, it is not known which standards will prevail, now that these countries are unified.

In some countries a single standard may cover all the fire tests normally covered by several standards in another country. Germany and Great Britain are examples. The standard or standards cited have many parts, and the methods of testing are embodied in these parts and subparts.

Identification Number	Sponsoring Organization	Name of Test
14 CFR Parts 25 and 121	US Government	Improved Flammability Standards for Materials Used in the Interiors of Transport Category Airplane Cabins
16 CFR 1610	CPSC	Standard for the Flammability of Clothing Textiles (General Wearing Apparel) (See also ASTM 1230)

Identification Number	Sponsoring Organization	Name of Test
16 CFR 1611	CPSC	Standard for the Flammability of Vinyl Plastic Films (General Wearing Apparel)
16 CFR 1615	CPSC	Standard for the Flammability of Children's Sleepwear: Sizes 0 Through 6X (FF3-71)
16 CFR 1616	CPSC	Standard for the Flammability of Children's Sleepwear: Sizes 7 Through 14 (FF5-74)
16 CFR 1630	CPSC	Standard for the Surface Flammability of Carpets and Rugs (FF1-70) (See also ASTM 2859)
16 CFR 1631	CPSC	Standard for the Surface Flammability of Small Carpets and Rugs (FF1-70) (See also ASTM 2859)
16 CFR 1632	CPSC	Standard for the Surface Flammability of Mattresses and Pads (FF4-72)
16 CFR 1633	CPSC	*Draft* Proposed Standard for the Flammability (Cigarette Ignition Resistance) of Upholstered Furniture (PFF 6-81)
Arapahoe Smoke Test	Arapahoe Chemical Company	This test is also known as the ASTM Standard Test D 4100 for Gravimetric Determination of Smoke Particulates from Combustion of Plastic Materials.
ANSI/UL 263	UL	Identical to ASTM E 119
Berkeley Test	University of California	Eight Foot Corner Test for Flame Spread
BFD IX-1	Boston Fire Department	Flammability of Upholstery, Curtains, Drapes, and Fabric Wall Coverings Used in Public Assemblies
BIFMA F-1-78	BIFMA	Flammability Standard for Business Markets. A. Small flame ignition B. Cigarette ignition
BNQ-7002-500 1982-05-31	Bureau de Normalisation du Québec	Textiles–Flame Resistance—Vertical Burning Test
BNQ-7002-510 1982-04-21	Bureau de Normalisation du Québec	Textiles-Flame Resistance—45° Angle Test
BNQ-7002-520 1982-09-17	Bureau de Normalisation du Québec	Textiles—Flame Resistance—Rate of Burning
BNQ-7002-530 1982-09-16	Bureau de Normalisation du Québec	Textiles—Flame Resistance—Ease of Ignition
BNQ-7002-580 1982-03-16	Bureau de Normalisation du Québec	Textiles—Flame Resistance—Selection of Materials

Identification Number	Sponsoring Organization	Name of Test
BNQ-7002-590 1982-09-16	Bureau de Normalisation du Québec	Textiles Burning Behaviour—Flame Resistance Classification
BNQ-7002-595 1982-04-19	Bureau de Normalisation du Québec	Textiles—Burning Behaviour, Determination of Oxygen Index
BS 476	British Standard	Fire Performance of Building Materials and Structures
BS 2782	British Standard	Test Methods for Plastics
BS 4066	British Standard	Tests on Electrical Cables Under Fire Conditions
BS 4735	British Standard	Horizontal Burning Characteristics of Small Specimens of Cellular Plastics and Cellular Rubber
BS 5111	British Standard	Smoke Generation Characteristics of Cellular Plastics and Cellular Rubber Materials
BS 6203	British Standard	Fire Characteristics and Fire Performance of Expanded Polystyrene Used in Building Applications
BS 6334	British Standard	Flammability of Solid Electrical Insulating Materials When Exposed to an Ignition Source
BS 6387	British Standard	Performance Requirements for Cables Required to Maintain Circuit Integrity Under Fire Conditions
BS 6401	British Standard	Laboratory Measurements of Specific Optical Density of Smoke Generated by Materials
BS 6458	British Standard	Fire Hazard and Testing for Electrochemical Products
BS 6336	British Standard	Development and Presentation of Fire Tests and Their Use in Hazard Assessment
BS PD 6520	British Standard	Guide to Fire Test Methods for Building Materials and Elements of Construction
CAN/CGSB 4.2 No. 27-M	Canada	Textile Test Methods, Flame Resistance—Selection of Methods
CAN/CGSB 4.2 No. 27.1-M87	Canada	Textile Test Methods, Flame Resistance—Vertical Burning Test
CAN/CGSB 4.2 No. 27.2-M87	Canada	Textile Test Methods, Flame Resistance—Surface Burning Test
CAN/CGSB 4.2 No. 27.3-M86/ ISO 6941-1984	Canada	Textile Test Methods, Textile Fabrics Burning Behavior—Measurement of Flame Spread Properties of Vertically Oriented Specimens
CAN/CGSB 4.2 No. 27.3-M86/ ISO 6940-1984	Canada	Textile Test Methods, Textile Fabrics Burning Behavior—Determination of Ease of Ignition of Vertically Oriented Specimens
CAN/CGSB 4.2 No. 27.5-M87	Canada	Textile Test Methods, Flame Resistance—45° Angle Test—One Second Flame Impingement

Identification Number	Sponsoring Organization	Name of Test
CAN/CGSB 4.2 No. 27.6-M84	Canada	Textile Test Methods, Flame Resistance—Methenamine Test Tablet for Textile Floor Coverings
CAN/CGSB 4.2 No. 27.7-M	Canada	Textile Test Methods—Combustion Resistance of Mattresses—Cigarette Test
CAN/CGSB 4.162 M80 (carpets)	Canada	Hospital Textiles—Flammability Performance Requirements
CAN/CGSB 4.162 M80 (apparel)	Canada	Hospital Textiles—Flammability Performance Requirements
CAN/CGSB 4.162 M80 (Bed Linen)	Canada	Hospital Textiles—Flammability Performance Requirements
CAN/CGSB 4.162 M80 (Drapes and Curtains)	Canada	Hospital Textiles—Flammability Performance Requirements
CAN/CGSB 4.162 M80 (Mattresses)	Canada	Hospital Textiles—Flammability Performance Requirements
CAN/CGSB-4.175 M87/ISO 4880-1984	Canada	Burning Behavior of Textiles and Textile Products—Vocabulary
CAN-S102-M83	Canada	Test Method for Surface Burning Characteristics of Building Materials and Assemblies
CAN 4-S102.2 M83	Canada	Test Method for Surface Burning Characteristics of Flooring, Floor Coverings, and Miscellaneous Materials and Assemblies
CAN 4 S109 M80	Canada	Standard for Flame Tests of Flame-Resistant Fabrics and Films (See also UL 214)
CAN 4-S117.1 M80	Canada	Test Method for Flame Resistance Methenamine Tablet Test for Textile Floor Coverings
Chrysler LP-463KC-13-01	Chrysler	Similar to Motor Vehicle Safety Standard #302
CPAI-84	IFAI	Specification for Flame-Resistant Camping Tentage Material
CSE RF 1/75/A	Italy	Edge Application of Flame: Small Burner Test
CSE RF 2/75/A	Italy	Surface Flame Application: Small Burner Test
CSE RF 3/77	Italy	Spread of Flame Test
CSE RF 4/83	Italy	Ignitability of Upholstered Furniture
CSN 73 0802	Czechoslovakia	Fire Protection of Buildings—General Regulations

Identification Number	Sponsoring Organization	Name of Test
CSN 73 0823	Czechoslovakia	Determination of Fire Classification for Building Materials
CSN 73 0851	Czechoslovakia	Determination for Fire Resistance of Building Construction
CSN 73 0852	Czechoslovakia	Determination of Fire Resistance of Fire Shutters
CSN 73 0855	Czechoslovakia	Determination of Fire Resistance of External Walls
CSN 73 0851	Czechoslovakia	Determination of Fire Resistance of Suspended Ceilings
CSN 73 0861	Czechoslovakia	Determination of Non-Combustibility of Building Materials
CSN 73 0862	Czechoslovakia	Determination of Flammability of Building Materials
CSN 73 0863	Czechoslovakia	Determination of Flame Spread of Building Materials
D 229	ASTM	Standard Method of Testing Rigid Sheet and Plate Materials Used for Electrical Insulation (contains two methods of fire testing)
D 470	ASTM	Thermosetting Wire and Cable Insulation
D 568	ASTM	Rate of Burning Flexible Plastics 0.050 inches and Thinner (Vertical)
D 635	ASTM	Rate of Burning Rigid Plastics 0.050 inches and Thicker (Horizontal)
D 757	ASTM	Incandescence Resistance of Rigid Plastics
D 876	ASTM	Nonrigid Vinyl Chloride Polymer Tubing Used for Electrical Insulation
D 1000	ASTM	Pressure-Sensitive Adhesive-Coated Tapes Used for Electrical and Electronic Applications
D 1230	ASTM	Test Method for Flammability of Clothing Textiles
D 1433	ASTM	Rate, Extent and Time of Burning of Flexible Thin Plastic Sheeting
D 1518	ASTM	Test Method For Thermal Transmittance of Textile Materials
D 1692	ASTM	Rate of Burning Cellular Plastics, Horizontal Position (Discontinued)
D 1929	ASTM	Ignition Properties of Plastics Procedure B
D 2584	ASTM	Test Method for Ignition Loss of Cured Reinforced Resins
D 2633	ASTM	Thermoplastic Wire and Cable Insulation
D 2671	ASTM	Heat-Shrinkable Tubing for Electrical Use
D 2843	ASTM	Smoke Density from Burning of Plastics
D 2859	ASTM	Flammability of Finished Textile Floor Covering Materials

Identification Number	Sponsoring Organization	Name of Test
D 2863	ASTM	Oxygen Index Flammability Test
D 3014	ASTM	Flame Height, Time of Burning, and Loss of Weight of Rigid Thermoset Cellular Plastics in a Vertical Position
D 3411	ASTM	Test Methods for Flammability of Textile Materials
D 3659	ASTM	Test Method for Flammability of Apparel Fabrics by Semi-Restraint Method
D 3675	ASTM	Radiant Panel Test for Surface Flammability of Cellular Plastics
D 3713	ASTM	Method for Measuring Response of Solid Plastics to Ignition by a Small Flame
D 3801	ASTM	Extinguishing Characteristics of Solid Plastics, Vertical Position
D 3874	ASTM	Ignition of Materials by Hot Wire Sources
D 3894	ASTM	Fire Response of Rigid Cellular Plastics Using a Small Corner Configuration
D 4100	ASTM	Method for Gravimetric Determination of Smoke Particulates from Combustion of Plastic Materials (Arapahoe Smoke Chamber)
D 4108	ASTM	Test Method for Thermal Protective Performance of Materials for Clothing by Open-Flame Method
D 4151	ASTM	Test Method for Flammability of Blankets
D 4205	ASTM	Flammability and Combustion Testing of Rubber and Rubber-Like Material
D 4372	ASTM	Specification for Flame-Resistant Materials Used in Camping Tentage
D 4391	ASTM	Terminology Related to the Burning Behavior of Textiles
D 4549	ASTM	Standard Specification for Polystyrene Molding and Extrusion Materials
D 4723	ASTM	Standard Index of and Descriptions of Textile Heat and Flammability Test Methods and Performance Specifications
D 4804	ASTM	Determining the Flammability Characteristics of Nonrigid Solid Plastics
D 4986	ASTM	Standard Test Method for Horizontal Burning Characteristics of Cellular Polymeric Materials
D 5048	ASTM	Standard Test Method for Measuring the Comparative Burning Characteristics and Resistance to Burn-Through of Solid Plastics Using a 125 mm Flame
DIN 0304	Germany	Thermal Properties of Electrical Insulating Materials; Flammability Under Action of an Ignition Source

Identification Number	Sponsoring Organization	Name of Test
DIN 0472	German Standard	Testing of Cables, Wires and Flexible Cords; Burning Behavior
DIN 4102	Germany	Fire Performance of Building Materials and Components
DIN 50 051	Germany	Burning Behavior of Materials
DIN 51 900	Germany	Test for Solid and Liquid Fuels
DIN 53 436	Germany	Toxicity (Test Also Used to Determine Smoke Density)
DIN 53 438	Germany	Testing of Combustible Materials; Response to Ignition by a Small Flame
DIN 54 332	Germany	Floor Covering Test (Small Burner Test)
DIN 54 336	Germany	Burning Behavior of Vertically Oriented Textile Specimens—Ignition at Lower Edge
DIN 57 472	Germany	Determination of Corrosivity of Combustion Gases (wires, cables and flexible cords)
DOC FF1-70	U.S. Department of Commerce	Standard FF-1 (Standard for Surface Flammability of Carpets and Rugs)
DOC FF1-70	U.S. Department of Commerce	Flammability Standard for Mattresses
DS 1051.1	Denmark	Elements of Building Construction: Fire Resistance (Equivalent to ISO 834 and NT Fire 005)
DS 1056.1	Denmark	Building Materials: Noncombustibility Test (Equivalent to ISO 1182 and NT Fire 001)
DS 1057.1	Denmark	Classification of Building Materials
DS 1058.3	Denmark	Test for Ignition Temperature and Flame Spread
DS 1063.2	Denmark	Floor Covering: Flame Spread
DS/Insta 410	Denmark	Building Products: Ignitability (Equivalent to NT Fire 002)
DS/Insta 411	Denmark	Coverings: Fire Protection Ability (Equivalent to NT Fire 003)
DS/Insta 412	Denmark	Building Products: Heat Release and Smoke Generation (Equivalent to NT Fire 004)
DS/Insta 413	Denmark	Roofing: Fire Spread (Equivalent to NT Fire 006)
DS/Insta 414	Denmark	Flooring: Fire Spread and Smoke Generation (Equivalent to NT Fire 007)
E 84	ASTM	Surface Burning Characteristics of Building Materials
E 108	ASTM	Response of Roof Covering to Flame
E 119	ASTM	Fire Tests of Building Construction and Materials
E 136	ASTM	Behavior of Materials in a Vertical Tube Furnace at 750°C
E 152	ASTM	Fire Tests of Door Assemblies

Identification Number	Sponsoring Organization	Name of Test
E 162	ASTM	Surface Flammability of Materials Using a Radiant Heat Energy Source
E 163	ASTM	Fire Tests of Window Assemblies
E 176	ASTM	Terminology of Fire Standards
E 286	ASTM	Surface Flammability of Building Materials Using an 8 foot (2.44 m) Tunnel Furnace
E 535	ASTM	Practice for Preparation of Fire Test Standards
E 603	ASTM	Guide to Room Fire Experiments: Time to Flashover, Total Rate of Smoke Production, Extent of Flame Spread for a Low Energy Ignition Source, Size of Ignition Source to Produce Flashover (Room Fires)
E 648	ASTM	Critical Radiant Flux of Floor-Covering Systems Using a Radiant Heat Source
E 662	ASTM	Specific Optical Density of Smoke
E 800	ASTM	Guide for Measurement of Gases Present or Generated During Fires
E 814	ASTM	Fire Tests of Through-Penetration Fire Stops
E 906	ASTM	Heat and Visible Smoke Release Rates for Materials and Products. The OSU Calorimeter.
E 970	ASTM	Critical Radiant Flux of Exposed Attic Floor Insulation Using a Radiant Heat Energy Source
E 1321	ASTM	Determining Material Ignition and Flame Spread Properties
E 1352	ASTM	Test Method for Cigarette Ignition Resistance of Mock-Up Upholstered Furniture Assemblies
E 1353	ASTM	Test Method for Cigarette Ignition Resistance of Components of Upholstered Furniture
E 1354	ASTM	Test Method for Heat and Visible Smoke Release Rates for Materials and Products Using an Oxygen Consumption Calorimeter
E 1355	ASTM	Guide for Evaluating the Predictive Capability of Fire Models
F 501	ASTM	Aerospace Materials Response to Flame with Vertical Test Specimen (for Aerospace Vehicles, Standard Conditions)
F 776	ASTM	Resistance of Materials to Horizontal Flame Propagation (for Aerospace Vehicles, Standard Conditions)
F 777	ASTM	Resistance of Electrical Wire Insulation Materials to Flame at an angle of 60° to the horizontal

Identification Number	Sponsoring Organization	Name of Test
F 828	ASTM	Radiant Heat Resistance of Aircraft Inflatable Evacuation Slide/Slide Raft Material
F 1103	ASTM	Standard Test Method for Materials Response to Flame (for Aerospace Vehicles)
FAA 25.853 (a) and (b) vertical	FAA-DOT	Part 25—Airworthiness Standards: U.S. Transport Category Airplanes
FAA 25.853 (b-2) and (b-3) horizontal	FAA-DOT	Part 25—Airworthiness Standards: U.S. Transport Category Airplanes
FAA 25.853 (a-1) 45° angle	FAA-DOT	Part 25—Airworthiness Standards: U.S. Transport Category Airplanes, Compartment Interiors
FAR 14CFR 25.853 (Paragraphs (a), (b), (c), and Appendix F	US Government	Airworthiness Standards—Transport Category Airplanes, Compartment Interiors
Fire Department Advisory Safety Provisions	New York Board of Standards and Appeals	Flammability Tests of Office Furniture and Furnishings [FTMS 191 Method 5903 (modified)]; ASTM E 84; or ASTM E 162 Radiant Panel Test (1974)
Fire Department Advisory Safety Provisions	New York Board of Standards and Appeals	Office Furniture and Furnishings (1981)
Fire Tube Test	Forest Products Laboratory	Same as ASTM E 286
Flooring Radiant Panel Test	ASTM and NFP	Flame Propagation of Flooring (See ASTM E 648 or NFP 253)
FM 4922	Factory Mutual	Flame Tests of Air and Fume Handling Ducts
FM Calorimeter	Factor Mutual	Heat Release Rate as a Function of Time (Composite Roof Assemblies)
FM Corner Test	Factory Mutual	Accepted If Material Does Not Produce a Self-Propagating Flame (Interior Finish Materials)
FPL Calorimeter	FPL	Similar to E 84 and NBS Calorimeter
FTMS 191A Method 5903	Federal Test Method Standard	Flame Resistance of Cloth: Vertical
FTMS 406	Federal Test Method Standard	The methods below are those that set up fire tests within FTMS 406
Method 1111		Standard Method of Testing Rigid Sheet and Plate Materials Used For Electrical Insulation (contains two methods of fire testing) (ASTM D 229)

Identification Number	Sponsoring Organization	Name of Test
Method 2021		Horizontal Burning Rate of Rigid Plastic Sheets or Bars (ASTM D 635)
Method 2022		Time For Film to Burn Completely or Extinguish Itself (ASTM D 568)
Method 2023		Flame Resistance of Difficult-to-Ignite Plastics (ASTM D 2863)
FTMS 501 Method 6411	Federal Test Method Standard	Combustion, Ignition Time, Char and Flame Length, and Smoke Density of Nontextile Floor Covering
Method 6421		Flame Spread Index of Nontextile Floor Covering (similar to E 162)
GM Test Method 32-12	GM Standard	Similar to Motor Vehicle Safety Standard #302
HUD (Proposed)	SRI	Fire Tests for Exterior Wall Materials
IEC 332	International Electrotechnical Commission	Tests on Electrical Cables Under Fire Conditions
IEC 512	International Electrotechnical Commission	Basic Testing Procedure for Fire Hazards, and Other Tests on Electrochemical Components
IEC 695	International Electrotechnical Commission	Fire Hazard Testing
IEC 707	International Electrotechnical Commission	Determination of the Flammability of Solid Insulating Materials When Exposed to Ignition Sources
IEC 829	International Electrotechnical Commission	Determination of the Flammability of Solid Insulating Materials When Exposed to Electrically Heated Wires
IEEE 383	Institute of Electrical and Electronics Engineers	Electrical Cables Mounted in a Tray
IMO A564(14)	IMO	Maximum Flame Spread
IPC TM-650 2.3.29	IPC	Flammability of Flexible Flat Cable
IPC TM-650 2.3.8	IPC	Flammability of Flexible Insulating Material
ISO 181	ISO	Determination of Flammability Characteristics of Rigid Plastics in the Form of Small Specimens in Contact with an Incandescent Rod
ISO 834	ISO	Fire Resistance Tests: Elements of Building Construction
ISO 871	ISO	Determination of Temperature of Evolution of Flammable Gases (decomposition temperature) from a Small Sample of Pulverized Material

Identification Number	Sponsoring Organization	Name of Test
ISO 1137	ISO	Plastics—Determination of Behavior in a Ventilated Tubular Oven
ISO 1172	ISO	Textile Glass–Reinforced Plastics—Determination of Loss on Ignition
ISO 1182	ISO	Combustibility or Noncombustibility of Building Materials
ISO 1210	ISO	Determination of Flammability of Plastics in Contact with a Small Flame Applied to a Horizontally Positioned Specimen
ISO 1326	ISO	Determination of Flammability and Burning Rate of Plastics in the Form of Film (ASTM 568)
ISO 1716	ISO	Determination of Calorific Potential
ISO 3008	ISO	Fire Resistance Tests: Door and Window Assemblies
ISO 3009	ISO	Fire Resistance Tests: Glazed Elements
ISO 3582	ISO	Laboratory Assessment of Horizontal Burning Characteristics of Small Specimens Subjected to a Small Flame (Cellular Rubber and Plastics)
ISO 3795	ISO	Determination of Burning Behavior of Interior Materials for Motor Vehicles (based on MVSS #302)
ISO/DTR 3814	ISO	Tests for Measuring Reaction of Fire to Building Materials
ISO 3941	ISO	Classification of Fires
ISO 4589	ISO	Determination of Flammability by Oxygen Index (ASTM D 2863)
ISO 4880	ISO	Burning Behavior of Textiles and Textile Products—Vocabulary
ISO 5657	ISO	Ignitability of Building Materials
ISO/DP 5658	ISO	Spread of Flame of Building Materials
ISO/DP 5659	ISO	NBS (NIST) Smoke Chamber
ISO/DP 5660	ISO	Rate of Heat Release Test from Building Products
ISO/DPR 5924	ISO	Test for Determination of Smoke Generated by Building Materials (ISO) Smoke Chamber
ISO/TR6160	ISO	Fire Resistance Tests—Protection of Steel Beams in Roofs and Floors by Suspended Ceilings
ISO/TR 6543	ISO	The Development of Tests for Measuring Toxic Hazards in Fires
ISO/TR 6585	ISO	Fire Hazard and the Design and Use of Fire Tests
ISO 6925	ISO	Textile Floor Coverings—Burning Behavior—Tablet Test at Ambient Temperatures

Identification Number	Sponsoring Organization	Name of Test
ISO 6940	ISO	Textile Fabrics—Burning Behavior—Determination of Ease of Ignition of Vertically Oriented Specimens
ISO 6941	ISO	Textile Fabrics—Measurement of Flame Spread Properties of Vertically Oriented Specimens
ISO 8421	ISO	Fire Protection Vocabulary
ISO 8191	ISO	Furniture—Assessment of the Ignitability of Upholstered Furniture—Ignition Source—Smoldering Cigarette
ISO/DIS 9239	ISO	Test Method for Flame Spread on Flooring Products
ISO/DP 9705	ISO	Corner Wall/Full Scale Room Fire Test for Surface Products
LIFT	NIST	Lateral Ignition and Flame Spread Test
Monsanto 2 foot tunnel	Monsanto Company	Flame Spread, Fuel Contribution, Afterflaming, Afterglowing, Smoke Contribution, Intumescence, and Insulating Value
Motor Vehicle Safety Standard #302	U.S. DOT	Flammability of Materials Used in the Interior of Automobiles
MSZ 595	Hungary	Fire Protection of Buildings
MSZ 1400	Hungary	Test Methods for Resistance to Fire
NBE-CPI Section 2.1.1	Spain	Regulations for Structural Fire Protection
NBN 713-010	Belgium	Fire Protection in Buildings, High Buildings (May be obsolete)
NBN 713-020	Belgium	Fire Resistance of Building Components
NBN S21-201	Belgium	Fire Protection in Buildings, Terminology
NBN S21-202	Belgium	Fire Protection in Buildings—High and Medium Height Buildings—General Conditions
NBN S21-203	Belgium	Fire Protection in Buildings—High and Medium Height Buildings—Fire Performance of Building Materials
NBN S21-205	Belgium	Fire Protection in Buildings—School Buildings, Boarding Schools, Student Homes
NBS Cone Calorimeter	NBS (NIST)	Heat release rates as a function of time
NBS TN 708	NBS (NIST)	Determination of Specific Optical Smoke Density
NBSIR 75-950	NBS (NIST)	Critical Radiant Flux of Floor Covering Systems Using a Radiant Heat Energy Source (ASTM E 648)
NBSIR 78-1436	NBS (NIST)	Flammability Testing for Carpet
NBSIR 82-2532	NBS (NIST)	Test Method for the Acute Inhalation Toxicity of Combustion Products

Identification Number	Sponsoring Organization	Name of Test
NEN 3881	Netherlands	Determination of Noncombustibility of Building Materials
NEN 3883	Netherlands	Determination of Combustibility of Building Materials
NF C 20-453	France	Determination of Corrosivity of Smoke
NF P 92-501	France	Determination of Fire Performance of Rigid Samples of All Thicknesses and Flexible Samples Thicker Than 5mm (Epiradiateur Test)
NF P 92-503	France	Determination of Fire Performance of Flexible Samples (Electrical Burner Test)
NF P 92-504	France	Rate of Flame Spread Test
NF P 92-505	France	Burning Drop Test
NF P 92-506	France	Radiant Panel Test for Floor Coverings
NF P 92-510	France	Determination of Calorific Potential
NFPA 225	NFPA	Standard Test Method for Surface Burning Characteristics of Building Materials
NFPA 251	NFPA	Standard Method of Fire Tests for Building Construction and Materials
NFPA 252	NFPA	Standard Method of Fire Tests for Door Assemblies
NFPA 253	NFPA	Critical Radiant Flux of Floor Covering Systems Using a Radiant Heat Energy Source (ASTM E 648)
NFPA 255	NFPA	Standard Method of Test of Surface Burning Characteristics of Building Materials
NFPA 256	NFPA	Standard Method of Fire Tests for Roof Deck Construction
NFPA 257	NFPA	Standard Method of Fire Tests for Window Assemblies
NFPA 258	NFPA	Maximum Optical Smoke Density Under Flaming and Smoldering Conditions Using a Radiant Heat Flux of 2.5 W/cm^2 (NBS Smoke Test)
NFPA 259	NFPA	Standard Test Method for Potential Heat of Building Materials
NFPA 260	NFPA	Standard Method of Test and Classification System for Cigarette Ignition Resistance of Components of Upholstered Furniture
NFPA 261	NFPA	Standard Method of Test for Determining Resistance of Mock-up of Upholstered Furniture Material Assemblies to Ignition by Smoldering Cigarettes
NFPA 262	NFPA	Standard Method of Test for Fire and Smoke Characteristics of Wires and Cables

Identification Number	Sponsoring Organization	Name of Test
NFPA 263	NFPA	Standard Method of Test for Heat and Visible Smoke Release Rates for Materials and Products
NFPA 264	NFPA	Standard Method of Test for Heat Release Rates for Upholstered Furniture Components or Composites and Mattresses Using an Oxygen Consumption Calorimeter
NFPA 701	NFPA	Flame resistance of textiles and film used in interior furnishings, protective clothing, tarpaulins and tents.
NFPA 702	NFPA	Standard for Classification of the Flammability of Wearing Apparel (as Applied to Tents, Tarpaulins and Other Protective Coverings
NFPA 1971	NFPA	Protective Clothing for Structural Firefighting
NFL Full Scale Test	National Fire Laboratory (Canada)	Combustibility of Vinyl Siding National Research Council of Canada (NRCC)
NS 3904	Norway	Elements of Building Construction: Fire Resistance (Equivalent to ISO 834 and NT Fire 005)
NS/Insta 410	Norway	Building Products: Ignitability (Equivalent to NT Fire 002)
NS/Insta 411	Norway	Coverings: Fire Protection Ability (Equivalent to NT Fire 003)
NS/Insta 412	Norway	Building Products: Heat Release and Smoke Generation (Equivalent to NT Fire 004)
NS/Insta 413	Norway	Roofing: Fire Spread (Equivalent to NT Fire 006)
NS/Insta 414	Norway	Flooring: Fire Spread and Smoke Generation (Equivalent to NT Fire 007)
NT Fire 001 to 036	Nordic countries	Nordtest Methods for Building Construction Materials
ÖNORM B 3800	Austria	Fire Performance of Building Materials and Components
ÖNORM B 3805	Austria	Fire Proofing Substances for Wood and Wooden Materials
ÖNORM B 3810	Austria	Fire Behaviour of Floor Coverings (Radiant Panel)
ÖNORM B 3820	Austria	Fire Behaviour of Curtains
ÖNORM B 3822	Austria	Behaviour of Equipment Materials in Fire, Decoration Articles
ÖNORM B 3825	Austria	Behaviour of Equipment Materials in Fire, Testing of Upholstery
ÖNORM B 3836	Austria	Building Components, Cable Penetration Seals
ÖNORM B 3850	Austria	Fire Resisting Doors

Identification Number	Sponsoring Organization	Name of Test
OSU Cone Calorimeter	Ohio State University	Rate of Heat Release Test Similar to E-84 and NBS Cone Calorimeter
P 190	ASTM	Proposed Test Method for Heat and Visible Smoke Release Rates for Materials and Products Using an Oxygen Consumption Calorimeter
Port Authority Tests	Port Authority of New York and New Jersey	Flammability Tests of Upholstery Material and Plastic Furniture [FTMS 191 Method 5903 (modified)]; ASTM E 84; or ASTM E 162 Radiant Panel Test
Port Authority Tests	Port Authority of New York and New Jersey	Curtains and Drapes. Uses FTMS 191 Method 5903
SABS Method 963 Surface Fire Index of Materials Used in Buildings	South Africa	Same as ASTM E 286
SAE-AMS 3851A	SAE	Fire Resistance Properties for Aircraft Materials
SAE-AMS 3852A	SAE	Fire Resistance Properties for Aircraft Materials
SAE J-369	SAE	Flammability of Automotive Trim Materials
SAE J-558	SAE	Low Tension Cable Flame Test
Schlyter Test	Austria	Determination of Low Combustibility of Building Materials (Derived from Nordtest NT Fire 002)
SFS 4190:E	Finland	Building Products: Ignitability (Equivalent to NT Fire 002)
SFS 4191:E	Finland	Coverings: Fire Protection Ability (Equivalent to NT Fire 003)
SFS 4192:E	Finland	Building Products: Heat Release and Smoke Generation (Equivalent to NT Fire 004)
SFS 4193	Finland	Elements of Building Construction: Fire Resistance (Equivalent to ISO 834 and NT Fire 005)
SFS 4194:E	Finland	Roofing: Fire Spread (Equivalent to NT Fire 006)
SFS 4195:E	Finland	Flooring: Fire Spread and Smoke Generation (Equivalent to NT Fire 007)
SIA 183 and Supplements /1,/2, and /3	Switzerland	Structural Fire Protection, Underground Garages, Use of Combustible Building Materials, Testing of Building Materials and Components, Signposting of Escape Routes and Exits, and Emergency Lighting
SIS 650082	Sweden	Test for Materials <3 mm Thick for Low Flammability

Identification Number	Sponsoring Organization	Name of Test
SNV 198 897	Switzerland	Floor Covering Text (Similar to DIN 54 332)
SNV 520 183/2	Switzerland	Use of Combustible Materials in Buildings
SRI Calorimeter	Southwest Research Institute	Similar to E 84 and NBS Calorimeter
SS 024820	Sweden	Elements of Building Construction: Fire Resistance (Equivalent to ISO 834 and NT Fire 005)
SS 024821	Sweden	Building Products: Ignitability (Equivalent to NT Fire 002)
SS 024822	Sweden	Coverings: Fire Protection Ability (Equivalent to NT Fire 003)
SS 024823	Sweden	Building Products: Heat Release and Smoke Generation (Equivalent to NT Fire 004)
SS 024824	Sweden	Roofing: Fire Spread (Equivalent to NT Fire 006)
SS 024825	Sweden	Flooring: Fire Spread and Smoke Generation (Equivalent to NT Fire 007)
ST CMEA 382-76	GDR	Determination of Noncombustibility
ST CMEA 2347-80	GDR	Determination of Low Combustibility
ST CMEA 5966-87	GDR	Propagation of Fire in Walls and Wall Linings
ST CMEA 5967-87	GDR	Propagation of Fire in Roofing
Technical Bulletin No. 116	California	Smoldering Test for Flexible Polyurethane Foams Used in Upholstered Furniture
Technical Bulletin No. 117	California	Requirements, Test Procedure and Apparatus for Testing the Flame Retardance of Resilient Filling Materials Used in Upholstered Furniture
Technical Bulletin No. 121	California	Flammability Test Procedures for Mattresses for Use in High Risk Occupancies
Technical Bulletin No. 133	California	Flammability Test Procedure for Seating Furniture for Use in High Risk and Public Occupancies
TGL 10685/11	GDR	Determination of the Combustibility of Building Materials
TGL 10685/12	GDR	Determination of the Fire Propagation of Building Components, Interior Structures, and the Suitability Group of Structures
TGL 10685/13	GDR	Determination of the Combustibility of Components, Firestops and Dust Seals
Title 19	California	Flame-Retardant Chemicals and Fabrics
UBC 17-5	ICBO	Room Fire Test Standard
UBC 17-6	ICBO	Wall Assembly Test

Identification Number	Sponsoring Organization	Name of Test
UBC 42-1	ICBO	Tunnel Test
UBC 42-2	ICBO	Standard Test Method for Evaluating Room Fire Growth Contribution of Textile Wall Covering
UBC 43-1	ICBO	Identical to ASTM 119
UBC 52-3	ICBO	ASTM D 1929
UFAC 83	UFAC	Six individual tests: similar to NFPA 260A
UL 9	UL	Fire Tests of Window Assemblies
UL 10	UL	Fire Tests of Door Assemblies
UL 44	UL	Flame Spread of Rubber-Insulated Wires and Cables
UL 83	UL	Flame Spread and Resistance of Thermoplastic Insulated Wire
UL 94	UL	Tests for Flammability of Plastic Materials for Parts in Devices and Appliances
UL 94V-0	UL	Flammability of Plastic Materials, Vertical
UL 94 HB	UL	Flammability of Plastic Materials, Horizontal
UL 214	UL	Vertical Flame Propagation of Film and Fiber
UL 224	UL	Flame-Retardant Properties of FR-1 Tubing
UL 263	UL	Fire Tests of Building Contruction and Materials Identical to ASTM E 119 (Also called ANSI/UL 263)
UL 723	UL	Surface Burning Characteristics of Building Materials
UL 790	UL	Sets limits for E 108
UL 910	UL	Smoke and Flame Production of Wire Insulation
UL 964	UL	Safety Standard for Electrically Heated Bedding
UL 992	UL	Test Method for Measuring the Surface Flame Propagation Characteristics of Flooring and Floor Covering Materials
UL 1056	UL	Standard for Safety for Fire Test of Upholstered Furniture
UL 1256	UL	Fire Test of Roof Deck Construction
UL 1479	UL	Fire Tests of Through Penetration of Fire Stops
UL 1666	UL	Test for Flame Propagation Height of Electrical Optical-Fiber Cables Installed Vertically in Shafts
UL 1820	UL	Fire Test of Pneumatic Tubing for Flame and Smoke Characteristics
UL 1887	UL	Fire Tests of Plastic Sprinkler Pipes for Flame and Smoke Characteristics
UL 1975	UL	Fire Tests for Foamed Plastics Used for Decorative Purposes

Identification Number	Sponsoring Organization	Name of Test
UL 2023	UL	Test Method for Flame and Smoke Characteristics of Nonmetallic Wiring Systems (Raceway and Conduits) Used in Environmental Air-Handling Spaces
UL 2024	UL	Test Method for Flame and Smoke Characteristics of Nonmetallic Raceways Used in Environmental Air-Handling Spaces
ULC-CAN4-S101-M82	ULC	Standard Methods of Fire Endurance Tests of Building Construction
ULC-CAN4-S104-M80	ULC	Standard Method for Fire Tests of Door Assemblies
ULC-CAN4-S105-79	ULC	Standard Specification for Fire Door Frames
ULC-CAN4-S106-M80	ULC	Standard Method for Fire Tests of Window and Glass Block Assemblies
ULC-CAN4-S114-M80	ULC	Noncombustibility of Building Materials (Similar to ASTM E 136)
ULC-S 102-M83	ULC	Test of Fire Performance of Interior Finishes (Similar to ASTM E 84)
ULC-S 102.2-M83	ULC	Test for the Surface Burning Characteristics of Floor Coverings and Miscellaneous Materials (Modified Tunnel Test with the flame applied downwards)
ULC-S-102.3-1982	ULC	Modified Tunnel Test (Light diffusers and lenses)
ULC-S107-M1980	ULC	Standard Method of Test for Fire Resistance of Roof Covering Materials
ULC-S 109-M 1980	ULC	Flame Tests of Flame-Resistant Fabrics and Films
ULC-S 127-1982	ULC	Corner Wall Test for Flammability Characteristics of Nonmelting Building Materials
Union Carbide 4 foot tunnel	Union Carbide Company	Flame spread
Upjohn small corner test	Upjohn Company	Designed to predict results of the Factory Mutual Full Scale Corner test
VDE-Standards	Germany	Distinguish Between Burning Behavior of Insulating Materials and the Fire Characteristics of Electrotechnical Equipment and Installations
Vertical Channel Test	NRCC (Canada)	Fire Characteristics of Exterior Wall Assemblies

Appendix E

References

ABS Plastics; Basdekis, author; Reinhold Publishing Corporation: New York, NY (1964)

Annual Book of ASTM Standards, Vol. 04.07, Building Seals and Sealants; Fire Standards, Building Constructions; ASTM: 1916 Race Street, Philadelphia, PA 19103 (1989)

Annual Book of ASTM Standards, Vol. 08.04, Plastic Pipe and Building Products; ASTM: 1916 Race Street Philadelphia, PA 19103 (1989)

Annual Book of ASTM Standards, Vol. 08.01, Plastics: ASTM: 1916 Race Street, Philadelphia, PA 19103 (1983)

Assessment of Test Methods to Determine Combustibility/Toxicity of All Materials; A Report to the California State Legislature by The Office of the State Fire Marshal; California, April 30, 1983

Behavior of Polymeric Materials in Fire; Schaffer, editor; ASTM: 1916 Race Street, Philadelphia, PA 19103 (1982)

Chemical Additives for the Plastics Industry; Radian Corporation, Noyes Data Corporation: Park Ridge, NJ 07656 (1987)

Coloring of Plastics; Ahmed, author; Van Nostrand Reinhold: 115 5th Avenue, New York, NY 10003 (1979)

Combustion Toxicology, Principles and Test Methods; Kaplan, author; Grand & Hartzell, Technomic Publishing Company: 851 New Holland Avenue, Box 3535, Lancaster, PA 17604 (1983)

Conference Proceedings, 13th International Combustibility Symposium; Society of the Plastics Industry: 1275 K Street, Washington, DC 20005 (1985)

Development and Use of Polyester Products; E. N. Doyle, author; McGraw-Hill, Incorporated: 1221 Avenue of the Americas, New York, NY 10020 (1969)

Engineering Thermpolastics: Properties and Applications; James Margolis, editor; Marcel Dekker, Incorporated: 270 Madison Avenue, New York, NY 10016 (1985)
Facts and Figures of the U.S. Plastics Industry; Society of the Plastics Industry: 1275 K Street, Washington, DC 20005 (1988)
Fire and Flammability Handbook; Schultz, author; Van Nostrand Reinhold: 135 West 50th Street, New York, NY 10020 (1985)
Fire and Polymers; Nelson, editor; American Chemical Society: 1155 16th Street, NW, Washington DC 20036 (1990)
Fire and Smoke: Understanding the Hazards; Committee on Fire Toxicology, National Academy Press: Washington, DC 1986
Fire Risk Assessment; Castino and Harmathy, editors; ASTM: 1916 Race Street, Philadelphia, PA 19103 (1980)
Fire Safety Aspects of Polymeric Materials, Vol. 1, Materials: State of the Art; Technomic Publishing Company: 851 New Holland Avenue, Lancaster, PA 17604 (1977)
Fire Safety Aspects of Polymeric Materials, Vol. 2, Test Methods, Specifications, and Standards; Technomic Publishing Company: 851 New Holland Avenue, Lancaster, PA 17604 (1979)
Fire Safety Aspects of Polymeric Materials, Vol. 3, Smoke and Toxicity (Combustion Toxicology of Polymers); Technomic Publishing Company: 851 New Holland Avenue, Lancaster, PA 17604 (1978)
Fire Safety Aspects of Polymeric Materials, Vol. 4, Fire Dynamics and Scenarios; Technomic Publishing Company: 851 New Holland Avenue, Lancaster, PA 17604 (1978)
Fire Safety Aspects of Polymeric Materials, Vol. 5, Elements of Polymer Fire Safety and Guide to the Designer; Technomic Publishing Company: 851 New Holland Avenue, Lancaster, PA 17604 (1979)
Fire Safety Aspects of Polymeric Materials, Vol. 7, Buildings; Technomic Publishing Company: 851 New Holland Avenue, Lancaster, PA 17604 (1979)
Fire Safety, Science and Engineering; T. Z. Harmathy, editor; ASTM: 1916 Race Street, Philadelphia, PA 19103 (1984)
Fire Test Standards, third edition; ASTM: 1916 Race Street, Philadelphia, PA 19103 (1990)
Flammability Handbook for Plastics, third edition; Hilado, author; Technomic Publishing Company: 851 New Holland Avenue, Lancaster, PA 17604 (1982)
Flammability of Solid Plastics, Part 2, Volume 17; Hilado, editor; Technomic Publishing Company: 851 New Holland Avenue, Lancaster, PA 17604 (1983)
Handbook of Plastics Testing Technology; Shaw, author; John Wiley & Sons, Incorporated: 605 Third Avenue, New York, NY 10036 (1984)
Identification and Analysis of Plastics, second edition, third printing; Haslam,

Willis and Squirrel, authors; Heyden and Son Incorporated; 247 S. 41st Street, Philadelphia, PA 19104 (1981)

Industry Standards and Engineering Data; Information Handling Services Incorporated: Inverness Business Park, 15 Inverness Way East, PO Box 1154, Englewood, CO 80150

International and Non-U.S. National Standards; Information Handling Services Incorporated: Inverness Business Park, 15 Inverness Way East, PO Box 1154, Englewood, CO 80150

International Plastics Flammability Handbook, second edition; Troitzsch, author; Carl Hanser, Verlag: Kolbergerstrasse 22, D-8000 Munchen 80 (1990)

International Plastics Handbook; Saechtling, author; Hanser Publications, Distributed in U.S. by Scientific and Technical Books, Macmillan Publishing Company, Incorporated: 866 Third Avenue, New York, NY 10022 (1983)

Kline Guide to the Plastics Industry, issued by Charles H. Kline and Company: Fairfield, NJ 07006 (1985)

Modern Industrial Plastics, issued by Richardson, Howard W. Sams and Company, Incorporated: Indianapolis, IN (1974)

Modern Plastics Encyclopedia; McGraw-Hill, Incorporated: 1221 Avenue of the Americas, New York, NY 10020 (1991)

Plastics Additives; Gachter and Muller, authors; Hanser Publishers, Distributed by Scientific and Technical Books, Macmillan Publishing Company, Incorporated: 866 Third Avenue, New York, NY 10022 (1985)

Plastics Engineering Handbook of the SPI; Frados, editor; Van Nostrand Reinhold: 115 Fifth Avenue, New York, NY 10003 (1976)

Plastics Industry Safety Handbook of the SPI; Cahners Publishing Company: 89 Franklin Street, Boston, MA (1973)

Plastics Technology Handbook; Chanda and Roy, authors; Marcel Dekker, Incorporated: 270 Madison Avenue, New York, NY 10016 (1987)

Plastics Technology Manufacturing Handbook and Buyers Guide; Bill Publications: 633 Third Avenue, New York, NY 10017-6743 (1990)

Plastics World; 1991 Plastics Yellow Pages Issue; Cahners: 275 Washington Street, Newton, MA 02158 (1990)

Polymer Degradation; Kelen, author; Van Nostrand Reinhold: 115 Fifth Avenue, New York, NY (1983)

Polymeric Materials; Alper and Nelson, authors; American Chemical Society: Washington, DC (1989)

Polymer Materials, second edition; Hall, author; Halsted Press, Division of John Wiley & Sons: 605 Third Avenue, New York, NY 10036 (1989)

Progress in Fire Safety; Fire Retardant Chemicals Association: 851 New Holland Avenue, Lancaster, PA 17604 (1976)

Pyrolysis of Polymers, Vol. 13; Hilado, editor; Technomic Publishing Company: 851 New Holland Avenue, Lancaster, PA 17604 (1976)

Rauch Guide to the U.S. Plastics Industry: Rauch Association, Incorporated: PO Box 6802, Bridgewater, NJ 08807 (1987)

Smoke and Products of Combustion, Part 2, Vol. 15; Hilado, editor; Technomic Publishing Company: 851 New Holland Avenue, Lancaster, PA 17604

Specifications and Standards for Plastics and Composites; Traceski, author; ASM International (1990)

Thermoplastic Polymer Additives; Lutz, editor; Marcel Dekker, Incorporated: 270 Madison Avenue, New York, NY 10016 (1989)

Thirteenth International Combustibility Symposuim; Society of the Plastics Industry: 1275 K Street, Washington, DC 20005 (1985)

"Toxicity of Combustion Products," *Fire Journal;* Frederic B. Clarke III, author; September 1983

Whittington's Dictionary of Plastics; Technomic Publishing Company: 851 New Holland Avenue, Lancaster, PA 17604 (1968)

Appendix F

Addresses of International Code and Standards Setting Organizations

American Association for Laboratory Accreditation (A2LA)
656 Quince Orchard Road, Suite 704
Gaithersburg, MD 20878

American National Standards Institute (ANSI)
1430 Broadway
New York, NY 10018

American Society for Testing and Materials (ASTM)
1916 Race Street
Philadelphia, PA 19103

Associación Española de Normalización y Certificación
Fernandez de la Hoz 52
Madrid, Spain

Association Française de Normalisation (AFNOR)
Tour Europe, Cedex 7
F-92080 Paris la Défense

Boligministeriets godkendelsessekretariat for Materialer og konstrukioner
Postbox 54
DK-2970 Hørsholm, Denmark

British Standards Institution
2 Park Street
London, W1A2BS, England

Brunamálastofnun ríkisins
Laugaveg 120
IS-105 Reykjvik, Iceland

Building Officials & Code Administrators International, Inc. (BOCA)
4051 W. Flossmoor Road
Country Club Hills, IL 60477-5795

Bureau de normalisation du Québec
Ministère de l'industrie du commerce et du tourisme
50 rue Saint-Joseph est
Québec, Canada, G1K 3A5

Canada General Standards Board
(CGSB)
Ottawa, Canada, K1A 1G6

Centro Studi ed Esperienze dei
Vigili del Fuoco (CSE)
I-00178 Capanelle
Rome, Italy

Deutsches Institut für Normung
(DIN)
Burggrafestrasse 4-10
Postfach 1107, D-1000
Berlin 30, Germany

General Services Administration
Seventh and D Streets, SW
Washington, DC 20407

Institut Belge de Normalisation
Avenue de la Brabançonne 29
B-1040 Brussels, Belgium

International Conference of Building
Officials (ICBO)
5360 South Workman Mill Road
Whittier, CA 90601

International Electrotechnical
Commission (IEC)
1-3, rue de Varembe, CH-1211
Geneva 20, Switzerland

International Organization for
Standardization (ISO)
1-3, rue de Varembe, CH-1211
Geneva 20, Switzerland

National Aeronautics and Space
Administration (NASA)
Langley Research Center
Hampton, VA 23665-5225

National Fire Protection Association
Batterymarch Park
Quincy, MA 02269

National Institute of Standards and
Technology (NIST) formerly the
National Bureau of Standards
(NBS)
Route I-279 and Quince Orchard
Boulevard
Gaithersburg, MD 20899

National Standards Association
1200 Quince Orchard Boulevard
Gaithersburg, MD 20878

Österreichisches Normungsinstitut
Leopoldgasse 4, Postfach 130
A-1021 Vienna, Austria

Prüfausschuß für Baustoffe und
Bautile
c/o Vereinigung Kantonaler
Feuerversicherungen Bundesgasse
20, CH-3011
Bern, Switzerland

Southern Building Code Congress
International, Inc. (SBCCI)
900 Montclair Road
Birmingham, AL 35213-1206

Statens bygningstekniske
Postboks 8150 Dep.
N-0034 Oslo 1, Norway

Statens planverk
Box 22027
S-120422 Stockholm, Sweden

Underwriters Laboratories, Inc.
333 Pfingsten Road
Northbrook, IL 60062

Underwriters Laboratories of Canada
7 Crouse Road
Scarborough
Ontario, Canada M1R 3A9

Ympäristöministeriö Kaavoitus-ja rakennusosasto
PL 399, SF-00121
Helsinki, Finland

Appendix G

Abbreviations for Code Bodies and Standards Setting Organizations

AA	The Aluminum Association
AATCC	American Assocation of Textile Chemists and Colorists
ACI	American Concrete Institute
AENOR	Associación Española de Normalización
AFNOR	Association Française de Normalisation
AIA	American Institute of Architects
AIA	American Insurance Association
AIA/NAS	Aerospace Industries Assocation of America
ASTM	American Society for Testing and Materials
BBC	Basic Building Code (BOCA)
BFD	Boston Fire Department
BIFMA	Business and Institutional Furniture Manufacturer's Association
BOCA	Building Officials and Code Administrators, USA
BS	British Standards Institution
CEE	Commission for Certification of Electrical Equipment
CEN	European Committee for Standardization
CENELEC	European Committee for Electrotechnical Standards
CMEA	Council for Mutual Economic Assistance
CNET	Centre Nationale d'Etudes des Télécommunications
CPSC	Consumer Product Safety Commission
CSE	Centro Studi ed Esperienze dei Vigili del Fuoco
DIN	Deutsches Institut für Normung
DOT	U.S. Department of Transportation

EEC	European Economic Community
EIA	Electronic Industries Association
FAA	Federal Aviation Administration
FAR	Federal Aviation Regulations
FM	Factory Mutual
FTC	Federal Trade Commission
FTMS	Federal Test Method Standards
FRG	Federal Republic of Germany
GDR	German Democratic Republic
GSA	General Services Administration
HHS	Department of Health and Human Services
HUD	Department of Housing and Urban Development
ICBO	International Conference of Building Officials, USA
IEC	International Electrotechnical Commission
IEEE	Institute of Electrical and Electronics Engineers
IFIA	Industrial Fabrics Association International
IPC	Institute for Interconnectivity and Packaging Electronic Circuits
ISO	International Organization for Standardization
MVSS	Motor Vehicle Safety Standards
NAMAS	National Measurement Accrreditation Service
NASA	National Aeronautics and Space Administration
NBC	National Building Code (AIA)
NBC	National Building Code of Canada
NBE-CPI	Norma Basica de la Edificación-Condiciones de Protección contra Incendio en los Edificios
NFPA	National Fire Protection Association
NKB	Nordiska Kommitten för byggbestämmelser
NMAB	National Materials Advisory Board
NRCC	National Research Council of Canada
SAA	Standards Association of Australia
SAE	Society of Automotive Engineers
SBC	Standard Building Code (SBCCI)
SBCCI	Southern Building Code Congreee International, USA
SFPE	Society of Fire Protection Engineers
SOLAS	International Convention for the Safety of Life at Sea
SRI	Southwest Research Institute
UBC	Uniform Building Code (ICBO)
UFAC	Upholstered Furniture Action Council
UIC	Union Internationale des Chemins de Fer
UL	Underwriters Laboratories Incorporated
ULC	Underwriter's Laboratories of Canada

Index

Page references in *italic* indicate illustrations or tables.

Acetaldehyde(s), 36, 130
Acetal(s), 50–52, *51*, 128–129, 134
Acetate, 77–78
Acetic acid, 137
Acids, 34, 119
Acrolein, 36
Acrylate, 53, 128
Acrylic(s), 44, 53–57, 97, 128, 135, 167
Acrylic–styrene–acrylonitrile (ASA), 102–103
Acrylonitrile, 48, 57, 81–82, 103
Acrylonitrile–butadiene–styrene (ABS), 48, *57*, 57–59, 76, 83, 103, 130, 134, 156, 167
Acrylonitrile–chlorinated polyethylene–styrene, 103
Additive(s), 16–17, 19, 52, 56, 58, 63, 65, 89, 92, 94, 98, 110–121
Air, 13
Alcohol(s), 34, 37, 48, 119
Aldehydes, 34, 36, 140–141, 149, 165
Alkenes, 44, 87
Alkyds, 44, 49, 106, 150
Alloy(s), 13, 22, 50, 69, 99, 105, 110, 130, 165
Allyls, 106, 150
Alpha cellulose, 113
Alumina hydrate, 112, 161
Aluminum hydrate, 112, 161
Aluminum powder, 116, 121, 123
Aluminum trihydrate, 112, 119, 161–164
American Society for Testing and Materials (ASTM), 132–133
  Fire Tests, 182–215
Amide(s), 80, 119
Amine(s), 34, 139
Amino(s), 106, 108, 131
Ammonia, 139, 151
Antimony oxide, 112, 119, 161, 162
Argon, 28
Aromatic polymer(s), 67, 69
Arsenic, 28
Astatine, 28
ASTM Fire Tests
  D 229, 182–183
  D 470, 183–184
  D 568, 184–185
  D 635, 185–186
  D 757, 186
  D 876, 186–187
  D 1000, 187
  D 1230, 187–188
  D 1433, 188
  D 1929, 188–189
  D 2633, 189
  D 2671, 189–190
  D 2843, 190–191
  D 2859, 191
  D 2863, 191–192
  D 3014, 192–193
  D 3675, 193–194
  D 3713, 194–195
  D 3801, 195–196
  D 3874, 196–197

ASTM Fire Tests *(Cont.)*
D 3894, 197
D 4100, 197–198
D 4205, 198
D 4549, 198–199
D 4723, 199
D 4804, 199–200
E 84, 200–201
E 108, 201–202
E 119, 202–203
E 136, 103
E 152, 203–204
E 162, 204–205
E 163, 205–206
E 286, 206–207
E 603, 207–208
E 648, 208–209
E 662, 209–210
E 800, 210–212
E 814, 212
E 906, 212–213
E 970, 213–214
E 1321, 214
F 777, 215
ATH, 112, 119
Atom, 7, 10–12, 14–16, 19, 27–28, 30–31, 44
Atomic mass unit (a.m.u.), 10, 18
Atomic weight, 10–11

Backdraft, 38
Barytes, 112
Benzene, 82, 97
Berkeley Test, 182
Bismaleimide, 107
Blend(s), 22, 50, 110, 130
BLEVE, 46
Blood, 13
Boiling point, 35
Bond, 16, 18
Bond energies, 18, *19,* 28
Bonding, 13
Boron, 28, 164
Brightener, 100
Brominated, 159, 162
Bromine, 21, 28, 37, 120, 131, 159–161
Butadiene, 48, 53, 57–58, *58,* 76, 139
Butane, 36
Butene, 48, 82, 87, 140
Butyl acrylate, 53
Butylene, 48, 76, 82, *83,* 87, 140

Calcium carbonate, 111
Calcium metasilicate, 112
Calcium sulfate, 112
Calories, 18
Carbon, 7, 15–17, 28–29, 31–33, 35–37, 44, 55, 78, 79, 88, 91, 98, 108, 112, 115, 120, 127–128, 130–131, 134–152, 159, 224
Carbon black, 113, 115, 117
Carbon dioxide, 13, 16, 26, 29–30, 32, 36, 120, 127, 130–131, 134–152, 159, 224
Carbon monoxide, 13, 16, 29–30, 32, 36–37, 120, 127, 130–131, 134–152, 159, 165, 224–225
Carboxyhemoglobin, 224
Carboxymethyl cellulose, 63, 139
Cardboard, 42
Catalyst, 47–48
Cellophane, 61, 62, 136
Celluloid, 61, 136
Cellulose, 35–38, 42, 60, 130
Cellulose acetate (CA), 61, 137, 168
Cellulose acetate butyrate (CAB), 61, 137, 168
Cellulose acetate phthalate, 62, 139
Cellulose acetate propionate (CAP), 61–62, 137, 168
Cellulose nitrate (CN), 60, 128–131, 137–139, 154, 167
Cellulose propionate, 63, 138
Cellulose triacetate, 62, 138
Cellulose xanthate, 63, 139
Cellulosic(s), 44, 60–63, 118, 128, 139, 168
Cellulosic fibers and flours, 113
Ceramic fibers, 113
Chain reaction of burning, 27, 33, 39
Char, 163
Chemical energy, 24
Chemical bond(s), 15–16
Chemical compound, 12–17, 19, 34, 98, 119
Chemical equation, 29
Chemical formula, 12, 87

Chemical reaction, 5, 11–12, 14, 16, 18–19, 21–22, 29, 40, 45–46, 49
Chemical symbol, 7
Chemistry, 5, 14–15, 47
Chloride, 12, 96
Chlorinated, 159, 162
Chlorinated polyethylene (CPE), 64, 71, 87, 131, 169
Chlorinated polyvinyl chloride (CPVC), 77, 97, 131
Chlorine, 10, 12, 17, 21, 28, 37, 64, 71, 73, 77, 82, 97–98, 120, 127, 131, 159–161, 224
Chlorofluoropolymer, 73
Chlorotrifluoroethylene (CTFE), 71, 131
Class A fires, 26
Class A materials, 26, 40, 152, 154
Class B fires, 26, 154
Class B materials, 154
Class C fires, 26
Class D fires, 26
Clay, 111
Code and Standards Setting Organizations, 314–318
Colorants, 59, 114–117
Combustible, 26, 34–35
Combustion, 16, 18, 21, 24, 29, 33, 36–39, 48, 53, 99, 120–123, 130
Compounding, 98
Compound(s), 6, 12–14, 22, 36, 50, 110
Copolyester, 64
Copolymerizing, 13, 71–72, 104
Copolymer(s), 13, 50–52, 54, 71–72, 78, 79, 82, 92–94, 104, 130
Cotton, 25
Covalent, 12–13, 18
Covalent bond, 14–18, 27–29, 33–35, 38–39, 44–46, 49, 88
Covalent compound, 16–18, 27, 33–34
Crosslinking, 48–49, 106
Cyanide, 82, 131, 151

Decomposition products, 98, 160, 164–165, 225
Deflagration, 21–22
Depolymerization, 134, 136, 144, 146, 225
Dilution, 26
Dry chemical, 33
Dye(s), 100, 114

Elastomeric, 43, 95
Elastomer(s), 64–65, 103
Electronic configuration, 11
Electronic structure, 11
Electron(s), 10–12, 15, 16, 19, 27–31, 45–47, 88
Element, 6, 7, *9*, 10–12, 16–17, 19, 27–28, 30, 44
Endothermic, 163
Energy, 17–18, 21, 23–26, 29–31, 35, 37–39, 46
Engineering plastics, polymers, or resins, 47, 50, 57, 64, 67–68, 70, 75, 79, 80, 83, 103
Engineering thermoplastic elastomers (ETE), 64–65
Epoxy, 44, 49, 107, 128, 150, 168, 173
Esters, 34, 119, 139
Ethane, 36
Ethene, 44, 47, 82, 140
Ethers, 34
Ethyl acrylate, 53
Ethyl cellulose, 62, 139
Ethylene, 16, *44*, *45*, 47–48, 71–72, 78, 82, *87*, 87–88, 91–92, 96–97, 140
Ethylene acrylic acid (EAA), 54
Ethylene butyl acetate, 55
Ethylene–chlorotrifluoroethylene (ECTFE), *71*, 71–72, 131
Ethylene–ethyl acrylate (EEA), 55
Ethylene methacrylate (EMA), 55
Ethylene methacrylic acid (EMMA), 55–56
Ethylene–tetrafluoroethylene, 72, *72*, 131
Ethylene–vinyl acetate (EVA), 55, 77–78
Ethylene–vinyl alcohol (EVOH), 78
Evaporation, 35
Exothermic, 48
Explosion(s), 21–22, 34–35, 38, 45–46
Explosive range, 24
Extender(s), 99, 110, 119

Feldspar, 112
Filler(s), 99, 110, 119
Fire, 21–25, 29, 34–35, 37, 40
  contrived, 124–125
  extinguishment, 25–27, 152–155
  load, 152
  parameters, 122
  real world, 124–125
Fire test(s), 124–125, 180–220, *291–309*
  ASTM fire tests, 182–215
  FTMS 191A, Method 5903.1, 215
  FTMS 406, Method 2021, 216
  IEC 707, 216
  IEC 829, 216
  ISO 871, 216
  ISO 1182, 216
  ISO 1326, 216
  ISO 5657, 216
  Motor Vehicle Safety Standard 302, 217
  NFL Full Scale Test, 217
  NFPA 701, 217
  NFPA 702, 218
  NFPA 1971, 218
  UL 94, 218–220
  UL 723, 220
Fire triangle, *23,* 23–25, 27, 40
Flame retardant(s), 99, 119–121, 156–179
  additive, 156–157
  reactive, 156, 160
  Table 7.1, 172–173
  Table 7.2, 174–178
Flaming, 21, 24, 26
Flammable, 32, 34–35
Flammable range, 24, 29
Flashover, 38
Flash point, 26, 35
Fluorinated, 75
Fluorinated ethylene propylene (FEP), 72–73, 76, 131
Fluorine, 10, 17, 21, 28, 37, 71, 73, 82, 127, 131, 145, 160, 224
Fluoropolymers, 44, 71–76, 145
Formaldehyde, 51, *51,* 107–108
Fractional effective dose (FED), 221
Free radical quenchant, 33
Free radical(s), 27–30, 32–36, 38–39, 45, 88, 162
Free radical trap, 33, 37
Fuel, 17–18, 21–25, 33–38, 40, 127
Functional group(s), 96–97

Gases, 34–35, 37–38, 46
Gasoline, 13
Glass, 112
Glossary, 267–288
Glowing, 37

Halogenated, 33, 34, 64, 87, 93, 119–120, 161–163, 165
Halogen(s), 21, 24, 37, 120, 127, 131–132, 154, 160, 224
Halon™, 162
Hardener, 48
Hazardous materials, 47
Heat, 18, 21, 23–25, 29, 34, 38–40, 46, 49
Heat of combustion, 33, *129,* 130, 135, 147
  non-plastic materials, *290*
Heat sink, 38, 39
Heat stabilizer, 98–99
Helium, 28
Hexane, 36
Homopolymer(s), 50–52, 79, 92, 94
Hydrocarbon derivative(s), 34–36, 51, 115, 147, 159, 162, 165
Hydrocarbon(s), 29, 33–34, 36–38, 44, 77, 87–88, 91, 119–120, 127, 140, 144, 147, 159, 161–163, 165
Hydrochloric acid, 98
Hydrogen, 10, 16–17, 19, 28, 30–36, 44, 55, 71, 73, 77–79, 82, 91, 96–98, 115, 127–128, 130–132, 224
Hydrogen bromide, 120
Hydrogen chloride, 98, 120, 131–132, 147–149, 154, 224–225
Hydrogen cyanide, 130, 139, 151, 225
Hydrogen fluoride, 131–132, 145, 224
Hydroxyalkyl cellulose, 63
Hydroxyl, 37, 162–164

Ignitability, 25
Ignitable mixture, 26

Ignition source, 24–25, 33, 35
Ignition temperature, 18, 24–26, 33, 35, 39, 134–152, 163
Imide, 67
Impact modifier, 99, 119, 121
Inhibitor, 45, 47
Initiator, 47
Inorganic, 12, 14, 19, 161, 163
Inorganic chemistry, 14–15
International Standards Organization (ISO), 133
Iodine, 21, 28, 37, 131, 160
Ion, 12, 14, 15
Ionic, 12, 13, 15, 17
Ionic bond, 15
Ionic compound, 15, 17
Ionomer(s), 78–79
Isocyanate(s), 48, 95
Isotope, 10

Kerosene, 13
Ketones, 34, 139
Krypton, 28

Leather, 42
Light, 18, 21, 24, 29
Light stabilizer, 99
Limiting Oxygen Index (LOI), 132, 164, *290–291*
Liquids, 26, 34–35
Lower flammable limit, 24
Lubricant, 99, 119

Magnesium carbonate, 164
Magnesium hydroxide, 164
Matter, 5
Maleic anhydride, 104
Mechanical energy, 24
Melamine(s), 44, 49, 107, 130, 150–151, 173
Melamine–formaldehyde, 106–107
Mer(s), 87–88
Metal fibers, flakes, powders, 113
Metallic, 15, 24
Metals, 7, 12
Methane, 16, 29–33, *29, 30,* 36, 151
Methyl, 31, 97
Methyl acrylate, 53
Methyl cellulose, 63, 139
Methyl methacrylate (MMA), 53–54
Mica, 111–112
Milk, 13
Mineral, 17
Mineral fillers, 111–112
Mixtures, 6, 13
Mole, 19
Molecular formula, 12, 16, 51, 55, 75, 91–92
Molecule, 13, 16, 28–31, 33–35, 37, 41, 44–47, 49, 55, 67, 71, 73, 77–78, 82, 88, 127
Molybdenum disulfide, 112
Monomer, 13–14, 16–17, 41, 43–49, 51, 53, 56–57, 71–72, 77, 79–82, 87, 91, 95, 98, 104

Natural polymers, 41–42, 49
Nephiline syenite, 112
Neon, 28
Neutron, 10–11, 19
Nitrile(s), 81, 146
Nitrogen, 10, 16–17, 28, 34–36, 127, 130–131, 151
Nitrogen dioxide, 16
Nitrogen oxides, 35–36, 137–139, 151
Nonmetallic, 15–16, 28
Nonmetal(s), 7, 12, 17, 27
Nuclear energy, 24
Nucleus, 7, 10–11, 19
Nylon(s), 80–81, 118, 130–131, 139, 168

Olefin-modified styrene–acrylonitrile, 103
Olefin(s) or olefin(ic), 82, 87–88, 91, 103
Orbit, 11
Orbital, 11
Organic, 12, 14–15, 17–19, 24, 34, 161, 165
Organic chemistry, 14
Oxidation, 18, 21–22, 29, 39, 131
Oxidized, 16
Oxidizer, 21, 23–26, 39
Oxidizing agents, 21

Oxygen, 10–13, 16–17, 21, 23–26, 28–30, *30,* 32, 34–38, 54, 78, 122, 127–132, 225
Oxygen index, 164
Oxymethylene, 51

Paper, 25, 35, 37, 42
Pentane, 36
Perfluoroalkoxy (PFA), 73, 76, 131
Periodic Table, 6–7, *8,* 19
Peroxides, 34
Phenol, 107, 141
Phenol–formaldehyde, 151
Phenolic(s), 44, 49, 107–108, 128, 151, 173
Phosphate ester(s), 119, 162
Phosphoric acid, 161, 162
Phosphorus, 28, 37
Phthalic anhydride, 62
Pigment(s), 13, 59, 100, 114–117, 119
Plastic compound, 13–14, 17, 19
Plasticizer(s), 99, 118–119, 132–133, 147
Plastic(s), 18, 21–22, 26, 39, 40–43, 48–50, 53–55, 98, 123–128
  abbreviations, 236–242
  ignition temperatures, *289*
  trade names, 243–266
Polyacrylonitrile (PAN), *81,* 81–82, 130–131
Polyamide–imide (PAI), 67–68
Polyamide(s) (PA), 44, 80, 108–109, 139
Polyarylate, 65
Polyarylsulfone, 65–66
Polybutadiene, 76–77, 127, 139
Polybutylene (PB), 44, 48, 82–83, 117, 127, 140
Polybutylene terephthalate (PBT), 83–85, 128, 140, 153
Polycarbonate (PC), 58, 83, 85, 128, 131, 141, 168
Polycarbonate/ABS, 168
Polycarbonate/high impact polystyrene, 168
Polychlorotrifluoroethylene (PCTFE), *73,* 73–74, 131
Polyester/polyether, 105
Polyester(s), 43, 48–49, 65, 84–87, 106, 108, 128, 151–152, 169, 173
Polyetheretherketone (PEEK), 66
Polyetherimide, 68
Polyetherketone (PEK), 66
Polyether/polyamide, 105
Polyethersulfone (PES), 66–67
Polyethylene (PE), 13, 16, 44, *46,* 46–48, 53, 55, 64, 76–79, 82, 87–92, 96, 117, 127, 140–142, 153, 169
Polyethylene ether, 69
Polyethylene glycol, 48
Polyethylene terephthalate (PET), 48, 82, 83, 85–86, 128, 140, 142–143
Polyformaldehyde, 51, 134
Polyimide, 67, 143
Polyisocyanate, 95, 146
Polyisocyanurate, 95–96, 146
Polymerization, 14, 17, 41–49, 51, 56, 64, 78, 88, 98, 106, 160
Polymerization reactor, 14, 45
Polymer(s), 13–19, 22, 35, 41–43, 45–50, 53–54, 64–65, 67–68, 71, 74, 76, 78, 80–82, 88, 91–92, 95, 97, 100, 108, 110, 127, 129, 130
Polymethylmethacrylate (PMMA), 54, 135, 165, 225
Polymethylpentene (PMP), 79, 117
Polyol(s), 48, 95
Polyolefin, 55, 64, 91, 141
Polyoxymethylene (POM), 51, 134
Polyphenylene oxide (PPO), 65, 69
Polyphenylene sulfide (PPS), 65, 69–70
Polyphenyl sulfone, 65
Polypropylene (PP), 44, 48, 82, 88, 91–92, 117, 127, 140, 143–144, 155, 169–170
Polystyrene (PS), 44, 48, 64, 69, 76, 82, 93–94, 97, 102, 120, 127, 144–145, 153, 156, 167, 170
Polysulfone (PSO), 58, 67, 70
Polytetrafluoroethylene (PTFE), 73–76, *74,* 131, 145–146
Polytetramethylene terephthalate, 84
Polyurethane(s) (PUR), 43–44, 48–49, 95–96, 130–131, 146, 171, 225
Polyvinyl acetate, 84, 97
Polyvinyl butyral, 84, 97

Polyvinyl chloride (PVC), 42, 44, 48, 58, 64, 71, 77, 82, 96–102, 117–118, 121, 131–133, 146–149, 153–155, 167, 171, 225
Polyvinyl chloride acetate, 84, 118
Polyvinyl fluoride (PVF), 75, *75,* 82, 131
Polyvinyl formal, 84, 97
Polyvinylidene chloride (PVDC), 71, 75, 97, 102, 131, 149
Polyvinylidene fluoride (PVDF), 75–76, *76,* 131
Polyvinyl(s), 96
Propane, 36
Propene, 82, 87, 140
Propylene, 48, 82, 87, *91,* 92, 140
Proton, 10–11, 19
Pure substances, 6, 12–14
Pyrolysis, 18, 32, 34–35, 37–40, 120, 122, 136, 139, 163, 165
Pyroxylin, 61, 136

Radical, 31–32, 36–38, 72, 77, 82, 88, 162
Radon, 28
Rayon, 62–63, 149
Reactants, 6, 18
Resin(s), 13, 22, 48, 50, 53, 110
Rubber, 76, 105, 139, 170
Runaway polymerization, 45

"Safety Film," 60
Saran™, 102, 149
Science, 5
Selenium, 28
Silica, 112
Silicon, 15, 17, 28, 108
Silicone(s), 14, 108, *108,* 164
Silk, 25, 42
Smoke, 32
Smoldering, 21
Sodium, 12
Sodium chloride, 12–13
Soot, 32
Space burning, 21, 37
Spontaneous ignition (spontaneous combustion), 39
Stabilizer(s), 117
Structural formula, 30, 51
Styrene, 48, 57–58, *58,* 93, *93,* 97, 103–104, 144
Styrene–acrylonitrile (SAN), 57, 104, 171
Styrene–butadiene (SB), 93, 104
Styrene/butadiene/styrene (SBS), 105
Styrene/ethylene–butylene/styrene, 105
Styrene–maleic anhydride (SMA), 58, 104
Styrenic(s), 57, 76, 93, 97, 102
Subatomic particle, 7
Sulfur, 17, 28, 34
Surface burning, 21, 37
Symbol, *9,* 12, 19, 30
Synergism, 162
Synergist, 162
Synthesize, 14
Synthetic polymers, 41, 43, 49, 60, 62

Talc, 112
Teflon™, 74
Tellurium, 28
Temperature, 24, 33, 39, 47
Terpolymer(s), 57, 102, 105
Tetrafluoroethylene, 72
Tetrahedron of fire, 27–28, *28,* 32, 35, 38, 40
Thermoplastic, 18, 39, 43–44, 47–50, 52, 54, 57–58, 65–68, 78, 84, 86, 95, 102, 104, 130, 134
Thermoplastic elastomers (TPE), 64, 104
Thermoplastic polyolefins, 105
Thermoset(ting), 18, 39, 43–44, 48–49, 95, 105–108, 130, 171–173, 226
Titanium dioxide, 117
Toxic, 32
Toxicity, 123–124
Toxicity testing, 221–235
Toxicity tests, 227–235
  Center de Recherches Toxicologiques Studies, 234
  DIN 53 436, 227–228
  Dome Chamber Toxicity Test, 228
  Dow Test Method, 228–229
  Federal Aviation Administration Method, 229
  GE/Southern Research Institute Lab and Full Scale Comparisons, 234

Toxicity tests *(Cont.)*
Harvard Medical School Test Method, 229
Hazard I, 234
Huntington Research Centre Primate Studies, 234
International Electrotechnical Commission, 234
ISO/TR 6543—The development of Tests for Measuring Toxic Hazards in Fires, 234
ISO/DTR 9122—Toxicity of Fire Effluents, 234
Japanese Combustion Toxicity Tests, 234
McDonnell-Douglas Test Methods, 229–230
National Fire Protection Research Foundation, 234
National Institute for Standards and Technology Tests, 230, 231
National Institute of Building Sciences Test Protocol, 230–231
Olin Corporation Studies, 234
PITT Test, 231
Radiant Heat Test Method, 231–232
Southwest Research Institute Primate Studies, 232
Stanford Research Institute International Test Method, 232
University of Michigan Test Methods, 235
University of Pittsburgh/PPG Industries Studies, 235
University of San Francisco Method, 232
University of Tennessee Test Method, 232–233
University of Utah Test Method, 233
U.S. Testing Company Test Method, 234
Vesicol/U.S. Testing Study, 235
Trimerization, 95

Ultraviolet absorber(s), 13, 117
Underwriters Laboratory (UL), 133
Upper flammable limit, 24
Urea-formaldehyde, 44, 49, 106, 108, 128, 130, 152
Urea(s), 108, 173

Vapors, 35
Vinyl acetate, 78
Vinyl acrylate, 53
Vinyl benzene, 93
Vinyl chloride, 48, 77, 97
Vinyl cyanide, 81
Vinyl dichloride, 97
Vinylidene chloride, 97
Vinyl radical, *96–97*
Vinyl(s), 42, 76, 82, 93, 96

Water, 13, 16, 29–30, 32, 120, 127, 130–131, 134–153, 224
Wollastonite, 112
Wood, 13, 25, 35, 37, 39, 42, 123, 130, 152, 165, 225
Wool, 25, 42

Xenon, 28